T0324797

DYNAMICALLY CONSOLIDATED COMPOSITES: MANUFACTURE AND PROPERTIES

DYNAMICALLY CONSOLIDATED COMPOSITES: MANUFACTURE AND PROPERTIES

T. Z. BLAZYNSKI

Formerly Reader in Applied Plasticity,
Department of Mechanical Engineering,
University of Leeds, UK

ELSEVIER APPLIED SCIENCE
LONDON and NEW YORK

ELSEVIER SCIENCE PUBLISHERS LTD
Crown House, Linton Road, Barking, Essex IG11 8JU, England

Sole distributor in the USA and Canada
ELSEVIER SCIENCE PUBLISHING CO., INC.
655 Avenue of the Americas, New York, NY 10010, USA

WITH 26 TABLES AND 288 ILLUSTRATIONS

British Library Cataloguing in Publication Data

Blazynski, T. Z. (Tadeusz Zdzislaw), *1924-*
Dynamically consolidated composites: manufacture
and properties.
I. Title
620.118

ISBN 1851666095

Library of Congress Cataloging-in-Publication Data

Blazynski, T. Z.
Dynamically consolidated composites:
manufacture and properties/T. Z. Blazynski.
p. cm.
Includes bibliographical references and index.
ISBN 1-85166-609-5
1. Composite materials. I. Title.
TA418.9.C6B555 1992
602.1'18—dc20   91-23462
                                                      CIP

No responsibility is assumed by the Publisher for any injury and/or damage to persons or property as a matter of products liability, negligence or otherwise, or from any use or operation of any methods, products, instructions or ideas contained in the material herein.

Special regulations for readers in the USA

This publication has been registered with the Copyright Clearance Center Inc. (CCC), Salem, Massachusetts. Information can be obtained from the CCC about conditions under which photocopies of parts of this publication may be made in the USA. All other copyright questions, including photocopying outside the USA, should be referred to the publisher.

To the memory of
Zofia and Witold
Blazynski

Preface

New composite materials and semi-fabricates, as disparate in their nature as solid multilaminates and powder compacts, have been steadily increasing in importance. Their application to a variety of industrial situations is being made easier by the considerable development of conventional manufacturing techniques which fulfil many of the requirements imposed on such materials. At the same time, however, the degree of their exploitation can be limited by, either the inadequate final product properties, or simply — as in the case of particulate matter — by the inability of these techniques to produce significant quantities of the composite.

For these reasons, combined with the ever increasing demand for highly sophisticated composites, attention has been focused on the dynamic manufacturing methods. Not only do they extend the range of the available routes, but they also offer the possibility of achieving chemical and/or structural syntheses of new materials from either the elemental or complex constituents. What is more, these techniques often tend to ensure integral bonding of the elements of the structure and they thus enhance the mechanical properties of the composite.

A considerable volume of information on these matters has been gathered in the last two decades or so, but it is contained in many separate papers, conference proceedings, and in only a few books. This makes it difficult for the practising engineer, material scientist, and researcher to have an easy, 'at a glance' access to the state of knowledge in this area. Since no specific, collated information on the dynamic methods of production and properties of composites appears to be available, this book has been written with the intention of bridging the various sources of information and providing, in consequence, an advanced review and a basis for further studies of these phenomena.

Dynamic processing involves the use of a number of very different techniques that range from explosive welding, through forming of bimetallic prefabricates, to shock treatment and, finally, shock compaction of particulate matter. The desired changes in material properties are effected by the stress and shock waves, but depend, in their effectiveness, on the processing method adopted and initial material conditions.

To place these parameters in their proper context, the first three chapters provide basic theoretical background to the problems normally encountered in this type of work. Here, the concepts of dynamic processing, stress and shock wave characteristics, and the material response are discussed. The methods of manufacture of solid composites are detailed in Chapters 3–5. Shock treatment and, where appropriate, consolidation and densification of particulate matter, are discussed in Chapter 6, whereas the specific conditions obtaining in the case of metallic composites are dealt with in Chapter 7. Ceramic and polymeric powders (including their various mixtures) form the bases of Chapters 8 and 9, with Chapter 10 dealing with the rather specialised composite compounds such as cermets, metallic glasses and super-conductors. All of these chapters are fully supplemented by lists of up-to-date references which enable the reader to delve deeper into any specific topic.

A book such as this could not have been written satisfactorily without the provision of adequate illustrative matter. In this respect, my grateful thanks go to all those researchers in the field who supplied me with the relevant illustrations and, in particular, to Dr R. A. Graham, Sandia National Laboratories, Albuquerque, New Mexico, Professor O. T. Inal, New Mexico Tech, Socorro, Professor L. E. Murr, University of Texas, El-Paso, ICI Nobel's Explosive Co. Ltd, Stevenston, Ayrshire, Scotland, Professor R. A. Pruemmer, Fraunhofer Institut, Freiburg i. Br., Germany, Dr D. Raybould, Allied Signal Corp., Morristown, New Jersey, and Dr M. L. Wilkins, Lawrence Livermore National Laboratories, University of California.

T. Z. BLAZYNSKI

Contents

ix

Chapter 1

Concepts of Dynamic Processing

1.1 INTRODUCTION

The last two decades have seen the ever increasing demand for economically viable and industrially acceptable new materials whose high strength and structural properties would be combined with highly sophisticated physical and/or chemical features. Many of the characteristics required can not be realised in a single material, but are possible to produce in a composite aggregate. It is for this reason that the concept of a composite material has dominated, of late, the thinking of engineers and material scientists. What is even more significant, it has been acquiring a new, wider definition and meaning. It is now recognised, and accepted, that the standard, narrow definition of what constitutes a composite material — usually associated with a chemical or metallurgical process of some kind — is inappropriate in the constantly changing and improving world of new technologies. The enlargement of the scope of manufacturing techniques, proceeding since the late 1960s, has made it possible to consider mechanical and electrical means of producing composites as well as the traditional routes. The development of the metallic powder compaction, in particular, was one step in the substantial changes that have taken place. This area of activity was recently strengthened by the introduction into its orbit of non-metallic materials such as ceramics and polymers. Ceramic/polymeric, ceramic/metallic and metallic/polymeric blends are now being produced thus adding to the wide range of potentially useful new materials. Explosive welding of solid aggregates, on the other hand, has opened the way to the production of an ever increasing range of composites that could not be produced otherwise.

The success, or otherwise, of any such production method is reflected, naturally, in the quality of the final product. It is precisely in this area that the conventional manufacturing techniques do not always pass examination because the degree of consolidation, homogenisation and/or solidification that they offer is not necessarily of a sufficiently high degree.

On the solid composite side, the conventional rolling, extrusion and drawing processes fulfil, up to a point, the requirements necessary for the production of very simple components, but are inherently incapable, as primary operations, to produce geometrically complex multimetallic or multimaterial composites.

The limitations associated with the conventional 'static' powder compacting, and metal-forming processes have led to the development, on an industrial scale, of a variety of dynamic techniques which not only considerably increase the scope of the possible, classical operations, but also introduce entirely new elements by changing the microstructure of the processed material, releasing the lattice energy, synthesising new compounds, and bonding particles. Normally incompatible solids can thus be joined together and materially and geometrically complex composites can be manufactured by explosive techniques to form well 'homogenised' structures. These, in turn, act as precursors to further, more conventional, operations.

The essential strength, resistance to wear, and uniformity of structure element of bonding, whether of individual grains or particles in a particulate aggregate, or of arrangements of solid sheets, plates, tubes, rods and wires, is always present in dynamically, particularly explosively, produced composites. The bonding of material elements is not, however, always present, or sometimes even possible, in many conventional processes.

To produce a comprehensive solution to the various manufacturing problems arising, a most general approach must be adopted — such as is used in this book — by defining a composite as a material element, obtained by processing solid or particulate matter, whose structure or composition has been so altered as to introduce a substantial degree of differentiation between its original and current properties. The change must be sufficiently significant to make the response of the composite to the working conditions in which it will operate considerably better than that expected from the dynamically unprocessed, constituent material(s).

Depending on the materials involved, their sizes, geometry, and, naturally, final 'destination' of the composite itself, different methods of building-in the required properties exist and should be explored from both the point of view of the economics and that of the product quality optimisation. The aim of this book therefore is to give a comprehensive review of the available dynamic tools of manufacture of the widest range of solid and particulate composites, to consider their mechanical, physical and chemical properties, and to indicate their respective applicability to modern industrial needs.

1.2 DYNAMIC PROCESSING TECHNIQUES

The three basic processing techniques — in order of their importance — are:

- explosive (chemical),
- mechanical,
- electrical,

Very high levels of energy, available in a chemical explosive charge, are easily accessible and can therefore be utilised cheaply in welding, forming, compaction and shock treatment processes. These are used either to alter the characteristic features of the material structure, as a preliminary procedure, prior to further processing, or as an end in itself. Since these techniques incorporate a wide range of variants, a detailed discussion of them is provided separately in the following chapters.

Various mechanical devices, such as HERF (High Energy Rate Fabrication) machines (Dynapak, Petroforge, etc.) or gas guns are used in forming and compaction with the latter also employed in the investigations of dynamic welding. The gas guns (Fig. 1.1) are particularly useful in the manufacture of solid or hollow cylindrical, particulate material composites. They are simple to design and operate, and, because of their basic simplicity, are mechanically reliable.

In a typical arrangement, an impactor is accelerated in the gun barrel by releasing a high pressure air or nitrogen charge. The powder to be compacted is contained, at tap density, in a cylindrical die and, on impact, is subjected to shock conditions. Impact velocities of at least 500 m/s are needed for most powders.

When the gun is used in the investigations of high velocity impact welding, a target plate (inclined at a suitable angle) takes the place of the powder container and the impactor is then used as a projectile.

The less frequently used, partly because of the cost, and partly because of the size limitations, are the hydro-electric and electromagnetic systems.

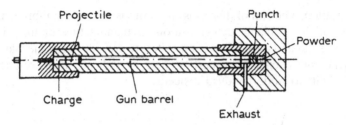

FIG. 1.1   An explosively activated gas-gun compacting system.

In hydro-electric devices, the principle of operation is based on the rapid dissipation and transmission of energy evolved when an electrostatic field is suddenly discharged. A basic circuit used is shown in Fig. 1.2.

Two different techniques of utilising the energy stored in a bank of condensers are employed, namely (a) underwater discharge, and (b) exploding wire.

In the former case, the discharge across two submerged electrodes produces a shock wave in the transmitting medium, accompanied by heating and vaporising of the adjacent layers of the medium. The plasma created by the spark expands as a gas bubble, transmitting the force of explosion to the workpiece. The efficiency of the operation depends on the conductor material, losses of energy in the circuit, and on the geometry and surface conditions of the electrodes.

The second method consists in connecting submerged electrodes with an initiating wire. The transmitted energy vaporises the wire and converts it into plasma, creating a pressure wave. The increase in volume of the vaporised wire is of the order of 25 000 times its original volume. The method possesses certain distinct advantages over the spark discharge in that the process can be more rigidly controlled. The shape of the shock wave can be determined by the shape of the wire, and a long arc discharge is obtained, as opposed to a point source one. The amount of energy can be controlled by the dimensions and material of the wire. It is found, for instance, that tungsten produces more energy than tantalum, niobium, molybdenum, titanium, nickel and aluminium, taken in that order.

The principle of electromagnetic forming is, basically, the same as that of the hydro-electric operation. The energy stored in a bank of condensers is rapidly discharged through a magnetic coil, surrounding, placed inside, or in the proximity of the workpiece (Fig. 1.3). A high intensity magnetic field is thus created and providing that the material

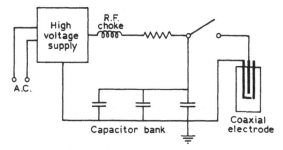

FIG. 1.2   A capacitor-discharge activated hydro-electric forming system.

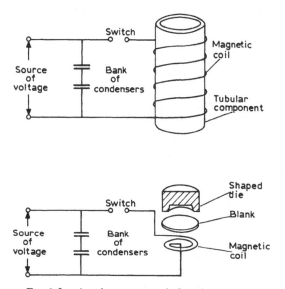

FIG. 1.3   An electromagnetic forming system.

of the workpiece is conductive, electric current is induced in the specimen. The current interacts with the coil field and produces high
transient forces. The specimen thus acts as a secondary-circuited coil.
The energy level produced by the magnetic field depends on the conductive properties of the formed metal, its shape and mass, and the
duration of the initial current pulse. These factors can be quite easily
controlled by a suitable choice of materials and geometry of the coils.
Materials of low conductivity are sometimes lightly coated with copper.
The shape of the impulse wave can be modified by using field shapers
which consist of shaped beryllium copper pieces of metal inserted in

the coils. Electrically, they help to depress or concentrate the intensity of magnetic flux in those sections of the workpiece which may require a lower or higher degree of deformation.

The life of a coil depends on the magnitude of the force to which it is subjected, and, this being equal to the force generated on the surface of the workpiece, the pressures can be very high. The usual practice consists therefore of using one-shot disposable coils for more complicated operations that involve only a few parts, and of limiting the pressures to the value of the compressive strength of the coil material for mass produced parts.

Most of the techniques described are likely to produce shock effects in the processed materials. In turn, the generation of a shock wave, whether in solid or particulate matter, may well produce structural and/ or chemical synthesis of the material(s) involved and will thus constitute an additional manufacturing process of its own.

1.3 RANGE OF APPLICATIONS

Although a more detailed indication of the applicability of the basic systems to specific industrial needs is given in the following chapters, a brief discussion of the sub-division of their operational areas (Fig. 1.4(a)) may be useful at this stage.

The widest range of possibilities is offered by the explosive type of operations (Fig. 1.4(b)). On the welding side, these include the manufacture of multimetallic 'sandwich' plates, bi- and trimetallic, axially clad tubes and cylindrical pressure vessels, welding and plugging of heat exchanger tube plates, production of planar and tubular transition joints for electrical and cryogenic applications, and bonding of anticorrosive and heat resistant liners to the working surfaces of chemical apparatus and plant.

Aircraft and aerospace components such as exhaust and thrust reverse parts, manifolds, airfoil, heat shields and panels are often fabricated out of explosively prewelded bimetallic sheet by explosive forming in open or closed dies. Of special interest here are cupronickel clad steels, and composites comprising welded lead/copper and lead/ steel plates offering antiradiation protection.

Light weight, but structurally strong pressure vessels made of integrally explosively welded foil/mesh laminated layers, geometrically complex axisymmetrical arrangements of arrays of rods and tubes

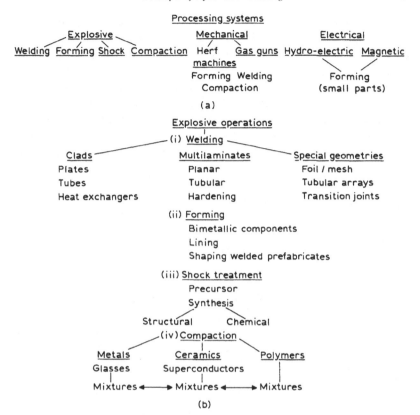

FIG. 1.4   Classification of dynamic processing. (a) General classification; (b) explosive/implosive systems.

encapsulated in tubular sheaths — suitable for sophisticated heat exchangers, or, at the other extreme, for remote control rods working in chemically or radiation hostile environment — are easily produced.

Particularly advantageous, in this context, is the possibility of bonding materials which are normally incompatible — from the welding point of view — and which are to be used in geometrical arrangements that, again, in normal conditions preclude full interface welding.

Compaction and intergranular bonding of particulate matter enables the introduction of metallic and non-metallic mixtures into the area of machine part design. High wear resistant and anticorrosive properties, combined with the structural strength of some of these combinations make them very attractive in a wide range of applications.

Metallic/polymeric blends find their uses in the electronic industry, as do explosively produced metglasses. A new generation of super-conducting materials is now being manufactured by these methods out of ceramic powders.

Shock treatment of both solid and particulate matter often changes the material structure and is therefore beneficial in itself as, say, a precursor to sintering, or an instrument of phase change. Chemical or structural syntheses are thus possible and can effectively produce composite compounds in reasonable quantities and in a matter of a few milli- or microseconds.

Mechanical means of generating dynamic effects, such as the already mentioned HERF machines of various types, are utilised mostly in the forming of small, possibly bimetallic, components, and in powder compaction, but together with the hydro-electric and electromagnetic devices they suffer from product size limitations.

1.4 FUTURE DEVELOPMENTS

In some areas of dynamic processing, notably in welding, a sufficiently high level of understanding of the processes, and of their utilisation has been reached to preclude any 'dramatic' new developments in the future. Rather, it is a matter of consolidation and adaptation of the existing techniques and 'know-how' to some specific applications that arise or may arise in a given manufacturing route, in particular in the car, marine, aircraft and space industries.

It is, however, in the areas of powder compaction and bonding, and fibre/wire or mesh reinforced particulate matrix compacts that considerable and, as yet, unexplored possibilities exist. A large variety of polymeric materials, which because of their thermal properties are unsuitable for conventional processing, can be easily incorporated into blends with metallic or ceramic powders, to produce composites of enhanced mechanical, physical and even chemical properties. Entirely new, industrially viable compounds may thus be made available at a relatively low cost and in sufficient tonnage to either offer an attractive substitute for the existing metallic alloys, or to become a welcome addition to the presently used range of engineering materials.

Chapter 2

Stress Waves and Material Response

Since the main objective of the present chapter is a consideration of the effects of the passage of pressure pulses through the body of a material component, the usually complex mathematical analyses of the actual nature of stress waves are excluded. A detailed evaluation of the existing theories can be found in the reviews of the dynamic material processing, by, among others, Myers and Murr,[1] Johnson and Yu,[2] and Johnson.[3] Consequently, in what follows, only general concepts and definitions of the basic wave properties are summarised with a view to facilitating the understanding of the developed arguments.

2.1 BASIC CHARACTERISTICS OF GENERATED WAVES

The application, in a dynamic mode, of an external agent to the surface of a material body results in the formation of a high pressure wave which, in turn, is transmitted through the medium in the form of a transient stress wave. The wave is capable of deforming and consolidating the structure, but, if it exceeds a certain magnitude — critical for the given material — of producing a fracture.

Depending on the level of energy dissipated, and the nature of the transmission generated, waves can be either elastic or plastic, but, again, with high pressures usually applied, the elastic component may well be negligible. Generally, a disturbance within the material results in the formation and propagation of two basic types of waves: (a) longitudinal — causing changes in the volume of the material, and (b) transverse, or shear — causing distortion of the structure. The rate of propagation depends on the properties and the geometry of the body; the longitudinal wave velocity being roughly twice that of the shear wave.

2.1.1 Elastic Waves

Although, from a practical point of view, the elastic waves are of a limited interest, it is nevertheless worth noting that in addition to the longitudinal and shear modes, surface or Raleigh waves can also be generated. These, as the name implies, are associated either with the surface and/or a shallow adjacent substratum of the material only and are sometimes observed in the explosive welding of foil-type materials.

2.1.2 Plastic Waves

If the uniaxial state of stress prevails in the processed material, when its critical elastic limit is reached, a plastic wave will be formed. However, if the material is subjected to uniaxial strain, a plastic shock wave will be generated. Plastic deformation propagates through the body of the material as a disturbance generated by a high pressure (stress) pulse of short duration and is similar, in its progression, to the elastic deformation. Microscopically, plastic deformation is inhomogeneous and will consequently give rise to a differential type of flow, and to the micro-defects associated with its microdefects. On a macroscopic level, however, it is possible to regard plastic strain as being uniformly distributed across the solid material body, providing, of course, that the rate of application of the external agent is comparatively low in comparison with that of the propagation of the disturbance. In a uniaxial stress field, the rate of stress $(d\sigma/dt)$ decreases with strain and so the shape of the wave front departs significantly from the abrupt shape of a discontinuity that is needed to generate a pure shock wave.

Pure plastic, non-shock waves are usually associated with explosive forming, and some explosive welding configurations in which the generated pressure pulse travels along the surface of the material but is prevented from passing through its body. The mode of these waves can be — as in the elastic regimes — either direct, longitudinal, tensile or compressive in nature, or transverse, shearing. The latter, because of its distortional properties, is of a particular significance in the manufacture of composites.

2.1.2.1 Shear waves

It is very seldom that in practical applications pure shear waves are likely to be generated. Their normal incidence is generally conditioned by the presence or absence of the deforming longitudinal waves. Both are generated by the impact of a projectile or by detonation of an explosive charge. Shear waves propagate at right-angles to the surface on which a pulse impulse is generated and produce lateral disturbances which are associated with the gradually increasing wave front. The rate of propagation is generally about half that of the pressure pulse.[4] This relationship resembles closely the situation that obtains in the case of longitudinal and shear elastic waves.

2.1.2.2 Adiabatic shear bands

The problem of adiabatic shear bands has recently received very considerable attention since, in the more extreme cases, the presence of the bands can lead to material failure. In its simplest form, the formation of a band is associated with intense localised shearing that, in turn, raises the temperature. If the heat generated can not be easily conducted away, near adiabatic conditions are created locally, and a reduction in the flow stress of the material follows. A nucleus of weakness is thus created and results in a concentration of deformation in this area. Naturally, this produces a further plastic flow and temperature rise and, so, may eventually result in local melting and mechanical failure of the material.

Two basic types of the bands exist, namely those of deformation and transformation. The former originate at high shear strains, of up to 100, and contain highly distorted grains which have not, however, undergone any structural changes. These bands are usually found in the ferrous, aluminium, titanium and uranium alloys and are characterised by their hardness that increases towards the band centre to a level approximating to that expected in a highly cold-worked material.[5]

The basic features of the transformed bands are the crystallographic phase changes, extremely high hardness, and very small thickness — not exceeding 50 μm. In the case of steels, the hardness is in excess of what would be expected in a conventional heat treatment and appears to be independent of that of the original material. As in the deformed bands, its maximum value is reached in the band core.

Generally, materials that are most susceptible to the incidence of either type of the adiabatic shear band, are those which exhibit low strain hardening, low specific heat, and high thermal softening.[6] The type of loading which is most likely to produce conditions favouring the generation of the bands, is that of a predominantly high rate compressive stress.[7]

The greatest potential danger of material failure can be expected in the fragmentation of exploding cylinders, high rate forming and welding, plugging and high velocity plate penetration.[7,8]

The importance of producing a mathematical model, capable of predicting the incidence and magnitude of the bands, is considerable and has consequently exercised the attention of a number of investigators. A full description of the adiabatic shear band phenomena, together with a discussion of their theoretical and experimental backgrounds is provided by, for instance, Dormeval,[6] Stelly and Dormeval[9] and Klepaczko et al.[10]

2.1.3 Shock Waves

As already pointed out, before a purely plastic wave can be converted to a plastic-shock one, the existence of a state of uniaxial strain is required to allow the build-up of the hydrostatic stress component to a level far exceeding that of the dynamic yield stress of the material. It is only in this situation that sharp discontinuities can be created in the pressure front, propagating through the material, and that a differentiality in the particle velocities can be produced. Displacements of considerable magnitude will take place as the particles accelerate and acquire high velocities. This is related to the fact that the rate of propagation of the disturbance behind the front is usually higher than that of the shock front. Under these circumstances, it is legitimate to assume — with a fair degree of accuracy — that the processed material has no resistance to shear, or, in other words, that its modulus of rigidity $G = 0$. In turn, this assumption allows the Rankine–Hugoniot hydrodynamic approach to be applied to the analysis of the conditions prevailing when very high pressures are induced. It is, of course, necessary for the velocity of the front to increase with the increase in pressure if the shock wave is to be maintained.

Irrespective of the method of initiation of the front (impact, detonation of an explosive charge, etc.), the physical state of the medium is altered, through compression, from its original density ρ_0, pressure p_0, and temperature T_0, to one of a higher density ρ, pressure p and temperature T. Further, particles of the medium acquire a translational velocity u, i.e. the velocity with which a particle moves in the compressed medium behind the shock front. The velocity U of the shock front, and the internal energies before (E_0) and after (E) compression, can be assessed by means of the standard equations of conservation of mass, momentum and energy.

Hence,

- Mass $\rho(U - u) = \rho_0 U$ (2.1)

- Momentum $p - p_0 = \rho_0 u U$ (2.2)

- Energy $E - E_0 = \frac{1}{2}(p + p_0[1/\rho_0 - 1/\rho])$

$$= \frac{1}{2}(p + p_0)(V_0 - V) \tag{2.3}$$

where V is the specific volume.

The solution of these equations is possible only if an additional relationship is established. This is normally obtained experimentally and relates the shock and particle velocities,

$$U = C_0 + S_1 u + S_2 u^2 \tag{2.4}$$

where C_0 is the velocity of sound at zero pressure, and S_1 and S_2 are specific material constants. For most metals, $S_2 = 0$, and the equation simplifies accordingly.

Since $E = f(p, V)$, eqn (2.3) yields a relationship between pressure and volume that is usually referred to as the Hugoniot curve. In effect, for a given material, the curve constitutes the locus of all the points that may be attained through the medium of shock transition.

The Hugoniot is usually a smooth curve, unless the shocked material is subject to pressure-induced phase transformation. In this case, the pressure-relative volume (V/V_0) Hugoniot curve will display a discontinuity on unloading and a hysteresis loop will be present.[11] The resulting change in the slopes of the, so-called, Raleigh lines (the lines passing through the $(p, V/V_0)$ and $(0, 1)$ points of a Hugoniot) will lead to the decomposition of the wave into two.

The hydrodynamic theory approach ignores the fact that, in reality, a uniaxial state of strain is associated with the presence of shearing. This fact, combined with the presence of the discontinuities in particle velocities, densities, temperatures and pressures across the shock front, as well as the rarefaction and attenuation of the wave, calls for a more sophisticated treatment if the proposed mathematical model is to give reasonably accurate numerical data. Computer codes now exist that take into account the non-linearity of the differential equations representing the variations in these parameters. A fuller and more detailed discussion of the solutions proposed will be found in Ref. 1.

From the more conceptual and visualisation points of view, the mechanical effect of the passage of a shock front through a solid is provided by Fig. 2.1.

Since, as indicated, the pressure front can not be planar, the shock wave propagated through a material can be visualised, in terms of pressure and time, as shown in the figure. The increase in pressure, up to its peak value, is associated with the indicated shock front. The peak shock pressure p operates in time Δt. This interval is followed by a

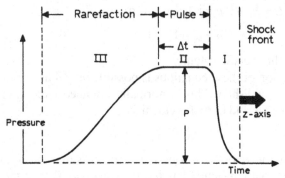

FIG. 2.1   Idealised representation of the effect of passage of a shock pulse through a solid (after Refs 1 and 12).

reduction in the pressure, or wave attenuation, and by the formation of a rarefaction wave. In consequence, compression of the solid takes place in Zone I, no overall volume change ($dV = 0$) is evident in Zone II, but a relief of the pressure is present in the rarefaction Zone III. Permanent, residual microstructural phenomena are therefore produced[1] in Zones I and III.

It is normally assumed in calculations of the internal energies across the shock front, and in the rarefaction zone, that adiabatic and isentropic conditions obtain respectively.

2.2 MATERIAL RESPONSE

2.2.1 General comments

The passage of a suitably profiled shock wave through a material results not only in the consolidation of its structure, but can also effect physical, mechanical and chemical changes in the initial body properties. Depending on the characteristics of the substance used, i.e. whether solid or particulate matter, synthesis of a new material, a new phase, possibly enhancement of the degree of reactivity, changes in the electrical and magnetic properties, and a general increase in the incidence, level and type of microfaults can be expected. In more complex materials and situations, many of these features may be present simultaneously, whereas in simpler cases only one or two may materialise. Irrespective, however, of the possible combinations actually present, the passage of a shock wave offers the means of creating a 'composite' material within the body of a single compound.

From a practical point of view, the generation and transmission of shock waves increases in its importance if it is aimed at either compaction of particulate blends, or consolidation and bonding of solid matter of differing basic properties.

The ease with which different particulate elements can be mixed and 'homogenised' into an acceptable aggregate, or — as with fibres and powder matrices — prearranged to form specific geometrical patterns, assumes very considerable practical importance. Skilful selection and manipulation of both the materials and their relative geometrical dispositions, can create a field of optimal stress distribution in the component when it is subjected to normal working conditions.

Since the passage of the shock wave through a particulate aggregate is likely to result in interparticle bonding, not only will the density of the blend increase, but also its strength. This effect is independent of the characteristics of the mixture and thus opens the way to the manufacture of aggregates of blends of metallic, ceramic and organic mixtures.

In the solid matter composites, the situation is slightly different. These are generally associated with metallic materials which are either shaped or prewelded and shaped and which consist of a number of material combinations. For reasons given later, explosive welding is more likely to lead to the avoidance of the generation of shock waves — taking attenuation into account — and will rely more on the effects of plastic non-shock pulses. Attenuation of the effect of shock — through the medium of protective buffers — is particularly important in the manufacture of metglasses.

The individual micro- and macroeffects of the passage of a shock wave are discussed in the following sections.

2.2.2 Metallurgical Microdefect Generation

A number of phenomena of basically metallurgical nature are likely to result from the passage of a shock wave and the associated stress system. Although these are microlevel occurrences, they affect, ultimately, the mechanical properties of the material.

The main microdefects are subdivided into four groups of linear, interfacial, point and volumetric and are known as:

- dislocations,
- twins,
- point defects,
- displacive/diffusionless phase transformations.

Their incidence, mode and magnitude depend on the prevailing levels of the deviatoric and hydrostatic stresses, as indicated schematically in Fig. 2.2. A model for the generation of dislocations in a crystal lattice, combined with an idealised shock front-relaxation model of Fig. 2.1, was proposed by Myers[1,8] and is shown in Fig. 2.3. Figure 2.1 should be regarded as being plotted at right-angles to each individual segment of Fig. 2.3.

The basis of the Myers' model[1] is that dislocations are generated at or near the shock front by the deviatoric stresses and the latter are then relieved by the presence of the formed dislocations. The dislocations are considered to be moving through short distances at subsonic speeds, with new dislocation interfaces being generated in the material as the shock wave propagates through its body. The effect of the pulse is felt throughout Δt time interval.

Figure 2.3(a) shows the first stage of this process, with the initial changes taking place in the lattice. In Fig. 2.3(b), the front coincides with the first dislocation interface; with the density of dislocations depending on the difference between the specific volumes of the two adjacent lattices. As the front moves forward (Fig. 2.3(c)) deviatoric stresses increase again and further layers of dislocations are formed (Fig. 2.3(d)). The process then continues throughout the bulk of the affected material.

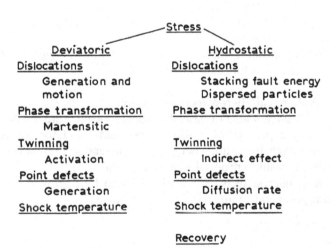

FIG. 2.2   Schematic representation of the influence of the deviatoric and hydrostatic stress components on the incidence and magnitude of microfaults.

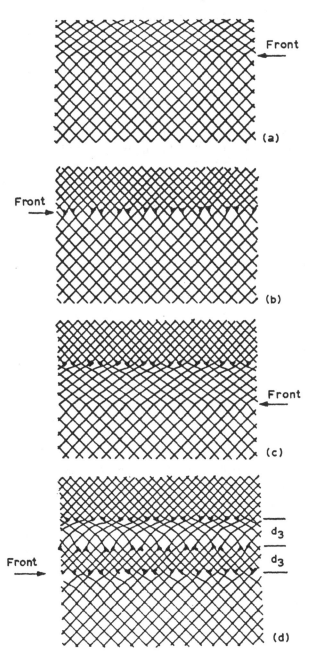

FIG. 2.3   Myers' model of the generation of dislocations by a shock wave.

The rarefaction wave appears to contribute very little to the generation of dislocations since it passes through the already dislocated material, nevertheless it is capable of, and often does produce some rearrangement and even destruction of the structure.

It has been shown[1,13] that an already shocked material does not display any pronounced increase in dislocations when it is subjected to further shock pulses.

Generally, the generation of this particular microfault is affected by the magnitude of the pressure pulse applied, and by the initial grain size, as implied by a Hall–Peck type of expression in which the residual stress σ is given in terms of the applied stress σ_0, a material constant K, and the average grain diameter D

$$\sigma = \sigma_0 + KD^{-1/2} \qquad (2.5)$$

However, the final result, in any particular situation, will depend on the mechanism of slip. Of the three possible slip systems, glide and climb are affected by the temperature, etc., but cross-slip depends on the stacking-fault free energy. The region of stacking faults, that separates partial dislocations, increases with a decrease in energy and thus determines the mode and extent of their propagation.

The intersection of mixed dislocations generates jogs whose motion, in turn, produces large concentrations of point defects such as vacancies and interstitial concentrations.[14] These are three to four times higher than defects generated by cold-forming of metals.[15] However, from a practical point of view, the presence of twinning is of much greater importance.

Twinning appears normally in a crystalline metallic matter at the critical, for the material, pressure, but it is present, under shock conditions, in materials such as, for instance, nickel and molybdenum which are not prone to it in other physical circumstances. The likelihood of its incidence is increased in low-stacking fault free energy metals in which thin twins, known as twin-faults, are formed. In the FCC metals and alloys, twinning occurs initially in (001) orientations. It is, however, dependent on the duration of the pulse. The threshold time for optimal twinning has to be established for each individual material. This is of particular interest in the case of Hadfield[16] and stainless[17] steels.

It should be pointed out, however, that twinning also occurs as a result of the passage of the elastic precursor wave, as, for instance, in Armco iron.[18]

The importance of the incidence and level of twinning lies in its contribution to the macro deformation of the shocked material. The four principal deformation mechanisms are slip, twinning, and, ultimately, brittle and ductile fractures. Of these, both slip and twinning are affected by the material properties, temperature and stress and strain rate. The structure of the metal affects slip insofar as the number of planes of the greatest atomic density is concerned. Although the number does not determine the relative ability of the metal to deform, this parameter, in conjunction with the component of shear stress along the slip surfaces, is indicative of the tendency of the metal to deform. Increased strain rates, alloying and strain hardening increase the critical shear stress, while an increase in the temperature produces a reduction in its magnitude. Twinning is less susceptible to the effect of temperature and although it accounts for only small strains, of the order of 5-10%, it re-orientates grains into a more favourable position for slip. This feature is particularly useful in the case of the materials characterised by a small number of slip planes.

Failure of the material can be caused by either ductile shear or brittle fracture or a combination of both. In the latter case, the initiation of failure is due to brittle separation of the grain boundaries which, in turn, appears to be caused by the critical normal stress acting on the planes of cleavage. Brittle fracture is very little affected by temperature, but it is very susceptible to high rates of strain since these affect, proportionally, the magnitude of the critical normal stress.

A useful model of the combined effect of these parameters is given in Fig. 2.4. The general inference that can be drawn is that most metals deform by slip, or fracture by shear at high temperatures and low strain rates, but that brittle fracture is characteristic of the mode of failure at low temperatures and high strain rates. In the high velocity (shock) systems, the deformation is caused primarily by twinning and only secondarily by slip. It seems, however, that the proportion is about equal in very ductile materials, such as stainless steel, and the failure in these cases is certainly of a ductile nature.

Displacive/diffusionless transformations are affected by pressure, temperature and shear stress field imposed. They result in volumetric and phase changes and can thus form a useful adjunct to the shock created effects.

The lattice distortion transformations are usually sub-divided into the dilation and deviatoric dominant.[19] The latter, often described as 'martensitic' refers to FCC material transformations, to either BCC or

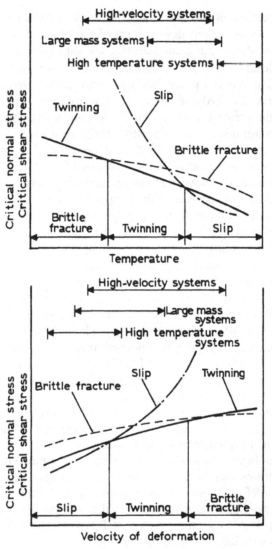

FIG. 2.4   Effect of process parameters on the mechanism of metal deformation.

HCP structures, and is thus associated mainly with the ferrous-type materials. A typical example of this is provided by Fig. 2.5 which illustrates the changes in 304 stainless steel that result in the formation of martensite at the intersection of twin-faults.[19] Martensite is created

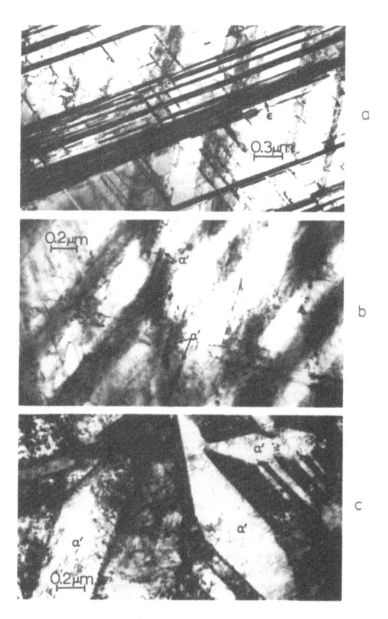

FIG. 2.5 Shock induced transformation products in 304 stainless steel. (a) ε-martensitic bundles mixed with stacking faults and microtwins — shock loading 25 GPa for 2 μs; (b) α'-martensite embryos nucleated at fault-band intersections after shock loading at 30 GPa, 2 μs; (c) α'-martensite platelets formed after shock loading at 45 GPa, 6 μs (courtesy L. E. Murr and M. A. Myers, Refs 1 and 20).

when Shockley partial dislocations are produced on every other (111) plane[1] (Fig. 2.5(a)). However, martensite α' is created within the volume of the intersection when more complex arrays of faults are present (Fig. 2.5(b)). Coalescence of the strain-induced martensite is further increased by the passage of a repeated shock pulse and can therefore reach considerably high levels (Fig. 2.5(c)).

As already indicated, the effect of the passage of a shock wave on the metallurgical properties of a material depends on its initial condition, as much as on the magnitude and characteristics of the pulse itself and will be therefore specific to the material and its original state. The existing volume of information, about a wide range of engineering materials, has increased in recent years to such an extent that the inclusion of it in this book is impossible. For this reason, the reader is referred to Refs 1 and 5-8 for detailed reviews.

2.2.3 Physical Properties

The nature of the changes in the physical properties depends, to a certain extent, on whether the processed material is in the solid or particulate state. In the former case, it is mainly the mechanical properties (discussed later) that are affected, whereas in the latter situation, in addition to the compacting effects, the powder may well experience changes in its magnetic resonance characteristics.

Depending on the material, isotropic and anisotropic paramagnetic defects are often noted in ceramics as, for instance, with silicon nitride, and, say, rutile.[21-23] In the first instance, two isotropic g-factor paramagnetic effects are present. A narrow line resonance with a g-factor of 2·0028, absent in the unshocked powder, appears after the application of a shock pulse, and a broader line resonance with a g-factor of 2·0038, already present in the original material, increases in concentration by a factor of 50 at 20 GPa, and by 100 at 27 GPa. In the case of the rutile powder, a shock pulse of 20 GPa produces two paramagnetic resonances; an isotropic one with a g-factor of 2·0029 and a high concentration of 3×10^{16} cm^{-3}, and an anisotropic one with g-factors of 1·937 and 1·969, and a high concentration of 3×10^{19} cm^{-3}.

In all such cases, considerable plastic deformation of individual grains is present, and lattice distortion, accompanied by a release of energy (up to 3·35 J/g) occurs. This, in turn, predisposes the material favourably to further sintering, by lowering the necessary pressure and temperature; a feature of particular practical significance in the case of non-oxide ceramics whose strong covalent properties make them difficult to sinter in normal conditions because of their low rate of diffusivity and low levels of crystal vacancies.

In organic materials, the response can be of a different kind in that molecular bonds may be broken (destruction reaction) leaving stable or active molecules. Isomerisation may result when the bonds combine to form slightly different molecules, or when they form larger ones this may lead to polymerisation.

Generally therefore, shocked materials are likely to acquire different properties which, if effected in a controlled operation, may well enlarge the range of their practical exploitation.

2.2.4 Chemical Properties

Shock modification of materials, leading to their synthesis, is a particularly attractive feature of the shock treatment to matter. The change in the lattice of the given material results in purely structural modification and is correspondingly known as structural synthesis. However, the synthesis of a compound from its basic constituents — whether in particulate or precipitate form — is regarded as chemical synthesis.

While structural modification is not necessarily dependent on the initial particle size and may occur in a solid, crystalline material subjected to a sufficiently high pressure, chemical synthesis is best conducted with initially small particles. The reason for this is that the rate of a chemical reaction is limited by the activation energy for diffusion and this increases with the increasing size of the reacting particle. In any individual case, the reaction rate must be matched to the rate of propagation of the shock front. Although it is sufficient to equalise the two quantities, success is better ensured if the rate of the reaction exceeds that of the shock front.

A pure chemical synthesis of two materials may be accompanied by changes in their crystalline structure or by the formation of a phase that differs in its unit cell parameters from a phase formed in a conventional process.

Both the purely structural and chemical syntheses, combined sometimes with compaction, are used in industrial processes, because not only are new materials thus produced, but also economically attractive bulk processing can be achieved in a very short period of time. Also, the fact that the heat generated is mostly adiabatic, ensures that thermal degradation of materials like thermoplastics is practically avoided.

Examples of chemical shock synthesis of metallic or ceramic compounds are provided by, for instance, syntheses of mechanically blended aluminium/titanium and aluminium/nickel composites (titanium and nickel aluminides), or, on the powder side, of zinc ferrite obtained from mechanically blended stoichiometric mixtures of zinc oxide and haematite powders.

Shock wave synthesis of diamond from graphite (Fig. 2.6) is a typical example of the result of a change that can take place in a single material.[24,25]

Further characteristic chemical changes that may occur are an increased rate of chemical reactivity that, among other things, may lead to the initiation of synthesis accompanied, possibly, by an exothermic reaction, a change, positive or negative, in the specific surface are a developed, and an increased catalytic activity.

2.2.5 Mechanical Properties

The direct macroeffect of the passage of a shock front through metallic components is the strengthening and hardening of the material. The phenomenon of strain-hardening is associated with most of the engineering alloys and reflects the effect of the piling-up of dislocations that, in turn, blocks the free movement of others. Strain-hardening leads to a constant increase in the value of the flow stress. Effectively there-

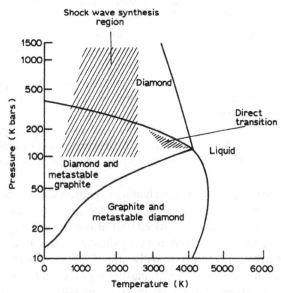

FIG. 2.6   Shock wave synthesis of polycrystalline diamond from graphite (after Ref. 25).

fore, it raises the level of the current yield stress and tends to affect the ductility of the material. The latter can be restored by annealing or a similar heat-treatment operation that will unblock the paths of the dislocations but, if this is not possible, the effect of strain-hardening has to be accommodated in the appropriate constitutive equation.

An arrested shock front produces a visible delineation within the body of the material (Fig. 2.7) leaving part of it unaffected (elastic regime, lower part of the figure) and part severely plastically deformed and strain-hardened. An increase in the rate of straining of the material will increase the strain-hardening effect and therefore raise the level of the flow stress even further.

Consideration of the general pattern of the true stress–natural strain curves indicates that the engineering alloys fall, broadly, into two types: (a) those having stress–strain curves that show that an increase in strain rate has initially no noticeable effect on the strain to fracture, but a further increase reduces that strain, and (b) those in which the true strain to fracture increases slightly with increasing strain rate and then decreases rapidly. Figure 2.8 shows these tendencies diagrammatically and it also indicates the existence of an optimal rate beyond which a further increase would not serve any useful purpose.

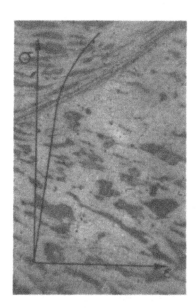

FIG. 2.7 The elastic–plastic material deformation boundary in a partially shock affected low carbon steel.

The ability of a metal to deform without fracturing is bound up with its ability to absorb any energy imparted. This property, known as toughness, can be assessed from the stress–strain curve because the area under the curve represents the amount of work done. Theoretical toughness of the two groups of metals of Fig. 2.8 is shown in Fig. 2.9. Again, the inadvisability of exceeding the optimal rate is clearly indicated.

A complete determination of the strain rate operating in an industrial process presupposes knowledge of the total strain involved, that is, both the homogeneous and the inhomogeneous. This is not always feasible and therefore the influence of the strain rate on the processed material may have to be gauged on the basis of the magnitude of the homogeneous deformation only. In general, however, interest centres

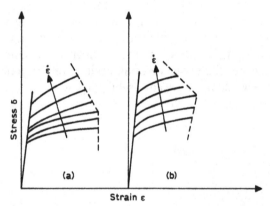

Fig. 2.8   Stress–strain to fracture curves for the basic groups of solid metallic materials.

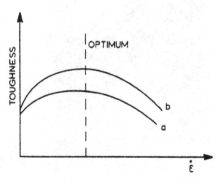

Fig. 2.9   Variation of toughness with strain rate for two groups of metals of Fig. 2.8.

primarily on two basic definitions of strain rate. Thus, reference is made either to the instantaneous rate, appropriate in material testing, or to the mean rate applicable to large plastic flow occurring in metal forming.

The instantaneous rate ε is defined simply as the ratio of the strain increment to that of the time in which the strain has occurred, or

$$\dot{\varepsilon} = d\varepsilon/dt \tag{2.6}$$

The definition of the mean strain rate raises certain problems since it is concerned only with the time required by a material element to travel through the zone in which the deformation takes place and it does not therefore reflect the variation in the velocity of the strained elements. The mean strain rate calculated on this 'time' basis will refer solely to the material element considered, but not to the total volume of the material undergoing deformation.

To obtain the mean strain rate for the whole zone of deformation, it is necessary to refer the strain rate to the distance travelled, rather than to the time taken.

These concepts are defined as follows:

(a) The time concept

$$\dot{\varepsilon}_{\mathrm{m}} = 1/t \int_{0}^{t} \varepsilon \, dt \tag{2.7}$$

(b) The distance concept

$$\dot{\varepsilon}_{\mathrm{m}} = 1/x \int_{0}^{x} \varepsilon \, dx \tag{2.8}$$

where $\dot{\varepsilon}$ is given by eqn (2.6), and t and x are the time and distance respectively.

The range of operational strain rates is large, extending from $10^{-8} \, \mathrm{s}^{-1}$ for slow creep, to $10^{6} \, \mathrm{s}^{-1}$ for hypervelocity impact. Table 2.1 is intended as a simplified guide to the classification of dynamic regimes. It provides time and strain-rate scales and indicates types of loading, the required testing facilities, and the basic considerations of the conduct of these tests.

Formability of metals reflects, naturally, their individual characteristics and the strain rate at which they are processed. A somewhat selective and simplified guide to the relative metal formability is provided by Fig. 2.10.

TABLE 2.1
Classification of Dynamic Regimes[40]

Characteristic time (s)	10^6	10^2 10^0	10^{-2} 10^{-4}	10^{-6}	10^{-8}
Strain rate (s⁻¹)	10^{-8}	10^{-4} 10^{-2}	10^0 10^2	10^4	10^6
	Creep rates	Quasistatic rates	Intermediate strain rates	High strain rates	Very high strain rates
Primary load environment	High or moderate temperatures	Slow deformation rates	Rapid loading or low velocity impact	High velocity impact or loading	Very high velocity or hypervelocity impact
Usual method of loading	Constant load or constant stress machines	Conventional hydraulic or mechanical machines	Fast-acting hydraulic or pneumatic machines, cam plastometers, low impact devices	High velocity impact devices, expanding-ring technique, high-speed metal-cutting	Light gas gun or explosively accelerated plate or projectile impact
Dynamic considerations in testing	Strain versus time	Constant strain-rate test	Machine stiffness wave effects in specimen and testing machine	Elastic–plastic wave propagation	Shock wave propagation, fluid like behaviour

←———— Isothermal ————→ ←———— Adiabatic ————→

←———— Inertia forces neglected ————→ ←———— Inertia forces important ————→

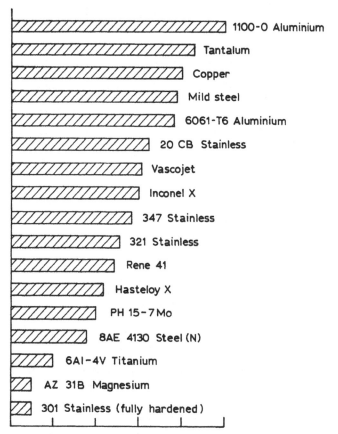

FIG. 2.10   Formability of a selection of solid metallic alloys, based on the deformation of 3·25 mm thick sheets subjected to the same shock pulse.

The dependence of material properties on the strain rate and, possibly, temperature is illustrated on the examples of an alloy steel (Fig. 2.11) and aluminium and titanium[26-28] (Fig. 2.12).

Low carbon steels constitute a problem of special interest since their ductile/brittle fracture transition temperature is easily affected by the high strain rates. This is indicated in, for instance, Fig. 2.13, obtained when explosively expanding, by increasing charges, an originally cold-drawn and annealed 0·1 C steel tubing, using a constant stand-off distance. The graph shows that the strains to fracture are much lower than those expected in a conventional bulge test of this material.

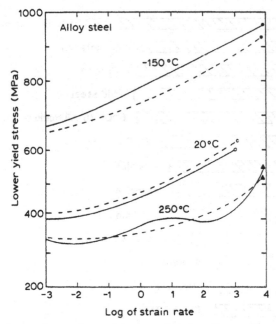

FIG. 2.11   Dependence of material properties of an alloy steel on strain rate.

Dynamic plastic deformation will produce a noticeable increase in the level of the flow stress in a variety of ordinary materials, when compared with the low strain rate deformation, as shown in, for instance, explosive welding (Fig. 2.14). More detailed data, referring to a wide range of materials, can be found in Refs 27–29.

Surface hardening of solid metallic alloys is utilised mainly in components made of Hadfield steels, e.g. railway points, stone crushers, etc., but, as will be shown later, it may introduce some practical problems when dealing with composites.

Compaction of organic and inorganic powders introduces special and rather individual effects best discussed on specific examples. It should be noted, however, that the density and hardness of many compacts are very high, the former can be of the order of 95%, and the latter can increase from some 400 Hv to 600 Hv in superalloys, and that the strength depends on the pressure applied. For well-bonded materials, values equivalent to those of the strengths of wrought materials are reached. Typically, a high strength of some 1·3 kPa — comparable with that of the wrought material — is obtained with AISI 9310 steel.

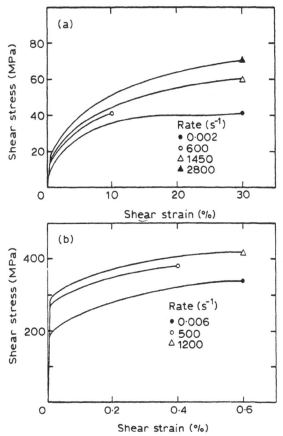

FIG. 2.12   Dependence of material properties of (a) aluminium and (b) titanium on shear strain rate.

The unique structure of dynamic compacts often gives them interesting fatigue properties. For aluminium compacts, for instance, the fatigue strength of 10^7 cycles is comparable, again, with that of the wrought material, and is equal to half the yield strength.

In consequence, mixtures of powders that would normally react together when processed at high temperatures, e.g. diamonds in steel, and, say, steel in aluminium, and alloys that would degrade at such temperatures, can be produced and will possess quite acceptable properties.

The rapid solidification of molten material on the interparticle weld boundary enables the consolidation of powder blends to take place

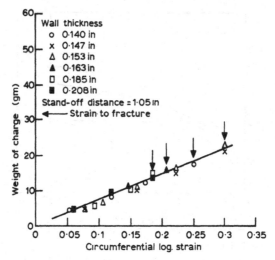

FIG. 2.13   The effect of the variation of the circumferential strain to fracture
with the level of shock pulse in 0·1 C steel.

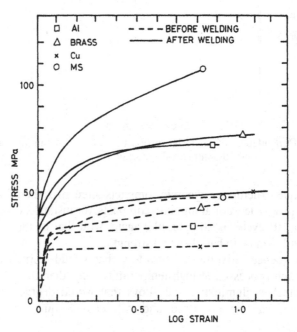

FIG. 2.14   Comparison between dynamic and static flow stresses in explosive
welding.

without, at the same time, resulting in degradation of metallurgical structures. Not only are therefore the advantageous properties of the initial powder mixture maintained, but they can also be easily enhanced. The effect of the rapid rate of quench and the short time scale in which the weld is molten, may mean that, on resolidification, a material with a very much finer structure can be formed. An illustration of this is provided by a water atomised Al 12%Si alloy (Fig. 2.15) in which the resolidified material of the weld zone possesses this characteristic.[30]

Constitutive equations of state have to be established in order that the functional stress/strain/strain-rate/temperature relationships may be used in mathematical assessments of material properties.

As suggested earlier on, the material properties at high strain rates depend not only on the strain actually imposed, but also on the strain-hardening effects and, therefore, on the thermal process control which will influence the movement of dislocations and the rate of their flow. The influence of these parameters varies with the material considered, or, at best, with the crystallographical type, and it therefore becomes

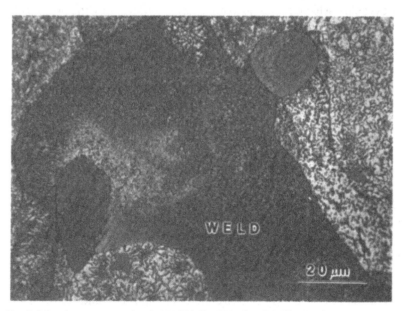

Fig. 2.15   A water atomised Al 12% Si alloy showing fine grain structure in the re-solidified weld zone (after Ref. 30).

extremely difficult to categorise metals neatly into the respective groups.

The research carried out in this field tends to support the idea that a more individual approach may be needed and the proposed semi-empirical constitutive equations are usually recommended for specific alloys and, often, ranges of the strain rate. For instance, it is found that at high rates of strain and low-to-moderate temperatures T, the steady-state relationship between the stress and strain rate for aluminium alloys is of the form

$$\dot{\varepsilon} = A[\sinh(\alpha\sigma)]^n \exp(-\Delta H/kT) \qquad (2.9)$$

where A, α and n are constants, H the enthalpy in the absence of stress, and σ the applied stress.

For alloy steel and, for example, a titanium alloy of the type Ti6Al4V the relationship takes the form

$$\sigma - \sigma_a = (\sigma_0)(\dot{\varepsilon}/\dot{\varepsilon}_0)^{\sqrt{3}\,kT/V(\sigma-\sigma_a)} \qquad (2.10)$$

where ε_0 is the nominal limiting strain rate, independent of both the temperature and the actual rate at which the test is conducted, σ_a is the stress below which thermal activation has no influence on the process of deformation, V is the activation volume involved, and k is the Boltzmann's constant. The degree of correlation between eqn (2.10) and the experimentally obtained data for steels, at a range of temperatures, is shown by dashed curves in Fig. 2.11.

Strain-rate history effects (stress–strain curves for constant rate and rate-jump situations) in FCC and BCC metals and alloys are summarised by Harding[28,31] in considerable detail.

Assessments of the validity of constitutive equations call for the development of material testing techniques, or of specially designed direct and torsional tests. At high rates of strain the application of a dynamic loading system will be responsible for the propagation of elastic stress waves in the testing machine itself and therefore the interpretation of results may prove difficult unless compensation is made either in the design of the apparatus or in the evaluation of the numerical data obtained.

The testing techniques can be sub-divided into two main groups: (a) those applicable to rapid or intermediate strain rates $(1-10^{-3}\ \mathrm{s}^{-1})$ at which the effect of wave propagation is unimportant, and (b) those at impact rates of strain (up to $10^3\ \mathrm{s}^{-1}$) for which the presence of elastic waves must be taken into account.

2.2.5.1 Intermediate strain rates
The testing apparatus used is either hydraulically or mechanically operated. The hydraulic apparatus includes biaxial tension–torsion, and tension-compression machines, whereas the mechanical testing techniques make use of the well-known Charpy pendulum and a variety of flywheel-powered equipment. Of these, the cam plastometer is of particular interest since a nearly constant, true strain rate of up to $2 \times 10^3 \, s^{-1}$ can be obtained in compression. Basically, the machine consists of a flywheel, capable of being rotated at requisite speeds, with an attached cam of a logarithmic profile. The cam is brought into contact with a platten on which the specimen rests and is thus compressed between the platten and a load cell. A similar arrangement has also been used in torsional tests on aluminium and its alloys for strain rates up to $10^3 \, s^{-1}$ and temperatures up to 600°C.

2.2.5.2 Impact strain rates
In this range of straining, wave propagation has to be accommodated. Of a number of testing techniques available, the split Hopkinson bar deserves a special mention because of its versatility. The bar, consisting of an input cylindrical element that can be activated according to the type of test performed and an output component that absorbs the impact, can be used for compressive, tensile and torsion testing. Both elements of the apparatus are fully strain gauged and the specimen to be tested is inserted between the ends of the two bars. On activating the input bar, a wave is created that is partly reflected from the surface of the specimen and partly transmitted through it. The transmitted wave is proportional to the force acting on the specimen and is recorded by the output bar; the change of the specimen dimensions and the velocity of the bar are monitored automatically. Repeated tests furnish a stress–strain curve at a given strain rate. A Hopkinson bar for compressive testing of cylindrical specimens was developed successfully at the Department of Mechanical Engineering of the University of Leeds.

A tensile test can be carried out on, for instance, a Harding and Campbell bar that uses a tubular input component with a freely sliding, inertia bar enclosed within it. Torsional split bars have also been developed. Strain rates attained in these tests are of the order of $3 \times 10^3 \, s^{-1}$.

2.2.5.3 Higher rates of straining
Strain rates of up to $4 \times 10^4 \, s^{-1}$ have been reached by introducing modifications of the specimens used in the Hopkinson arrangement and

to the apparatus itself. A solid cylindrical specimen of the preceding section is replaced either by a flat plate in which a circular hole will be punched out or by a thin strip with a double notch. The output bar is replaced by a tube and the shear strain rate is measured. In another version of the system, a projectile, fired by a gas-gun (see Chapter 1), impacts directly on a miniaturised specimen with the resulting strain rates being as high as 5×10^4 s^{-1}.

2.3 SHOCK WAVES IN COMPOSITES

2.3.1 Solid Materials

As the detonation front travels along the surface of a multilayer, possibly multimetallic, component with a velocity v_D, pressure and, in consequence, stress waves are propagated obliquely through the assembly. The waves are reflected at the interfaces and thus change in nature (tensile → compressive) and magnitude. Multireflections take place and result in a complex in its mode and magnitude, stress field.

The reflected wave intensity is dependent on the difference in the acoustic impedances of the materials involved (ρC) (where ρ is the density, and C is the velocity of propagation of the wave in the medium). With the particle velocity u in a longitudinal wave, the stress in the x-direction is

$$\sigma_x = \rho C_1 u \tag{2.11}$$

and the shear, transverse stress is

$$\tau = \rho C_2 v \tag{2.12}$$

where C_1 and C_2 are the velocities of the respective waves, and v is the particle velocity in the transverse direction,

The two velocities are given by

$$C_1 = [E(1 - \mu)/\rho(1 + \mu)(1 - 2\mu)]^{1/2} \tag{2.13}$$

and

$$C_2 = [E/2\rho(1 + \mu)]^{1/2} \tag{2.14}$$

where E is the Young's modulus, and μ is the Poisson's ratio.

Transmission and reflection of waves through a multilayer assembly of materials, varying in their densities (ρ), is shown schematically, for a single, incident compressive wave A_o, in Fig. 2.16.[32,33]

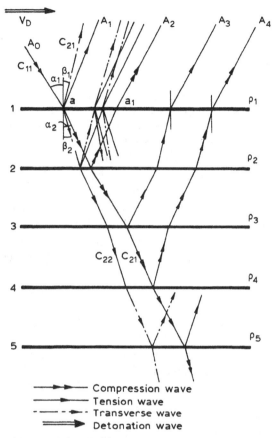

FIG. 2.16 Incidence and reflection of stress waves in a multilayer, multi-metallic assembly (after Refs 32, 33).

When reflection takes place, the relationship between stress σ_1 generated by the incident wave and that produced by the reflected wave (σ_r) is given by

$$\sigma_r = R\sigma_1 \qquad (2.15)$$

whereas that for the shear stress is

$$\tau_r = [(R+1)\cot 2\beta]\,\sigma_i \qquad (2.16)$$

R, the coefficient of reflection is defined as

$$R = [(\tan \beta \tan^2 2\beta - \tan \alpha)/(\tan \beta \tan^2 2\beta + \tan \alpha)] \qquad (2.17)$$

and, further

$$\sin \alpha / C_1 = \sin \beta / C_2 \qquad (2.18)$$

Equation (2.15) can also be expressed as

$$\sigma_r = [(\rho_2 C_{12} - \rho_1 C_{11})/(\rho_2 C_{12} + \rho_1 C_{11})] \sigma_i \qquad (2.19)$$

The reflected wave is thus seen to be dependent on the interplay between the acoustic impedances of the two materials, with R being a function of the angle of incidence. The reflected wave changes signs alternately and is either compressive or tensile.

Assuming, for the sake of argument, that $\rho_1 C_{11} > \rho_2 C_{12}$ and considering, for simplicity, the longitudinal compressive wave of general intensity A_0 only, it is seen that A_1 and A_2 are tensile and compressive respectively. It is clear that the angle of incidence constitutes the only factor that governs displacements between the points of intersection with the surfaces. The distance aa_1 between the points of incidence and reflection is then

$$aa_1 = 2h \tan \alpha \qquad (2.20)$$

where h is the thickness of the layer. The velocity of the point of reflection is $C_1 \sin \alpha$.

Two distinct possibilities can arise here. If the collision point between, say, two initially separated layers (stand-off distance) and the incident wave travel at the same velocity ($v_D = C_1 \sin \alpha$) the interference will occur between the tensile reflection of A_0 and the compressive reflection of A_1. A similar interference, in reverse order, will be produced at the next interface if $\rho_2 C_{12} > \rho_1 C_{11}$. However, if $v_D > C_1 \sin \alpha$, a situation may arise in which a reflected wave behind the collision point produces a high stress intensity which, in turn, may result in damage to the interface.

The optimal condition, suggested by Blazynski and El-Sobky[33,34] is

$$v_D < C_1 \sin \alpha \qquad (2.21)$$

2.3.2 Particulate Matter
Although, as already indicated, the transmission of a shock pulse through a material element can produce a wide variety of effects, the most important one, insofar as particulate composite matter is concerned, is that of consolidation and compaction.

The mode of passage of a shock wave through an aggregate of particulate matter differs from that in a solid material in that the pulse is not plane in its configuration, but, because of the variation in particle size and spacing, is rather complex and it progresses in jumps from particle to particle. In consequence, the degree and nature of consolidation of the material vary across its section and depend on particle size, shape and morphology, on its initial density, the pattern of grain or particle distribution in the composite, and, finally, on the geometry of the shock front.

The individual and/or combined effects of these parameters increase with decreasing metallic content of the mixture, because the response of the individual particles is conditioned by their mechanical and geometrical properties, as well as by the characteristics of the compacting system. These are reflected in the disposition of the particles with respect to each other, and, further, by the material and geometry of the container in which compaction takes place.[35,36]

The relative movement of the elements of the assembly is caused and characterised by the degrees of their individual accelerations — and therefore their individual stand-off distances — the nature of their collisions, randomness of the directionality of impacts, and by their different surface morphologies. The shape of the shock wave and the associated physical properties of the front will determine the final response of any of the sections of the composite affected by the passage of the wave.

It can be said therefore that the three basic parameters determining the quality of shock consolidation of a powder are (a) the explosive to powder mass ratio (defining the level of energy available and the mode of its dissipation), (b) the velocity of detonation and the associated acoustic impedance, and (c) the rate of dissipation of the energy supplied.

Three possible situations arise with respect to the geometry of the shock wave. If a uniform pressure distribution across a section of a composite aggregate is achieved, a homogenised structure will result. However, either under- or overcompaction will occur if this condition is not realised. The three cases are illustrated in Fig. 2.17 on an example of a compact of PVC polymeric powders. Here, a hyperbolic shock front (Fig. 2.17(a)), associated with a reduction in pressure across the compact, produces undercompaction in the centre that is reflected either in the presence of loose powder in the core or in a pipe (see cross-sectional micrograph). A parabolic type of the shock front, consistent with increase in the pressure in the core of the compact (Fig. 2.17(c)) results in two specific effects. On the one hand, the concentration of the pressure pulse in the core produces localised high tempera-

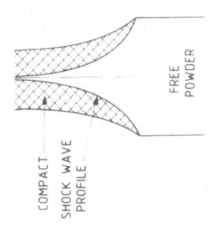

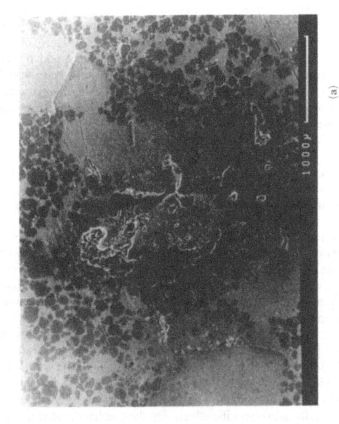

(a)

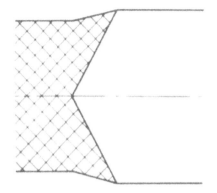

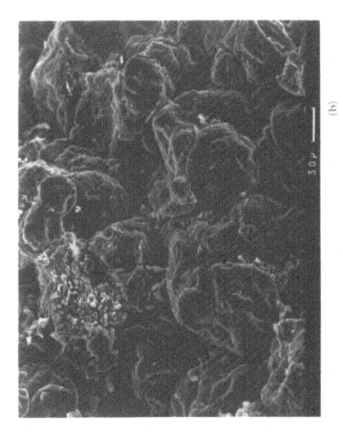

(b)

FIG. 2.17 The effect of the geometry of the shock front on the quality of a particulate compact of a blend of PVC homo- and copolymer powders. (a) Undercompaction (presence of pipe); (b) uniform pressure distribution; (c) overcompaction (melting).

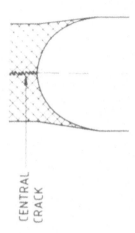

CENTRAL CRACK

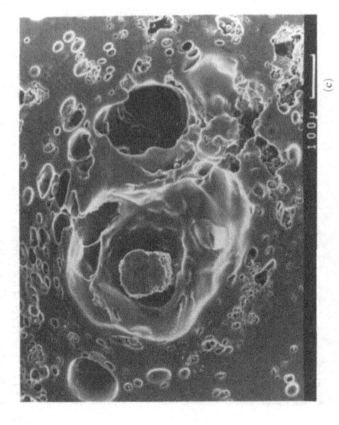

(c)

100μ

Fig. 2.17. — contd.

ture and the consequent melting (especially in organic materials), but, on the other, the presence of a high intensity tensile, release wave results in the cracking of the already compacted product (Fig. 2.18). Containment of the release wave is thus seen as being of primary importance and it is therefore under these circumstances that the material and geometry of the container can play a decisive role. However, if a conical shape of the shock wave (Fig. 2.17(b)) can be generated, a uniform

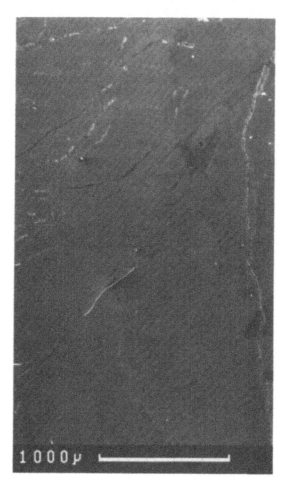

FIG. 2.18   Cracking of a SiO_2 consolidated compact caused by a tensile release wave.

pressure distribution across the specimen will produce a satisfactory degree of compaction throughout the body of the compact.

The presence of interparticle bonding constitutes the main and the most important difference between the shock compacted composites and the statically or isostatically consolidated materials. The occurrence, type and level of bonding depend entirely on the local effects of the passage of the shock front and on its transformation into a plastic deforming wave. A simplified model of transformation of the shock wave was proposed by Myers and Murr[1] (Fig. 2.19). A loose aggregate of particulate matter (Fig. 2.19(A)), when subjected to a shock front (Fig. 2.19(B)), experiences the closing of the interparticle voids by lateral flow of the material. However, with the resulting disappearance of the state of uniaxial strain, at least on the particle boundaries, the shock wave turns to a pure plastic one with shock conditions existing only in the cores of individual particles. The velocity of the shock wave is reduced and its energy is dissipated as heat in practically adiabatic conditions.

The bonding of particles is effected by a combination of three distinct mechanisms that come into play when the aggregate is shocked. In addition to the adiabatic heat already generated, high speed interparticle

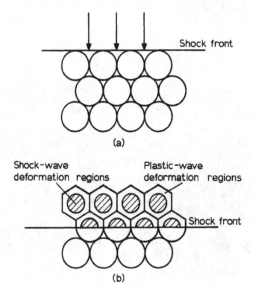

FIG. 2.19   An idealised model of the propagation of a shock wave through particulate matter (after Myers and Murr).[1]

collisions may create a sufficiently high temperature rise to produce local melting. Fusion welding will then be present (Fig. 2.20(a)). Equally, the high velocity collisions combined with considerable surface morphology variations of composite materials can result in the creation of 'hot-spots' that lead to the localised friction welding (Fig. 2.20(b)). High, local pressure pulses accelerating individual elements of the assembly and resulting in angled interparticle impacts, can exceed those of the propagating detonation front. Hydrodynamic flow that originates on the affected surfaces produces jetting and therefore the conditions which normally obtain in explosive welding[6,21,25] (Fig. 2.20(c)). Any or all of the bonding links may exist at any time as shown, for instance, in Fig. 2.20(b) where both fusion and friction welds are present.

The presence and concentration of the effects of the three mechanisms is random and therefore the properties of a compact, in any of its sections, will vary slightly, but the overall effect is that of considerably increased strength and density.

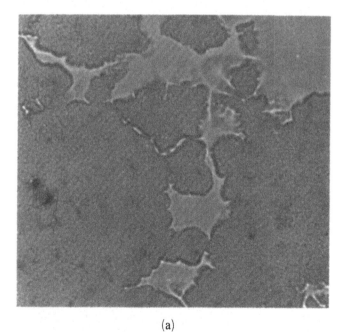

(a)

FIG. 2.20   Three mechanisms of interparticle shock bonding of polymeric powders. (a) Fusion welding; (b) friction and fusion welding; (c) explosive welding.

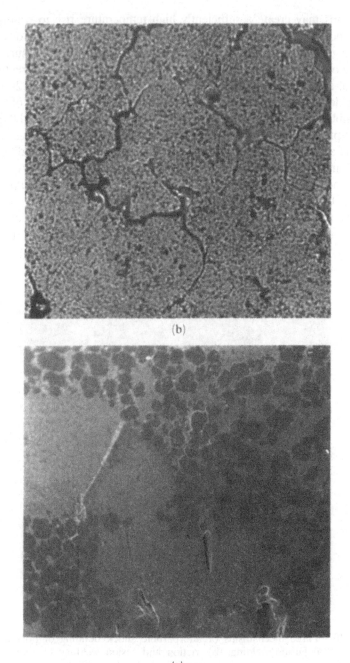

(b)

(c)

Fig. 2.20. — contd.

Any quantitative assessment of the conditions developed when a shock wave is transmitted through a particulate composite, is based essentially on the mass, momentum and energy conservation principles defined by eqns (2.1)–(2.4). However, a simplified approach, proposed by Herman[37] and further developed by Raybould,[38] can be used. Rather than determining the Hugoniot for a specific material by making use of the four equations, it is assumed that the pressure developed is not significantly higher than that needed for full densification. Since the pressure is lower than the value usually taken in a standard hydro-dynamic approach, it can be further assumed that no significant change in the internal energy occurs. In these circumstances, only the mass and momentum equations, in conjunction with the quasistatic pressure–density relationships (obtainable from the manufacturers), are required. The latter is in the form of

$$\rho = \rho_0[a + (1 - a)/(1 + b)]^{-1} \tag{2.22}$$

where a is the initial fractional density of the powder and b defines the powder stiffness. By combining eqns (2.1), (2.2) and (2.22), the shock velocity is obtained as

$$U = C_0(1 + abp_0)[(1 + bp)/(1 - bp_0)]^{1/2} \tag{2.23}$$

where $C_0 = [(1 - a) b\rho_0]^{-1/2}$.

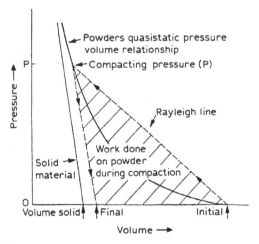

FIG. 2.21   Schematic pressure–volume relationship for powder (loading along the Rayleigh line, elastic unloading from the new density).

Schematically, the resulting Hugoniot is given by Fig. 2.21. The determination of the increase in the surface temperature ΔT_m and the average temperature rise T_A of a particle are of practical importance.

Taking into account a 95% efficiency factor, the rise is given[21] by

$$\Delta T_A = [0 \cdot 475 p (1 - \rho_o / \rho)] / a \rho_f c \qquad (2.24)$$

where c_P and ρ_f are the specific heat and density of the solid respectively.

For an approximately spherical particle of radius r, the heat Q stored after compaction is

$$Q = 4 \pi r^3 \rho_f c_P \Delta T_A / 3 \qquad (2.25)$$

The additional heat associated with the rise of ΔT_s is

$$Q_1 = \int_0^D R c_P \rho_f \Delta T_s (1 - x^2 / D) \, \mathrm{d} x \qquad (2.26)$$

where x is the distance from the surface, and

$$\Delta T_M = T_M (1 - x / D) \qquad (2.27)$$

and, further

$$D = (12 \, K t / \rho_f c_P)^{1/2}$$

t being the rise time and K the thermal conductivity.

If the work of compaction is associated primarily with the interface, the temperature on the surface is basically the same as the melting temperature.[21] In consequence, eqns (2.25) and (2.26) when equated, and a substitution for D is made, give the time t as

$$t^{1/2} = r \Delta T_A (\rho_f c_P / 12 K)^{1/2} / \Delta T_M \qquad (2.28)$$

The following approximate relationship links the time t with the particle diameter D and shock front velocity U

$$t \approx (3 D / U) \qquad (2.29)$$

On combining eqn (2.23) and (2.29) with (2.24) and (2.28), the following expression for the qualifying factor R is given

$$R = S [2 r (1 - a) (a^{2 \cdot 5} b^{2 \cdot 5} T_M (c_P K \rho_f^{1 \cdot 5})^{-1}] (b p)^2 (1 + b p)^{1/2} \qquad (2.30)$$

where T_M is the particle melting point, and S is the shape factor accounting for the irregularity of particle surfaces.[39] When $R = 1$ complete interparticle melting is expected.

REFERENCES

1. Myers, M. A. and Murr, L. E. In: *Explosive Welding, Forming and Compaction*, Ed. T. Z. Blazynski, Applied Science Publishers, London, 1983, p. 17.
2. Johnson, W. and Yu, T. X. In: *Plasticity and Modern Metal-Forming Technology*, Ed. T. Z. Blazynski, Elsevier Applied Science, London, New York, 1989, p. 73.
3. Johnson, W. *Impact Strength of Materials*, Edward Arnold, London, 1972.
4. Abu-Sayed, A. *Analytical and Experimental Investigation of Pressure-Shear Waves in Solids*, PhD thesis, Brown University, 1972.
5. Rogers, H. C. and Shastry, C. V. In: *Shock-Waves and High Strain-Rate Phenomena in Metals*, Eds E. L. Murr and M. A. Myers, Plenum Press, New York, 1981, p. 285.
6. Dormeval, R. In: *Materials at High Strain Rates*, Ed. T. Z. Blazynski, Elsevier Applied Science, London, New York, 1987, p. 47.
7. Shockey, D. A. In: *Metallurgical Applications of Shock-Wave and High Strain-Rate Phenomena*, Eds L. E. Murr, K. P. Staudhammer and M. A. Myers, Marcel Dekker, New York, Basel, 1986, p. 633.
8. Murr, L. E. and Staudhammer, K. P. In: *Shock Waves for Industrial Applications*, Ed. L. E. Murr, Noyes Publications, Park Ridge, NJ, 1988, p. 1.
9. Stelly, M. and Dormeval, R. In Ref. 7, p. 607.
10. Klepaczko, J. R., Lipinski, P. and Molinari, A. In: *Impact Loading and Dynamic Behaviour of Materials*, Eds C. Y. Chiem, H. D. Kunze and L. W. Myer, DGM Informationsgesellschaft Verlag, Oberursel, FRG, 1988, p. 695.
11. Duvall, G. E. and Graham, R. A., *Rev. Mod. Phys.*, **49** (1977), 523.
12. Murr, L. E. In Ref. 6, p. 1.
13. Myers, M. A. *Scripta Met.*, **12** (1978), 21.
14. Murr, L. E., Inal, O. T. and Morales, A. A. *Appl. Phys. Letters*, **28** (1976), 432.
15. Kressel, H. and Brown, N. J. *J. Appl. Phys.*, **38** (1967), 1618.
16. Champion, A. R. and Rhode, R. W. *J. Appl. Phys.*, **41** (1970), 2213.
17. Murr, L. E. and Staudhammer, K. P. *Mater. Sci. Engngr*, **20** (1974), 95.
18. Rhode, R. W. *Acta Met.*, **17** (1969), 353.
19. Staudhammer, K. P., Frantz, C. E., Hecker, S. S. and Murr, L. E. In Ref. 5, p. 91.
20. Murr, L. E. and Myers, M. A. Private communication, 1989.
21. Raybould, D. and Blazynski, T. Z. In Ref. 6, p. 71.
22. Graham, R. A., Morosin, B., Venturini, E. L. and Carr, M. J., *Ann. Rev. Mater. Sci.* (1986).
23. Davison, L. and Graham, R. A. *Phys. Rep.*, **55** (1979), 255.
24. Bergmann, O. R. In: *Proc. 7th HERF Conference*, Ed. T. Z. Blazynski, University of Leeds, 1981, p. 142.
25. Prümmer, R. In Ref. 1, p. 369.
26. Harding, J., Wood, E. O. and Campbell, J. D., *J. Mech. Engng Sci.*, **2** (1960), 88.

27. HARDING, J. In Ref. 1, p. 123.
28. HARDING, J. In Ref. 6, p. 133.
29. MURR, L. E. In Ref. 8, p. 60.
30. RAYBOULD, D. Private communication, 1989.
31. HARDING, J. In Ref. 10, p. 23.
32. EL-SOBKY, H. and BLAZYNSKI, T. Z. In: *Proc. 15th Int. MTDR Conference*, Eds S. A. Tobias and F. Koenigsberger, Macmillan Press, London, 1975, p. 399.
33. BLAZYNSKI, T. Z. In Ref. 1, p. 289.
34. BLAZYNSKI, T. Z. and EL-SOBKY, H. In: *Proc. 5th Int. HERF Conference*, University of Denver, Colorado, 1975, Paper 4.6.
35. HEGAZY, A. A. and BLAZYNSKI, T. Z. *Int. J. Imp. Engng*, **6** (1987), 63.
36. BLAZYNSKI, T. Z. and HEGAZY, A. A. In Ref. 10, p. 979.
37. HERMAN, W. *J. Appl. Phys.*, **40** (1969), 2490.
38. RAYBOULD, D., *Int. J. Powder Met. Powder Technol.*, **16** (1980), 9.
39. LEMCKE, B. and RAYBOULD, D. US Patent No. 4 255 374.
40. LINDHOLM, U. S., *Techn. Metals Res.*, **V** (1971), 199.

Chapter 3

Solid Matter Composites

3.1 INTRODUCTION

Of its very nature, the manufacture of integrally formed solid matter composites calls for the employment of explosive welding techniques which, by providing conditions of high velocity impact, result in solid-phase welded joints. The strength of the weld is enhanced by the mechanical locking produced by the, inherently, wavy form of the surface deformation (Fig. 3.1), caused by jetting, that originates at the collision point between the elements of the assembly collapsing onto each other, and by extensive diffusion at the explosion weld interface.

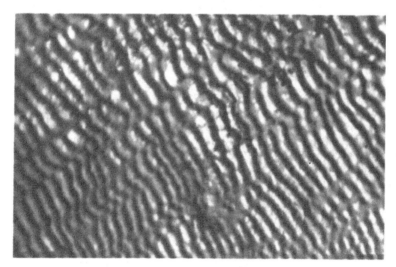

FIG. 3.1   Wavy interface of an explosively processed surface.

The enhancement of diffusion is attributed to the more thermally stable defect structure that results from the intense plastic deformation and flow at the metal interfaces.[1] It is generally held that subsonic plastic wave propagation velocities are preferable to the supersonic, if the quality of the weld is to be satisfactory.

The use of explosive systems, in preference to mechanical dynamic arrangements, is dictated by practical considerations. The high pressures and high levels of energy, associated with solid-phase welding, combined with the relatively low developed temperatures and the high rate of heat dissipation are the requirements which, in contrast to the complicated and very expensive mechanical equipment, are easily and cheaply satisfied by the explosive systems.

Explosive forming processes, on their own, are of limited interest in the manufacture of composites since they can only produce a material combination that, generally, is just mechanically assembled. In special situations, however, when the forming is combined with a welding or, at least a residual stress building operation, interesting composite structures can be produced (see Chapter 5).

A schematic classification of the basic processes, operated primarily in the explosive welding mode, is given in Fig. 3.2.

Both types of processes are, naturally, complex in their respective characteristics and material response and, consequently, a detailed study of their nature is outside the scope of this book. An extensive

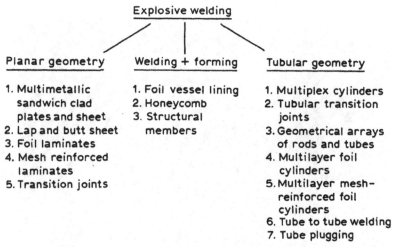

FIG. 3.2   Schematic classification of explosive welding processes.

volume of published information is in existence and the operational details and descriptions are provided in, for instance, Refs 2 and 3, both of which contain lists of further references.

The success of any operation, as well as the appreciation of the way in which a material is likely to respond to the dynamic plastic flow, depend on the understanding of the role of the basic parameters involved. This includes knowledge of the characteristic properties of explosives, of the corresponding levels of energy available, and of the mode and rate of its dissipation. A clear view of the known mechanisms of welding and of the basic geometrical configurations used is also needed together with an appreciation of the physical realities that follow from any particular system geometry employed. Similar observations apply also to the forming techniques.

Brief outlines of these concepts — insofar as they apply to and are helpful in following the arguments developed later in connection with composites — will be given in the following sections of this chapter.

3.2 EXPLOSIVES AND DETONATION PHENOMENA

3.2.1 Classification of Explosives
From a practical point of view, commercial chemical explosives will generally fulfil the requirements of the welding, forming or compacting operations. On detonation, the detonating front travels with a velocity v_D through the explosive converting its mass to a high temperature, high pressure/stress wave and, depending on the conditions, producing a shock and release wave. Chemical explosives are sub-divided into high and low (deflagrating) materials.

3.2.1.1 High explosives
High explosives are characterised by very high detonation pressures and high rates of reaction. This class can be further sub-divided into two groups:

(i) Primary high explosives (or detonating high explosives) which are very sensitive and may be detonated by slight impact, heat, flame, static electric charge, or simple ignition. They are usually used in detonators, but seldom as a source of energy in metal working operations.

(ii) Secondary high explosives are used mainly in metal working and other industrial applications. They usually require a detonator to initiate the reaction and sometimes a booster charge to reinforce the detonation wave. In terms of energy, they have a higher content than primary explosives (Table 3.1).

3.2.1.2 Deflagrating or low explosives

These materials burn rather than detonate on the initiation of the reaction and produce large amount of gases and much lower pressures. They usually contain their own oxygen supply, and burn easily, but in some types the reaction is difficult to initiate. Their low rate of burning makes them excellent propellants. Fire risk is considerable when handling such materials.

The most commonly used explosives are listed below:

Primary high explosives

* mercury fulminate,
* lead azide,
* diazodinitrophenol,
* lead styphnate,
* nitromannite.

Secondary high explosives

* TNT (trinitrotoluene),
* tetryl (trinitrophenyl-methylnitramine),
* RDX (cyclotrimethylenetrinitramine),

TABLE 3.1
Explosives

	High explosive	Low explosive
Method of initiation	Primary high explosives — ignition, spark, flame or impact. Secondary high explosives — detonator, or detonator and booster combination	Ignition
Conversion time	Microseconds	Milliseconds
Conversion rate	1830–8500 m/s	A few cm to a few m/s
Pressures	Up to 28 GPa	Up to 0·3 GPa

- PETN (pentaerythritol tetranitrate),
- ammonium picrate,
- picric acid,
- ammonium nitrate,
- DNT (dinitrotoluene),
- EDNA (ethylenediaminedinitrate),
- NG (nitroglycerine),
- nitro starch.

Low explosives

- smokeless powder,
- NC (nitrocotton),
- black powder (potassium nitrate, sulphur, charcoal),
- DNT (dinitrotoluene ingredient).

Non-explosive ingredients used in explosive mixtures

- aluminium,
- waxes,
- diphenylamine,
- metal nitrates,
- mononitrotoluene,
- metal nitrates
  metals (aluminium, ferrosilicon),
  wood pulps and other combustibles,
  paraffin and other hydrocarbons.

3.2.1.3 Detonators
Detonators are devices used to initiate an explosive charge. Electric detonators are widely used and are safe to handle, if reasonable precautions are taken. The detonators consist basically of a thin metallic container, usually copper, protecting their sensitive contents of an initiating or primary high explosive and a small quantity of a sensitive secondary explosive, e.g. PETN or tetryl, totalling about 1 g. Initiation is achieved electrically using an exploding bridge wire. Some types contain a slow burning material to provide a time delay when many charges have to be fired at different intervals.

3.2.1.4 Forms of available explosives
Explosive materials are available in different forms and some in more than one. Many of them can be melted, thus allowing other explosives to

be added in the form of slurries. For instance, cyclotol is made by slurrying the granular RDX in molten TNT or in Composition B. Powder, granular, solid, liquid and plastic explosives are available. One of the most useful types is the so-called 'Datasheet' or 'Metabel', which is essentially a PETN explosive combined with other ingredients to form a tough, flexible waterproof sheet that can be cut and shaped to the required size for contact and stand-off operations. This material is available in different thickness sheets which can be glued to each other or used as the shaping back-up material, if a shaped charge is required. Another very useful type is the powder explosive and particularly so the mixtures of TNT with aluminium powder which can be used to fill a container of any geometry and can then be compacted to attain higher densities. These do not require the use of sealed containers which gives them an advantage over the liquid explosives. 'Cord' explosives, e.g. Cordtex (ICI), are also available. These consist of a flexible cord containing an explosive core. They are very useful when continuous long charges are required, and provide a means of charge application with reasonable accuracy in a number of forming operations.

3.2.2 Effectiveness of the Charge

This depends on the characteristics of the explosive, as reflected by the pressure/time function, velocity of detonation, mass ratio of explosive to specimen, stand-off distance and the transmitting medium. The relevant properties of the more commonly used high explosives are given in Table 3.2.

Figure 3.3 shows the basic difference between the shapes of the pressure/time distribution curves for typical high and low explosives, whereas Fig. 3.4 indicates the effects of the explosive mass and that of the transmitting medium, as well as of the stand-off distance.

Contact operations, i.e. those in which the charge is in direct, or nearly direct (through the medium of a thin buffer) contact with the specimen are used in explosive welding operations. Stand-off systems, characterised by the charge(s) being applied at a distance from the target, are more appropriate in the forming processes. The compaction of particulate matter falls in-between these two operations.

3.2.3 The Detonation Process

3.2.3.1 Introduction

Detonation is a term used to describe the process in which an explosive charge undergoes a chemical reaction accompanied by a

TABLE 3.2
Properties of More Commonly Used High Explosives

Explosive	Melting or softening point (°C)	Max working temp. (°C)	Detonation		Products of explosion	
			Pressure (MPa)	V_D (m/s)	Spec. vol. (dm³/kg)	Spec. heat (MJ/kg)
Ammonium nitrate	170	125	—	2690	968	1·44
Picric acid	122	125	18·3	7345	668	4·16
PETN	139	100	22·0	8290	781	5·76
Tetryl	130	100	20·0	7833	753	4·66
TNT	80	115	15·5	6890	724	3·85
RDX	202	100	23·4	8380	900	5·40
Amatol	80	115	—	6400	850	4·07
Pentolite	80	100	19·3	7430	—	5·07

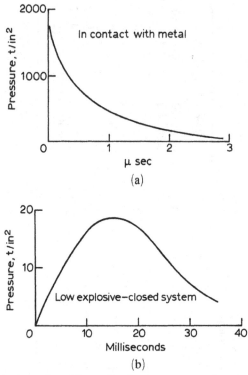

Fɪɢ. 3.3   Typical pressure/time distribution curves for the products of explosion: (a) high explosive; (b) low explosive.

characteristic type of shock wave (or detonation wave). Depending on the properties and type of explosive material, the velocity and intensity of the characteristic shock wave will vary, but will remain constant for a given type of explosive and for a charge of uniform geometry and density. This simplifies the mathematical solution of the hydrodynamic theory which applies to the process. The general behaviour of a primary explosive during reaction is characterised by a slow combustion process at the beginning, and then by deflagration, up to a point of a sudden transition to detonation. The whole process is completed in a few microseconds. However, the rate of the build-up of the reaction and of the transfer to detonation, when a secondary explosive is detonated without a detonator, is much slower and burning before detonation may occur.

Low explosives are characterised by the absence of the transition period. They react at rates which are proportional to the build-up of

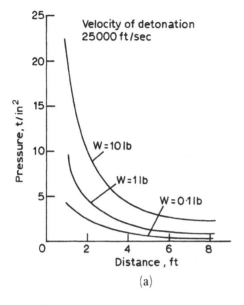

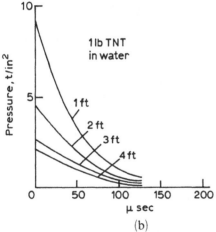

FIG. 3.4   The effect of the mass of explosive (a) and that of the stand-off distance (b) on the pressure generated.

pressure. This, in turn, increases as the chemical reaction increases. A cycle like this leads to an explosion within a small fraction of a second, but the rate of reaction is usually much slower than 1% of that in detonation, and peak pressures also attain lower levels. However, energy comparable with that obtained from high explosives can be generated by the low explosives when they are adequately confined or used in sufficient quantities. The pressure distribution is easily controllable.

3.2.3.2 Parameters of the detonation process

Insofar as metal working processes are concerned, the most important parameters of the detonation process are:

(i) The total amount of energy (E) released by the detonation of an explosive charge.
(ii) The detonation velocity (v_D), i.e. the velocity of propagation of the detonation front.
(iii) Pressure (P) or the transient pressure exerted by the gaseous products of detonation on a solid body.

Other aspects of the process, such as thermal stability, sensitivity of the explosive, temperature, heat generation and ionisation phenomena in the gaseous products and their composition are not of any importance in the manufacture of composites.

There exist a number of experimental methods and instruments employed in the various techniques associated with the determination of detonation velocity and pressure.

In a steady state detonation process, the wave travels at a constant velocity that is determined by the thermodynamic laws, in terms of the energy released, the rate of release, the density, and, finally, the size of the charge.

The manner of propagation of the detonation wave varies from a helical path, known as spinning or helical detonation, to straight or spherical propagation. There are two discrete velocities of propagation which may differ by a factor of five. The ideal velocity corresponds to the hydrodynamic value V_D^*, the theoretical maximum, and is a function of the explosive density ρ_1 only,

$$V_D^* = D(\rho_1) \tag{3.1}$$

A non-ideal detonation refers to a steady state wave propagation at $V_D < V_D^*$. If a boosting charge is used, a greater value of V_D is expected, but this is of a transient nature which will soon settle to the steady state value. The velocity of detonation may also increase gradually or discretely either because of the variation in the charge itself, or because of a discontinuity in the process or the operational system.

The parameters of the detonation process are related to the hydrodynamic equation and thermodynamic laws as follows (Fig. 3.5):

$$P_2 = \rho_1 V_D u_2 + P_1 \tag{3.2}$$

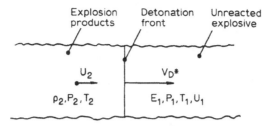

FIG. 3.5 Schematic representation of physical changes taking place on the detonation of an explosive charge.

On the other hand, on the Chapman–Jouguet (C–J) plane, which separates the reaction zone from the detonation products, the following relation obtains:

$$u_2 = V_D(1 - \rho_1/\rho_2) \qquad (3.3)$$

where P_1 is the initial pressure, P_2 is the detonation pressure, ρ_1 is the original density of the explosive, ρ_2 is the density immediately after detonation, V_D is the detonation velocity and u_2 is the particle velocity in the detonation plane.

Equation (3.2) can be used only by measuring either u_2 or ρ_2, but it can be written in the form

$$P_2 = V_D^2 \rho_1 (\rho_2 - \rho_1)/(\rho_2 + P_1) \qquad (3.3a)$$

It is thus seen that the pressure is more sensitive to the detonation velocity than to the density of the explosive.

One of the more accurate methods of measurement of ρ_2 is that due to Gehring and Dewesy, quoted by Cook[4] as follows. 'The surface of the metal behind the detonation front in an oblique shock wave configuration (Fig. 3.6) can be considered to be a stream line in the flow of the gases at a certain angle α between the detonation front and the normal to the original surface. The normal component of stress on the deformed interface equals the detonation pressure. This can be computed from the following equations:

$$V_D \tan \alpha = (V_D - u_2) \tan(\alpha + \delta) \qquad (3.4)$$

$$P_2 = (\rho_1 V_D^2 \sin \delta)/\cos \alpha \sin(\alpha + \delta) \qquad (3.5)$$

where α is the critical angle and δ is the angle of depression.

In the case of normal incidence on the metal surface, the direct reflection will cause a large increase in both the pressure and particle velocity, and so, the density of the detonation gases will increase

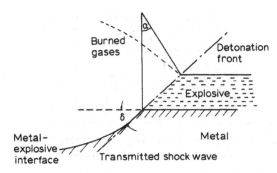

FIG. 3.6   Explosive metal interface (after Ref. 4).

proportionally. The other extreme situation is reached when the wave travels parallel to the surface of the metal. A rarefaction wave is then produced and a reduction in the average effective pressure is observed in contrast to the previous case.'

The critical value of α, at which the average effective pressure is a maximum with no reflection or rarefaction, can be determined by direct observation from, say, flash radiographs, or by plotting the pressure, as determined by the angle δ, against α. The resulting curve is smooth up to the required value of α. For some explosives, the error in the pressure estimated by this method is of the order of 5% as compared with pressures computed from the standard equation of state:

$$PV = nRT + \alpha(T, V)P \tag{3.6}$$

In many cases, approximate estimates of the detonation process parameters are quite sufficient. Under such circumstances, the following approach can be used:

(i) For most explosives the changes in the ratios ρ_2/ρ_1 and V_D/u_2 are so small that it can be assumed, with sufficient accuracy, that for condensed explosives, in air: $\rho_2/\rho_1 \simeq 4/3$, and $V_D/u_2 \simeq 4$, and that therefore

$$P \simeq 0{\cdot}0098\,\rho_1 V_D^2/4 \tag{3.7}$$

(ii) A better approximation is obtained in terms of the empirical quantities ρ_1 and V_D^*:

$$P_2^* \simeq 0{\cdot}00987\,V_D^{*2}(0{\cdot}380 - 1270\,\rho_1)/V_D^*\rho_C \tag{3.8}$$

where ρ_C is the crystal density of explosives, V_D^* is the ideal detonation velocity (m/s), and P_2^* is the ideal detonation pressure (atm).

The equation is based on the experimental observation that

$$V_2^*/V_1 = \rho_1/\rho_2^* \approx 0{\cdot}62 - (1270\,\rho_1/V_D^*\rho_C) \qquad (3.9)$$

where V_2^* is the ideal volume of detonation products. For non-ideal detonation, eqn (3.8) becomes:

$$P_2 = 0{\cdot}00987\,\rho_1\,V_D^{*2}[0{\cdot}380 - (1270\,\rho_1\,V_D/V_D^{*2}\rho_C)] \qquad (3.8a)$$

Equation (3.8) gives better results for large quantities of explosives, when the detonation velocity approaches the ideal value, i.e. the maximum theoretical value determined from the thermodynamic characteristics of the process. The choice between eqns (3.7)–(3.9) depends on the knowledge of the detonation velocity.

The pressure on the C–J plane has a characteristic value for each type of explosive. In the reaction zone, the pressure decays within a few microseconds and, hence, the mean effective pressure is lower than the peak value. However, for a given system, the two are often related.

The temperature of the detonation products is usually very high, and lies in the range of 2–6×10^3 K. Being of a highly transient nature, the duration of the peak temperature is of the order of a few microseconds. The shape of the function $\alpha(T, V)$, in eqn (3.6), cannot be determined from a comparison between the calculated and measured values of pressure and density, because they are too insensitive to the form of the function. An experimental determination of the temperature T_2 has to be carried out. Because of the short duration, the transient nature, and high magnitude of T_2, as well as of the high pressure of the detonation gases and their chemical composition, it is very difficult to measure the temperature experimentally. However, some estimates of the temperature in the C–J plane are available for a number of commercial and military explosives.

The C–J postulate is based on the observation that a complete solution to the detonation (shock) wave requires that the following relationship be established:

$$V_D = u + c \qquad (3.10)$$

For condensed explosives, in which P_1, P_2 and the variation of the exponent n with density, are small, consideration of eqns (2.1)–(2.4), (3.6), and of the Second Law of Thermodynamics, leads to the following relationships:

$$V_D = V_1(P_0 + 1)(nRT_2)/(V_1 - \alpha)\beta \qquad (3.11)$$

$$u = (nRT_2)^{0{\cdot}5}/\beta \qquad (3.12)$$

$$T_2 = (Q + T_1 c_v)\, \beta / (\beta c_v - 0{\cdot}5\, nR) \tag{3.13}$$

where c_v is the average specific heat, Q is the heat, α is the covolume and β is given by

$$\beta = (c_v - nR)/c_v - (\alpha/V_1)_s + (nR/c_v P_2)(E/V)_T \tag{3.14}$$

where the subscript s denotes static conditions.

The geometry of the charge, and the fact of whether it is confined or unconfined, have considerable effect on the shape of the detonation front and on the process of energy transfer to the surrounding medium. In the steady state phase, the detonation front is often assumed to be a plane wave. This configuration is widely used in the literature to facilitate the analysis. It has to be borne in mind, however, that this is an approximation because plane waves do not actually exist in condensed explosives.

The front of a wave emerging from the end of an unconfined cylindrical charge is, in general, spherical in shape in both ideal and non-ideal detonation. The radius of curvature of such a front increases at first, but soon settles to a steady value, the ratio of which to the diameter of the charge can vary between 0·5, at the critical diameter, and a maximum of 3–4 when the charge diameter is far greater than the critical. At a long distance from the initiation point, the wave is independent of the mode of initiation or of the initial wave profile. Although the detonation wave can be given almost any required shape by using 'shaping booster charges' (Fig. 3.7), or by initiating the detonation on a solid surface of any arbitrary shape, it tends to settle to the steady state profile at a distance far from the point of initiation. Since the wave front is immediately followed by a rarefaction, the reaction part forms a truncated zone, with rarefaction waves at the sides, and then, as the detonation continues and the steady state is reached, it gradually assumes a completely conical shape. When a fully developed cone is formed, the release, or the rarefaction waves reach the axis of the charge at a distance of about one diameter, after the wave has travelled a distance $L_M = 3{\cdot}5\ d$. At distances $L \geqslant L_M = 3{\cdot}5$, along the charge, the detonation head propagates in a condition of steady state, and the wave profile remains unchanged because, by then, all the effects of the point of initiation will have been eradicated by the influence of the lateral rarefaction.

The thickness of the reaction zone and the volume of the detonation head can be estimated for different geometries, other than the cylindrical (see Ref. 4).

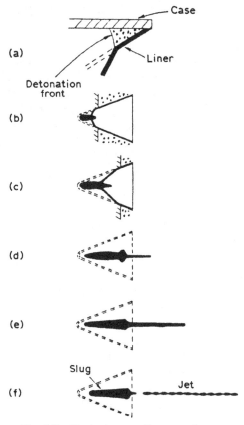

Fig. 3.7 Basic shapes of booster charges.

The ratio of the mass M_H of the detonation head to that of the whole charge M_C is a measure of the efficiency of the charge in accelerating a material element.

Since the surface (A) of the element affected by the detonation head has to be taken into account, the ratio M_H/A (in g/cm^3) represents the surface density of the material of optimal thickness h required for the maximum utilisation of energy.

Hence,

$$M_H/A = \rho h_S \qquad (3.15)$$

The maximum possible value of the ratio M_H/M_C is 0·5.

3.3 CHARACTERISTICS OF THE WELDING MECHANISMS

3.3.1 Introduction

In its simplest form, an explosive welding arrangement consists of a rigidly supported base element, a flyer element or plate initially separated from the base by an air gap (Fig. 3.8), and an explosive charge in direct, or via a buffer, contact with the surface of the flyer.

On exploding the charge, the detonation front will travel, in finite time, with a velocity V_D along the surface of the flyer causing the latter to collapse onto the base. Geometrically, the collision is oblique and takes place at an initial angle α. The flyer collapses with a velocity V_P, but at the point of collision the shearing stresses developed in the materials of both elements are so high that they can not be sustained. Liquefaction and jetting are present and a solid phase weld is formed.

The nature of the weld depends on the level of energy delivered, the stand-off distance, the respective acoustic impedances of the participating materials, the profile and property of the plastic wave generated, and the degree of rigidity of the base support.

Irrespective of the interplay between these parameters in any given situation, the following sequence of events will be observed. As the front is initiated, the flyer is accelerated towards the base and, initially, closes the stand-off gap, at the first point (line) of collision, it undergoes plastic deformation, via the medium of a plastic hinge. The pressure profile in the collision region, assessed by Taylor[5] (Fig. 3.9), is associated with the rigid to inviscid response of the material in this area.

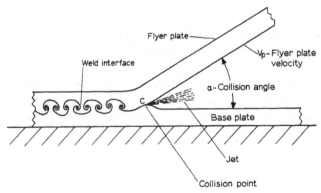

Fig. 3.8   Diagrammatic representation of an explosive welding system.

The variation is continuous and in terms of viscosity, μ, can be defined as

$$\mu = f(P) \tag{3.16}$$

Numerically, this attains infinity for rigid behaviour and is zero for the inviscid state. The viscosity distribution (Fig. 3.9) is a function of the distance, x, along the collision zone and can be represented as

$$\mu = f'(x) \tag{3.17}$$

Thus, there exist two viscous regions of length L_{v1} and L_{v2}, ahead and behind the stagnation point respectively, and an inviscid region l_i. The extent of these respective regions is defined, in each case, by pressures P_V and P_{CR} at which viscous and inviscid behaviours become significant.[6,7]

As the weld is formed, and the process continues along the interface of the flyer and the base, the now welded assembly undergoes a further elastic compression followed, after the passage of the front, by elastic recovery.

The bonding of the material elements is achieved by melting, of a very short duration, followed by extremely rapid solidification and cooling of a very thin layer along the interface. The short period of

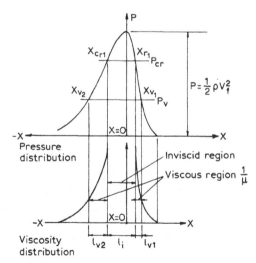

FIG. 3.9 Theoretical pressure and viscosity distributions in the collision region (after Refs 6 and 7).

annealing produces a higher resistance recrystallisation of the collision point region as compared with the regions farther away from the collision plane. The formation of a narrow submicroscopic interlayer, consisting of the materials of both the flyer and the base, has, generally, no detrimental effect. Since very small grains are present, the weld itself is usually stronger in shear than the weaker assembly material.[2,8,9]

3.3.2 Brief Review of the Theories of Wave Formation

It is generally agreed that the appearance of periodic, and fairly regular surface deformation on the interface of two explosively welded components can be legitimately regarded as part of the same general phenomenon which is associated with the formation of waves on the interface of any two continua subject to fluid flow. The similarity between hydrodynamic fluid behaviour and that of explosively welded metals arises out of the fact that when two metallic components come into contact at a high impact velocity, the very high shearing stresses in the collision region produce liquefaction. This clearly takes place not as a result of a temperature rise, but, as already indicated, only because of high pressures generated at that instant. In consequence of this physical fact, the explanations of the nature, characteristics and properties of the welded surfaces tend to be given in terms of fluid behaviour. However, because of the basic difficulties of observing the development of patterns of deformation, the validity of any of the four hypotheses of wave formation has not as yet been completely resolved. The indications are that, in at least some cases, a combination of the four mechanisms may be responsible, rather than just a single parameter.

The choice of the mechanisms lies between:

- indentation mechanism (jetting),
- flow instability mechanism,
- vortex mechanism,
- stress wave mechanism.

The indentation mechanism[10] is based on the theory of a two-dimensional, incompressible jet of fluid (originating at the point of collision) impinging on a flat plate. When an explosive weld is produced, the flyer plate collapsing on to the base can, in an appropriate reference frame, be regarded as the originator of such a jet. Qualitative modifications to this theory exist that include the deformation of the base plate near the successive collision points[7,10] (Fig. 3.10).

A further consideration to be taken into account is that two fluids, moving with different velocities parallel to their common surface, will

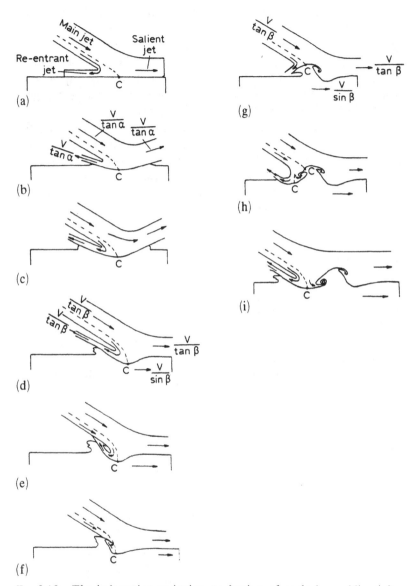

FIG. 3.10   The indentation or jetting mechanism of explosive welding (after Ref. 10). (a) Rigid elastic base; (b) stationary elastic base; (c) stationary elastic base with shear; (d) allowing for velocity of the parent plate; (e) hump interfering with jet; (f) formation of tail; (g) formation of forward trunk; (h) formation of front vortex; (i) completion of process.

experience an instability in which small perturbations in the interface grow in amplitude and eventually cause it to roll-up into vortices at regular intervals, although, in a real fluid, these features tend to be masked by the effects of viscosity and turbulence. The phenomenon is known as Helmholtz instability.[11] The mechanism does not explain the presence of the waves on the interface of symmetrical welding arrangements where, in theory, there can be no discontinuity.

The third hypothesis is that of a vortex mechanism[12,13] which suggests that the stagnation region acts as a solid obstacle in the flow, past the collision point, and that the waves develop behind that region in a manner similar to that proposed by von Karman with reference to the classical vortex street theory (Fig. 3.11).

The stress wave mechanism (Fig. 3.12) considers the situation in which compressive waves, initiated at the collision point, are reflected from the free surfaces of both the flyer and the base with the reflection taking place either in front or behind the collision region. The geometry of the system determines the actual physical situation. The possibility of forming disturbances on the free interface has been studied in some detail[7,14] and the optimal conditions for successful welding are stipulated by eqns (2.20) and (2.21).

Although the controversy about the actual mechanism of welding, in any given situation, is not yet settled, the indentation (jet) concept, in conjunction with that of the stress wave (for thin material layers) appears to be appropriate in the majority of practical cases involving composite materials.

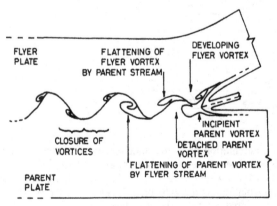

FIG. 3.11   The vortex street welding mechanism (after Ref. 16).

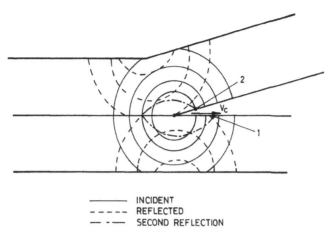

——— INCIDENT
— — — — REFLECTED
— · — SECOND REFLECTION

Fɪɢ. 3.12   The stress-wave welding mechanism (after Refs 7 and 15).

As the plastic hinge is formed, the 'static' angle α changes to a dynamic angle β (Fig. 3.10) and it is effectively the latter, together with the jet velocity, that influences the property of the system. This is characterised by four basic types of bonds that can be easily distinguished.

If the stand-off, energy level and acoustic impedance conditions are suitably chosen, a metal-to-metal, large amplitude, high frequency wave will be generated (Fig. 3.13(a)). In the same circumstances, a lower level of energy will reduce the wave amplitude (Fig. 3.13(b)) to, possibly, almost linear. Equally, excess of energy delivered will result in the interface melting and in the consequent formation of intermetallics which resolidify in either a wavy (Fig. 3.13(c)) or linear form (Fig. 3.13(d)). On the other hand, a high impact velocity may easily cause the intermetallic jet to be entrapped in the troughs of the waves (Fig. 3.13(e)) thus introducing an element of brittleness into the system.

For a given angle β, the use of a low energy level will result in a low jet velocity ($V \cot \beta$ in Fig. 3.10) and therefore in a shallow penetration of the base. The relatively small wave hump produced will be large enough to trap the low velocity re-entrant jet at an early stage (Fig. 3.13(e)) and consequently waves of low amplitudes will originate.[17]

Superimposition of the stress wave effect onto that of a 'normal' welding wave is particularly noticeable when layers of foil or similar thin elements are welded. The additional curling of the welded interface becomes then very pronounced (Fig. 3.13(f)).

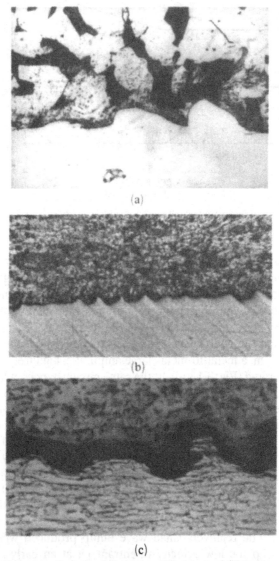

Fɪɢ. 3.13   Typical explosive welding bimetallic bonds. (a) Metal-to-metal low carbon/stainless steels ($\times$ 300); (b) metal-to-metal low carbon/stainless steels ($\times$ 200); (c) wavy form of the resolidified intermetallics, low carbon steel/brass ($\times$ 100); (d) linear form of the resolidified intermetallics, aluminium/low carbon steel ($\times$ 50); (e) trapped jet (dark interface areas) copper/low carbon steel ($\times$ 100); (f) superimposed stress wave effect, aluminium/brass foils ($\times$ 100).

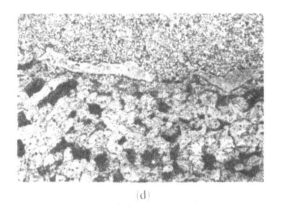

(d)

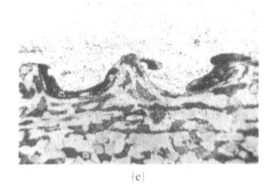

(e)

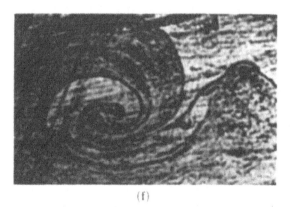

(f)

3.3.3 Simulation of Explosive Welding

3.3.3.1 Liquid analogue

The hydrodynamic nature of the wavy interfaces, even at low amplitudes, suggests that investigations of the welding processes can be simulated and valid observations made when using a suitable liquid analogue. This can lead to the acquisition of a better understanding of the mechanism of wave formation and therefore be helpful in selecting the correct hypothesis for the considered situation. It is really in this role that the simulation of explosive welding by a liquid analogue finds its place in both the theoretical investigation and in industrial application of the process.

Experimental investigations utilising liquids to simulate the behaviour of metals were carried out in the past using relatively simple apparatus.[18-20] The limitations inherent in these techniques have been removed, to a considerable degree, in the apparatus designed by El-Sobky[6] and used for an extensive investigation of the conditions of wave formation.[7,21,22]

The apparatus, shown in Fig. 3.14, consists of two tanks with transparent perspex walls, supported by two hinged steel frames positioned on top of each other. The top tank (1) contains a layer of liquid which represents the flyer plate whereas the lower (2) contains the base component. The bottom of the top tank (3) is made of thin steel sheet which can freely slide in guides (4) made of brass and fitted beneath the walls. The guides extend beyond the length of the tank along the entire length of the top steel frame (5). A drive unit is used to draw the bottom of the tank allowing the flyer liquid to impact on the base liquid plate. The unit consists of a DC variable speed electric motor (6) driving a pulley (7) via a speed reduction unit (8) and a 'one revolution clutch' (9). This engages the pulley, causing it to rotate for one revolution only and then come to a stop. A special triggering mechanism (10) was designed to be used in conjunction with the one revolution clutch to ensure instant engagement and to facilitate synchronisation with a high speed camera. The pulley is grooved and a steel wire (11) is attached to it (12) at one end, and to the thin steel sheet (3) at the other end (13). The diameter of the groove and the length of the wire are calculated in such a way that the thin sheet (3) is completely withdrawn within one revolution of the pulley.

The top frame supports the drive unit and the top tank and is directly hinged to the lower frame (17) at one end (14). At the other end, it is

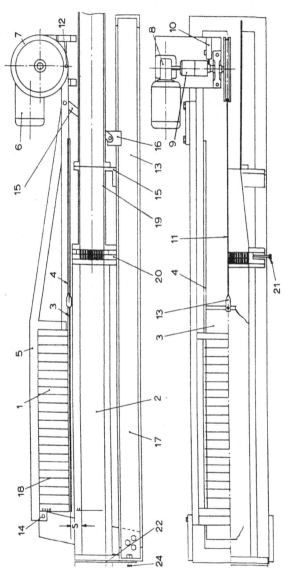

FIG. 3.14   Schematic diagram of a liquid analogue (after Ref. 7).

hinged to two supporting levers (15) which are, in turn, hinged on two blocks (16) that slide on the lower support (17). The two frames (5, 17), the levers (15) and the sliding blocks (16) constitute a four bar mechanism which allows tilting of the top tank, and thus simulates an inclined welding arrangement. In order to avoid accumulation of the flyer liquid, or the flow across the tank in the parallel system, the top tank is divided into smaller compartments by means of aluminium separators which can be inserted into vertical slots in the tank walls (18). The lower edge of each aluminium separator is rubber padded and fits tightly inside the grooves so that the liquid in every compartment is completely sealed.

Two further extensions exist in the lower tank for the purpose of creating a continuous steady flow, rather than using a stationary layer. A motion of the liquid in the direction opposite to that of withdrawal of the thin plate (3), can be imparted to increase the relative velocity of the collision point. The liquid is delivered to the tank by means of a tube (19) through a special valve (20) which consists of two plates, one sliding relative to the other (21); both having 320 uniformly arranged small holes, 4 mm in diameter. The flow is controlled by adjusting the amount of coincidence of the holes in the two plates. The holes have the additional function of ensuring a uniform velocity profile everywhere and at all flow rates. At the other end of the tank, another tube (22) is connected to guide the fluid to a circulating pump. The overall dimensions of the apparatus are 170 cm long by 40 cm high by 30 cm wide. The lower tank is 12·5 cm deep and 118 cm long and extends on both sides beyond the length of the upper tank to ensure that the collision begins and ends far enough from the lower tank end-plates. The top tank is 10 cm deep and 50 cm long.

The apparatus can simulate the mechanism of wave formation in the post collision region only. It cannot demonstrate the effect of reflection of waves from the surfaces since this is not apparent in fluids. The formation of a hump is however clearly reproducible.

The basic parameters of the process are identified in Fig. 3.15 as follows:

- a, the wave amplitude,
- V_C, the velocity of the collision point, equivalent to V_W the velocity of withdrawal of the membrane,
- V_p, the velocity of the flyer ($\sqrt{2gy}$)
- s, the stand-off distance,
- h', the thickness of the flyer layer,
- h, the thickness of the base layer,

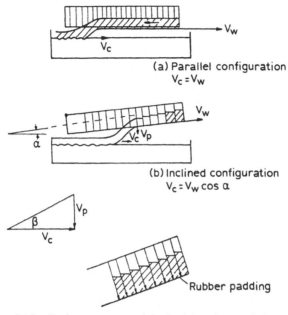

FIG. 3.15   Basic parameters of the liquid analogue of Fig. 3.14.

- y, the height of the centre of gravity of the flyer layer above the base surface ($y = s + 0.5h$),
- λ, the wave length,
- E, the energy per unit area delivered at the surface,
- β, the dynamic angle of collision ($\beta = \tan^{-1}(V_P/V_W)$),
- c_S, the surface velocity $\sqrt{gh'}$,
- N, the wave frequency.

The collision of the two liquid layers is recorded by means of a high speed cine camera operating at some 200 frames/s (Fig. 3.16).

The most suitable medium is found to be a mixture of kerosene with aluminium powder. The concentration of about 3 g/litre gives a chemically stable, long lasting suspension. The powder consists of minute flakes which act both as mirrors and as plates that align with streamlines, thus giving well-defined photographs of the flow pattern. Kerosene of density of 800 kg/m^3, and viscosity of 1.2×10^{-3} N s m^{-1} can be used successfully. To simulate the welding of composites, it is sufficient to either change one of the liquids, or to vary the density of a layer. This will reproduce the sought after relative properties of the analogue.

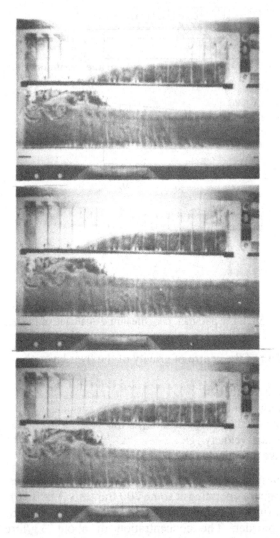

F<small>IG</small>. 3.16   An example of the development of a 'welding wave' after the collision of two layers of liquid in the liquid-analogue system.

3.3.3.2 Computer modelling

The explosive welding process has been successfully simulated for a
variety of material combinations of steels, aluminium and copper, by
using a Lagrangian finite difference computer code. The code enables

the determination of the criteria of weld formation and provides graphic visualisation of the relationship between basic process parameters.[23]

At the macrolevel, this particular simulation gives deformation patterns, velocity fields, pressures, temperatures, etc., whereas at the microlevel the formation of the wave hump can be studied.

A detailed approach to the building of a computer program for the determination of the optimal explosive loading and weld geometry, is offered by Stivers and Wittman[24] who have produced flow charts and have discussed the philosophy behind the steps taken to assess the relationship between, and the importance of basic process parameters.

3.3.4 Velocity Field Configurations

Depending on the geometry of the assembly to be welded, two basic welding systems can be employed. For axisymmetrical configurations, e.g. multiplex cylinders, arrays of tubular, etc., components, and, in

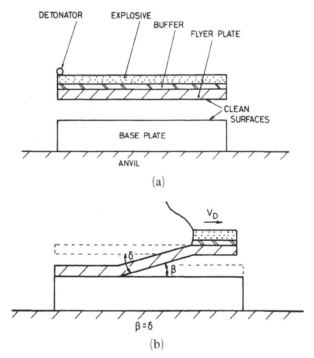

FIG. 3.17 A parallel welding system: (a) pre-detonation geometry; (b) detonation geometry.

some cases, for flat plate, the parallel welding system is used. This is shown in Fig. 3.17(a) and applies, in this particular case, to a simple flyer-base arrangement. The assumed geometry of deformation, during the passage of the detonation front, is indicated in Fig. 3.17(b).

In more complex planar and shape geometries, an inclined configuration (Fig. 3.18(a)) is used, in which both the initial, static angle α and the dynamic hinge angle β play, respectively, major roles (Fig. 3.18(b)). In this case, the angle δ is a measure of the magnitude of the hinge.

The three major parameters of the explosive welding process are:

- the characteristics of the charge, including the level of energy delivered and usefully dissipated,
- the deformation and impact velocity of the colliding elements,
- the physical nature and the geometry of the collision zone.

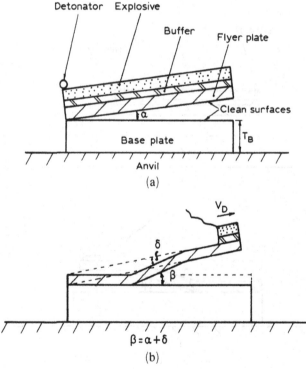

FIG. 3.18   An inclined welding system: (a) pre-detonation geometry; (b) detonation geometry.

In turn, these are influenced by the impact (flyer) velocity V_P, that of the collision point V_C, and the dynamic angle of collision β.

The geometry of the collision zone, i.e. in the vicinity of the collision point C (Fig. 3.8), presents considerable difficulty, in the sense that its definition depends on the assumptions made by the investigator.

The simplest definition of the velocities V_R of the re-entrant, and V_S of the salient jets (Fig. 3.19), is, in terms of fixed co-ordinates (Fig. 3.19(a)), as follows:

$$V_R = V_P(1 + \cos \beta)/\sin \beta, \quad V_S = V_P(1 + \cos \beta)/\sin \beta \qquad (3.18)$$

In transformed co-ordinates, moving with point C (Fig. 3.19(b)), V_C and V_F become:

$$V_C = V_P/\sin \beta, \quad V_F = V_P/\tan \beta \qquad (3.19)$$

The possible variations in the assumed velocity fields are shown in Fig. 3.20, and the relevant data, for some of the inclined configurations, are summarised in Table 3.3. According to Birkhoff *et al.*,[25] for instance, the velocity V_P bisects the angle between the already accelerated segment of the plate, behind the front, and the, as yet, undeformed

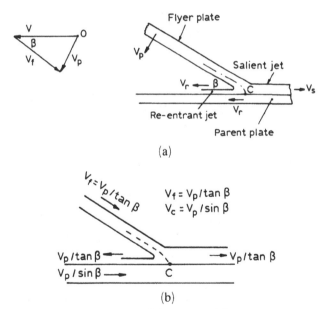

FIG. 3.19 Simple velocity fields in (a) fixed, and (b) transformed co-ordinates.

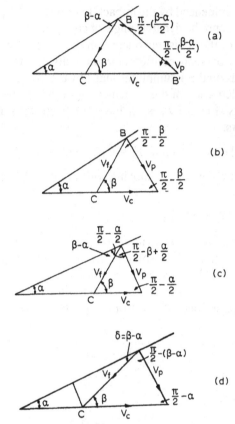

FIG. 3.20   Velocity field configurations with respect to the collision point C: (a) Birkhoff *et al.*;[25] (b) Crossland;[26] (c) Wylie *et al.*;[27] (d) Jouguet–Chapman postulate.

element (Fig. 3.20(a)). Crossland,[26] on the other hand, assumes that $V_C = V_F$ (Fig. 3.20(b)), or that the length of the flyer remains unchanged and that V_P is perpendicular to the original flyer position. Wylie *et al.*[27] assume, however, that V_P is perpendicular to a line bisecting the initial angle α (Fig. 3.20(c)). When this velocity is calculated from the consideration of the J–C plane (Fig. 3.20(d)), the direction of it is, however, taken to be perpendicular to the original position.

Naturally, the calculated values of the parameters will vary accordingly (Table 3.3) and the angle θ, defining the inclination of V_P, will

TABLE 3.3
Welding Parameters in Inclined Configurations

	Ref. 23	*Ref. 25*	*J–C postulate*
V_P	$2V_D \sin^{1/2}(\beta - \alpha)$	$2V_D \sin(\alpha/2)$	$V_D \tan(\beta - \alpha)$
V_C	$V_P \cos^{1/2}(\beta - \alpha)/\sin\beta$	$V_D V_P/[V_P + 2V_D \sin(\alpha/2)]$	$V_P \cos(\beta - \alpha)/\sin\beta$
V_f	$V_P \cos^{1/2}(\beta + \alpha)/\sin\beta$	$V_P \cos(\alpha/2)/\sin\beta$	$V_P \cos\alpha/\sin\beta$
β	$\tan^{-1} V_P \cos^{1/2}(\beta + \alpha)/$ $V_C - V_P \sin^{1/2}(\beta + \alpha)$	$\tan^{-1}[V_P \cos(\alpha/2)/$ $(V_C - V_P \sin(\alpha/2))$	$\tan^{-1}(V_P \cos\alpha)/$ $(V_C - V_P \sin\alpha)$
θ	$\frac{1}{2}(\pi - \beta + \alpha)$	$\frac{1}{2}(\pi + \alpha) - \beta$	$\frac{1}{2}\pi - (\beta - \alpha)$

differ in each case. For instance:

$$\theta_A > \theta_B > \theta_C \qquad (3.20)$$

with θ_D lying in the B–C range.

N.B. The equations of Table 3.3 are applicable to parallel configurations on substituting $\alpha = 0$.

To arrive at an estimate of the various process parameters, it is necessary to assess the value of V_P. A number of existing empirical expressions can be used for this purpose, on the understanding however that, strictly, they all apply to one-dimensional detonation in the plane perpendicular to that of the flyer and parallel to the direction of V_D.

The classic equation, due to Gurney,[28] gives

$$V_P\{2E[3r^2/(r^2 + 5r + 4)]\}^{0.5} \qquad (3.21)$$

where r is the ratio of the mass of charge per unit area to the flyer mass per unit area, and E is the available energy of the charge.

For the conditions of $1/r > 1$, Duvall and Erkman[29] suggest:

$$V_P = (4E)^{0.5}[1 + 27/16r(1 - [1 + 32r/27)^{0.5}] \qquad (3.22)$$

where E is the total energy released by the charge.

Deribas[30] has found that for granular, low detonation explosives, the following relationship is valid:

$$V_P = 1{\cdot}2\, V_D[(1 + 32r/27)^{0.5} - 1]/[(1 + 32r/27)^{0.5} + 1] \qquad (3.23)$$

and, in general, that consideration of gas dynamics gives:

$$V_P = 3\, V_D/4\{1 + 2/r[(r + 3)/(r + 6)](1 - [1 + 3r(r + 6)/(r + 3)^2]^{0.5})\} \qquad (3.24)$$

3.3.5 Multi-layer Composites

Unlike the relatively simple situation that obtains in the welding of a bimetallic composite (Fig. 3.21), the multi-layer, possibly multimetallic assemblies present a much more complex picture. A composite, multi-layer system results from the successive welding of an ever increasing, in thickness, flyer element to a free-to-deform base.

As the detonation front propagates along the first of the n layer system surface, collision takes place on successive interfaces from, say, $i = 1$ to $i = n$ (Fig. 3.22).

The momentum imparted to the first layer produces a velocity $V_{P(i=1)}$. The available momentum is then shared, upon collision, with

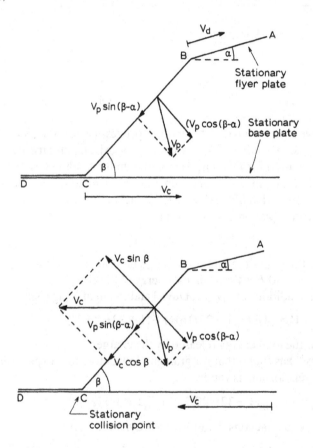

Fig. 3.21   Velocity diagrams for a bimetallic explosively welded composite: (a) for a moving collision point C; (b) for a stationary collision point C.

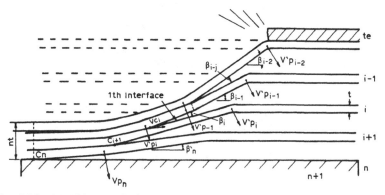

FIG. 3.22   Multi-layer collisions in the welding of a composite (t_e is the thickness of the explosive charge).

the next layer which then moves with the first, but at a reduced velocity V_{P2}.[31] This sequence is repeated until the collision of the last layer with the base has occurred. Referring the layer velocity to the base (V'_P), the relationship can be defined as:

$$V'_{P(i-1)} > V'_{Pi} > V'_{P(i+1)} \qquad (3.25)$$

Considering the collision point C_i on the ith interface (Fig. 3.23), the ith layer acts as a base for the preceding $(i-1)$th layer, ahead of C_i, and as part of the $(i+1)$ flyer-layer, behind C_i.

Hence, ahead of C_i, one has

$$V'_{F(i-1)} = V'_{P(i-1)} \cot \beta'_{(i-1)}$$

and, behind C_i:

$$V'_{F(i-1)} = V'_{Pi} \cot \beta'_i$$

Consequently,

$$V'_{P(i-1)} / V'_{Pi} = \cot \beta'_i / \cot \beta'_{(i-1)} \qquad (3.25a)$$

and therefore, considering eqn (3.25),

$$\beta'_{(i-1)} > \beta'_i > \beta'_{(i+1)} \cdots \qquad (3.26)$$

It follows from Fig. 3.23 that the following relationships exist:

the initial angle of collision $\alpha_i = \beta'_{(i-1)} - \beta'_i \qquad (3.27)$

the dynamic angle of collision $\beta'_i = \alpha_i - \delta_{(i-1)} = \beta''_{(i-1)} - \beta'_i \qquad (3.28)$

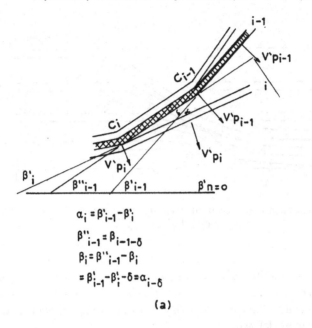

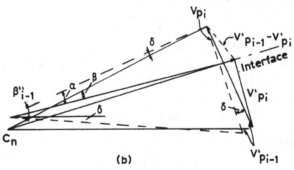

FIG. 3.23   Multi-layer collisions. (a) Geometry of collision at the ith interface;
(b) the appropriate velocity diagram.

$$\beta''_i = \beta'_i - \delta_i \qquad (3.28a)$$

and, further

$$\cot \beta'_i = V'_F / V'_{Pi} \qquad (3.28b)$$

and

$$\cot \beta_i = V_F / V_{Pi} \qquad (3.28c)$$

Further consideration of the relationships between the flyer velocity and the charge ratio R indicates[31] that with increasing i, the following sequence of velocity ratios will be observed:

$$V'_{P(i-1)}/V'_{Pi} > V'_{Pi}/V'_{P(i+1)} > V'_{P(n-1)}/V'_{Pn} > 1 \qquad (3.29)$$

Also,

$$\tan \alpha_i = (\tan \beta'_{(i-1)} - \tan \beta'_i)/(1 + \tan \beta'_{(i-1)} \tan \beta'_i) \qquad (3.30)$$

and

$$\tan \alpha_i = [1 - (V'_{Pi}/V'_{P(i-1)})]/(\cot \beta''_{(i-1)} + \tan \beta'_i) \qquad (3.31)$$

It is evident from these two equations that α_i will decrease with i. The change in the dynamic angle of collision β_i is governed by the change in the deflection angle $\delta_{(i-1)}$, and in α.

Consequently,

$$\beta_{(i+1)} - \beta_i = (\beta'_i - \beta'_{(i-1)}) - (\delta_i - \delta_{(i+1)}) - (\beta'_{(i+1)} - \beta'_i) = C_i - \Delta\delta_i \quad (3.32)$$

where C_i is always positive. β_i will change according to the value of $\Delta\delta_i$. As the loss of momentum continues, $\Delta\delta_i$ is expected to decrease and the difference $(\beta_{(i+1)} - \beta_i)$ should be positive.

The wave amplitudes should thus be greater as i increases, providing that the effect of thickness on the width of the collision region, and the increase in the dynamic angle of collision compensate for the effect of reduction in the collision velocity.

It follows from eqn (3.25a) that uniformity in the welding conditions can be enhanced artificially by using a driver plate. The number of layers can then be increased.

If V_{Pmax} and V_{Pmin} refer to the limits of impact velocity, within which welding is likely to occur, and, further, if they can be assumed to be approximately equal to V_{P1} and V_{Pn} respectively, then the condition for welding n layers of the assembly is[6]

$$V_{P1}/V_{Pn} \simeq [(m+n)/(m+1)]^2 \qquad (3.33)$$

where m is the number of layers equivalent to the thickness of the driver plate.

3.3.6 The Welding 'Window'

The success in the welding of solid composites depends on selection of suitable parameters that, on the one hand, would guarantee the existence of the weld on the impacting interfaces and, on the other, would ensure that the weld is both metallurgically and mechanically acceptable.

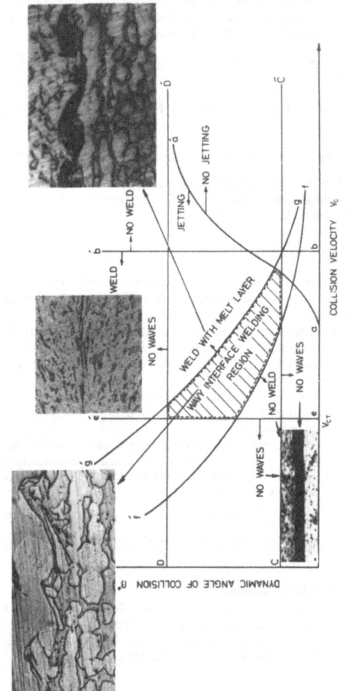

Fig. 3.24 Construction of a 'welding window'.

As has been indicated in the earlier discussion, the three basic parameters are the dynamic angle of collision β, the collision point velocity V_C, and the impact velocity V_p. The interplay between the respective sets of these parameters will determine the nature of the physical conditions developing in the system and consequently the response of the materials involved in the operation.

The variation in the values of the respective parameters, with the changing geometry, etc., of the welding system, has been established for a number of monometallic material combinations and is usually represented in the form of a 'welding window' (Fig. 3.24).

The early investigations of Abrahamson[32] and Cowan and Holtzman[33] have provided means of estimating critical, for the jet formation, values of β (aa' in Fig. 3.24), whereas the upper and lower values of the angle, necessary for the weld to exist, have been indicated by, among others, Bahrani and Crossland[34] (CC' and DD'). The lower, acceptable limit of V_C was provided by Cowan et al.,[18] and the upper by Deribas[30] and Christensen et al.[35] (bb' and ee' respectively in Fig. 3.24). The lower and upper limits of V_p, giving respectively a sufficiently high pressure to produce a weld and to avoid melting, were estimated by Wittman.[36]

As can be seen in the figure, the limits imposed on the values of the three parameters produce a shaded area within which, for the specific material considered, a wavy interface weld will be produced. Although

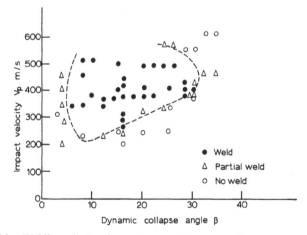

FIG. 3.25   Welding window for a monometallic copper/copper system (after Ref. 37).

the presence of a wave is not a necessary condition for the existence of a strong bond, many users of the process favour this configuration because it introduces an additional mechanical locking mechanism.

A selection of pictorial representations of the conditions that can exist on the interfaces is given in Fig. 3.24.

An additional example of a welding window is provided by Fig. 3.25 which refers to the conditions obtaining in a monometallic copper/copper welding system.[37]

When multimetallic composites are welded, the optimal conditions are established by superimposing monometallic 'windows' for the required metals onto each other and thus considerably narrowing the shaded area that represents the working range of parameters.

REFERENCES

1. NAGARKAR, M. D. and CARPENTER, S. H. Diffusion studies in explosion welded metals, Internal Report, University of Denver, Colorado, 1974.
2. BLAZYNSKI, T. Z. (Ed.) *Explosive Welding, Forming and Compaction*, Applied Science Publishers, London, New York, 1983.
3. CROSSLAND, B. *Explosive Welding of Metals and its Application*, Clarendon Press, Oxford, 1982.
4. COOK, M. A. *The Science of High Explosives*, Reinhold, New York, 1958.
5. TAYLOR, G. I. *Phil. Trans. Roy. Soc.*, **260** (1966), 96.
6. EL-SOBKY, H. PhD thesis, University of Leeds, 1979.
7. EL-SOBKY, H. In: *Explosive Welding, Forming and Compaction*, Ed. T. Z. Blazynski, Applied Science Publishers, London, New York, 1983, p. 189.
8. HAMMERSCHMIDT, M. and KREYE, H. In: *Shock Waves and High Strain-Rate Phenomena in Metals*, Eds M. A. Myers and L. E. Murr, Plenum Press, New York, 1981, p. 961.
9. HAMMERSCHMIDT, M. and KREYE, H. In: *Proc. 7th Int. Conference on High Energy Rate Fabrication*, Ed. T. Z. Blazynski, University of Leeds, Leeds, 1981, p. 60.
10. BAHRANI, A. S., BLACK, T. J. and CROSSLAND, B. *Proc. Roy. Soc.*, **A296** (1967), 123.
11. HUNT, J. N. *Phil. Mag.*, **17** (1968), 669.
12. KLEIN, W. In: *Proc. 3rd Int. HERF Conf.*, University of Denver, Colorado, 1971.
13. REID, S. R. *Int. J. Mech. Sci.*, **16** (1974), 399.
14. STIVERS, S. MSc thesis, University of Denver, Colorado, 1974.
15. BLAZYNSKI, T. Z. In Ref. 7, p. 289.
16. KOWALICK, J. F. and HAY, D. R. *Met. Trans.*, **2** (1971), 1953.
17. BLAZYNSKI, T. Z. and DARA, A. R. *Metals Mater.*, **6** (1972), 258.
18. COWAN, G. R., BERGMANN, O. R. and HOLTZMAN, A. H. *Met. Trans.*, **2** (1971), 3145.

19. WILSON, M. P. W. and BRUNTON, J. H. *Nature*, **226** (1970) (5), 538.
20. KUDINOV, V. M., PETUSHKOV, V. G. and FADEENKO, Yu. I. In Ref. 9, p. 224.
21. EL-SOBKY, H. and BLAZYNSKI, T. Z. In: *Proc. 5th Int. HERF Conf.*, University of Denver, Colorado, 1975, Paper 4.5.1.
22. BLAZYNSKI, T. Z., *Materialwissenschaften u. Wekstofftechnik*, **20** (1989), 262.
23. OBERG, A., SCHWEITZ, J. A. and OLOFSSON, H. *Proc. HERF-1984 Conf.*, Eds I. Berman and J. W. Schroeder, ASME, New York, 1984, p. 75.
24. STIVERS, S. W. and WITTMAN, R. H. In: *Proc. 5th Int. HERF Conf.*, University of Denver, Colorado, 1975, Paper 4.2.
25. BIRKHOFF, G., McDOUGALL, D. P., PUGH, E. M. and TAYLOR, G. *J. Appl. Phys.*, **9** (1948), 563.
26. CROSSLAND, B., *Metals Mater.*, **5** (1971), 401.
27. WYLIE, H. K., WILLIAMS, P. E. G. and CROSSLAND, B., In Ref. 12.
28. GURNEY, R. W., Report No. 405, Ballistic Research Laboratories, Aberdeen Proving Ground, Maryland, 1943.
29. DUVALL, G. E. and ERKMAN, J. O. Tech. Rep. No. 1, Stanford Research Institute, Project No. GU-2426, 1958.
30. DERIBAS, A. A. *Explosive Welding*, Siberian Academy of Sciences, USSR, 1967.
31. EL-SOBKY, H. and BLAZYNSKI, T. Z. In: *Proc. 7th Int. HERF Conf.*, Ed. T. Z. Blazynski, University of Leeds, Leeds, 1981, p. 100.
32. ABRAHAMSON, G. R. *J. Appl. Mech.*, **83** (1961), 519.
33. COWAN, G. R. and HOLTZMAN, A. H. *J. Appl. Phys.*, **34** (1963), 928.
34. BAHRANI, A. S. and CROSSLAND, B. *Proc. Inst. Mech. Engrs London*, **179** (1964), 264.
35. CHRISTENSEN, K. T., EGLY, N. S. and ALTING, L. *Proc. 4th Int. HERF Conf.*, University of Denver, Colorado, 1973, Paper 4.3.
36. WITTMAN, R. H. In: *2nd Symposium on The Use of Explosive Energy for Manufacturing Metallic Materials of New Properties*, Czechoslovak Scientific and Technical Society, Prague, 1973, p. 153.
37. MEYER, M. D. In Ref. 33, Paper 5.3.1.

Chapter 4

Explosive Welding Operations

4.1 WELDING SYSTEMS

4.1.1 Introduction

The methods of manufacture of welded composites fall into two main categories, depending on whether the geometry of the system is planar or tubular. A further sub-division occurs with the tubular components which can be produced either in axi- or asymmetric systems.

Irrespective of the system employed, the basic precept that has to be observed is that of providing a sufficient stand-off distance between the flyer and the base to allow the former to acquire a high impact velocity V_p. In multilayered systems, the individual elements of the assembly have to be separated from each other both initially and during the welding operation.

To ensure that the detonation front proceeds with a uniform velocity V_D over the whole area of the flyer element, it is necessary to initiate detonation in a manner compatible with the geometry of a plane wave. This involves arranging the initiating (low detonation) charge either in the form of an equilateral triangle (planar geometries) or a regular cone (tubular geometries). Figure 4.1 demonstrates both these possibilities and indicates also a basic charge arrangement.

Planar welding geometries are used in the aerospace, transportation (particularly ship building), chemical and petrochemical industries, in which 'sandwich' clad plate composites (accounting for some 75% of the total), electrical, structural and cryogenic transition joints, honeycomb structures, and sheet multilaminates in, perhaps, unusual metallic combinations, are often used. The actual components range from chemical vessels clad explosively with thin sheaths of anticorrosive, structurally strong and expensive materials, e.g. stainless steels, through aluminium/steel transition plates (in shipbuilding), clad fuel ducts in

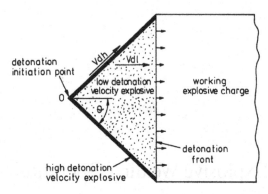

FIG. 4.1   Schematic charge arrangement producing a plane detonation front.

nuclear installations, to clad gun liners and honeycomb type components. The range of applications and material combinations is steadily increasing as shown in Table 4.1.

On the other hand, the various axisymmetrical tubular, explosively welded systems are concerned with providing the means of manufacturing multiplex cylinders, geometrically intricate arrays of rods and tubes for remote control mechanisms, tubeplates for heat exchangers, electrical and cryogenic tubular transition joints, and multilayered foil and foil/mesh reinforced cylinders.

4.1.2 Planar Systems
Three basic geometrical configurations are possible.[1]

In its simplest form, the parallel system (Fig. 4.2(a)) creates conditions in which the collision point velocity is of the same magnitude as the detonation velocity V_D. The systems of this type are used in commercial cladding of large plates, but depending on the size and kind of cladding, a further sub-division of the technique is noted.

A large standard bimetallic clad is processed horizontally by first providing a charge container and inserting stand-off spacers on what will eventually constitute the interface (Fig. 4.3(a)), and then filling the container with powdered explosive (Fig. 4.3(b)). After the detonation, the welded clad is retrieved (Fig. 4.3(c)), cleaned, straightened, etc.

Heavier plates, welded to relatively thin bases, require the use of a supporting anvil, as shown in the sequence of Fig. 4.4. When smaller bimetallic plates of a considerable degree of flatness are required, the vertical double, back-to-back configuration of Fig. 4.5 is often used.

TABLE 4.1
Commercially Available Bimetallic Clads

	CS	AS	SS	Al	Cu	Ni	Ti	Hastelloy	Ta	Columbium	Ag	Au	Pt	Stellite 6B	Mg	Zr	Pb	Mo	Inconel
Carbon steels (CS)	X	X	X	X	X	X	X	X	X	X	X	X			X	X	X		
Alloy steels (AS)	X	X	X	X	X	X	X	X							X	X	X		
Stainless steels (SS)	X	X	X	X	X	X	X	X	X	X	X		X			X			
Aluminium alloys	X	X	X	X	X	X					X	X	X	X					
Cu alloys	X	X	X	X	X	X					X	X	X						
Ni alloys	X	X	X	X	X	X	X	X	X				X	X					
Ti	X	X	X			X	X	X	X	X	X	X			X	X	X		
Hastelloy	X	X	X			X	X	X	X	X									
Ta	X		X			X	X	X	X	X	X	X	X	X					
Columbium	X		X				X	X	X	X	X	X	X						
Ag	X		X	X	X		X		X	X	X	X	X						
Au	X			X	X		X		X	X	X	X	X						
Pt			X	X	X	X			X	X	X	X	X						
Stellite 6B				X		X			X					X					
Mg	X	X					X								X				
Zr	X	X	X				X									X	X		
Pb	X						X									X	X		
Mo																		X	X
Inconel																		X	X

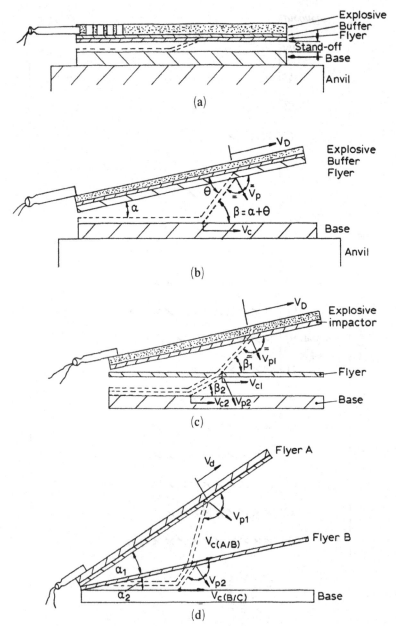

FIG. 4.2   Planar welding techniques. (a) Basic parallel system; (b) basic inclined system; (c) inclined system incorporating an impactor; (d) double inclined system.

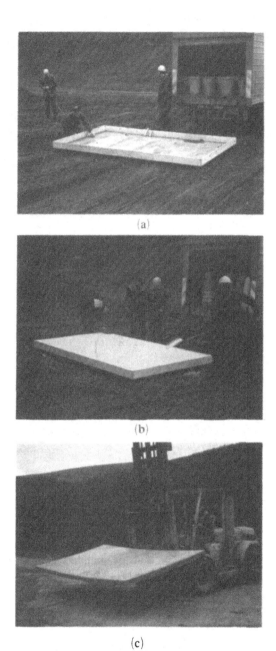

(a)

(b)

(c)

FIG. 4.3   Cladding a large standard bimetallic plate. (a) Insertion of stand-off spacers; (b) provision of powdered explosive charge; (c) recovery of the clad (courtesy ICI Nobel's Explosive Co. Ltd, Stevenston Scotland).

(a)

(b)

Fig. 4.4   Cladding a smaller plate to a relatively thin base requiring an anvil.
(a) Clad ready for firing; (b) clad and anvil after firing (courtesy ICI Nobel's
Explosive Co. Ltd, Stevenston, Scotland).

The second basic system is that of inclined (single or double) geometry. In a single inclined system (Fig. 4.2(b)), the initiation of the charge takes place simultaneously along the edge of the flyer, and a plane detonation front is generated. In this case, the relationship between V_C, V_P and V_D is as follows:

$$V_C = V_P/\sin \beta = V_P/[\sin \alpha - \sin^{-1}(V_P/V_D)] \qquad (4.1)$$

The advantage of an inclined system is its ability to adjust — through the medium of the static angle α — to the type of explosive available, so that supersonically detonating explosives can be used because their effect will be attenuated by the choice of a suitable α. In consequence, the impact and collision point velocities will still remain in the subsonic range.

The effectiveness of an inclined system can be enhanced by using an impactor (Fig. 4.2(c)) which converts the system to an inclined/parallel one. The possibility of varying the range of impact velocities is thus increased since it will depend on the geometry of the impactor as well as on the charge energy. The technique is used for the welding of long, narrow elements or obtaining small weld areas.[2] The collision point velocity is again given by eqn (4.1).

The third configuration used is that of double inclined geometry (Fig. 4.2(d)). This is employed in those cases where the relative acoustic impedance compatibility of two materials to be welded is low, and therefore an intermediate metallic sheet is provided between them to make the inequality less pronounced.

The appropriate velocity relationships are given by:

$$V_{C(AB)} = V_{P1}/[\sin(\alpha_1 + \sin^{-1}(V_{P1}/V_D))]$$

and

$$V_{C(BC)} = V_{P2}/[\sin \alpha_2 + \sin^{-1} V_{P2}/V_{C(AB)}] \qquad (4.2)$$

Special problems can arise in the manufacture of multilaminates, and especially so if these consist of relatively thin sheets. It follows from the analysis of eqns (3.25)–(3.32) that in an uncorrected multilayered system there exists a variation in both the respective impact velocities and the dynamic angle β. This is shown diagrammatically in Fig. 4.6(a). The variability in the values of these two parameters will result in the corresponding variations in the properties of the individual layers (particularly in the bond integrity) and it may make the laminate unsuitable for more demanding applications in, for instance, the aerospace industry.[3]

A solution to this problem is found in the use of a mass driver plate,[3] or an equivalent to an impactor of Fig. 4.2(c). When this is used, in the manner indicated in Fig. 4.6(b), the momentum generated by the driver plate is sufficiently high not to be noticeably reduced by the impact with the first layer of the assembly, or by the successive impacts. Similarly, the initial dynamic angle, associated with the plate will be imparted to

(a)

(b)

(c)

FIG. 4.5 Arrangement for simultaneous cladding of two elements, back-to-back (titanium flyers and aluminium bases). (a) Arrangement for initiation of the charges; (b) assembly prior to firing; (c) two clads after firing (courtesy ICI Nobel's Explosives Co. Ltd, Stevenston, Scotland).

the laminates. In consequence, the conditions become stabilised and easier to control, thus producing a better quality product.

For the specific conditions developing in planar configurations, the classic Gurney eqn (3.21) can be modified[1,4] to give:
For $R > 5$

$$V_P = V_D[3R/(40R^2 + 8R + 32/R)]^{0.5} \qquad (4.3a)$$

For $R > 2.5$

$$V_P = 0.162RV_D/(2 + R) \qquad (4.3b)$$

and, if $R \approx 0.1$ (small explosive/flyer mass ratios), eqn (4.3b) can be further modified to give

$$V_P = 0.587RV_D/(2 + R) \qquad (4.3c)$$

4.1.3 Axisymmetrical Systems
Axisymmetrical or tubular composites can be manufactured in two basic configurations of either explosive or implosive systems. An explosive system (Fig. 4.7(a)) consists of, in this case, two concentric,

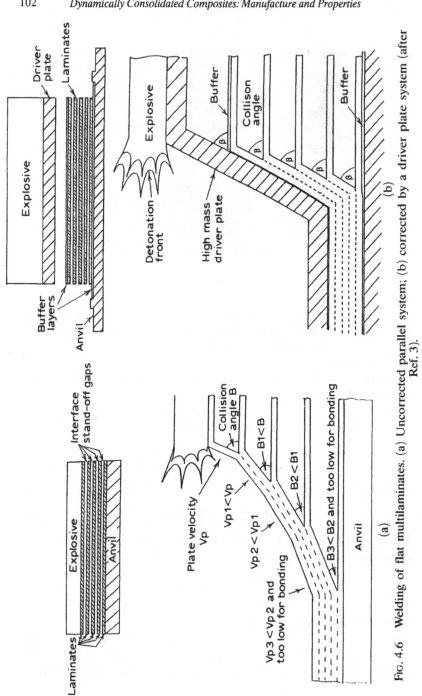

Fig. 4.6   Welding of flat multilaminates. (a) Uncorrected parallel system; (b) corrected by a driver plate system (after Ref. 3).

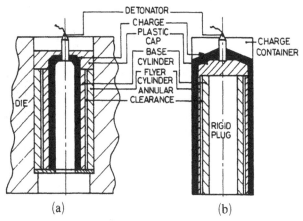

FIG. 4.7   Axisymmetrical welding systems: (a) explosive; (b) implosive.

tubular components to be welded together (completely) on their respectively outer and bore surfaces. An 'anvil' in the form of a heavy, thick-walled die is necessary. The charge, contained in the inner cylinder, is detonated in a configuration ensuring the generation of a plane wave front. The front then travels along the whole length of the cylinder causing it to deform and collapse onto the supported outer one in the same manner as that observed in a parallel plate system. For reasons of economy, inclined systems are not used in tube-type work.

The limitations of an explosive system lie in the relatively short clad cylinder that can be produced, again, because of the cost of very large dies and the associated operational difficulties, and in the low level of charge energy that can be allowed to be generated without causing the retaining die to fracture. Consequently, only thin-walled clad cylinders may be produced.

These disadvantages are eliminated in an implosive system (Fig. 4.7(b)) in which the charge is located on the outer surface of the outer tube. Although, apart from possibly unpleasant environmental effects, there is no apparent limitation on the specimen dimensions, the quality of the bonds does, however, tend to deteriorate if tube lengths greater than 4 m are processed. Both for the reasons of ease of access and therefore manipulation, and of the mechanical system properties, the implosive technique has been widely adopted.

Further examples of its use are provided by the bonding of multiplex cylinders, as in Fig. 4.8, and multilayered foil, and foil/mesh reinforced cylinders indicated in Fig. 4.9.

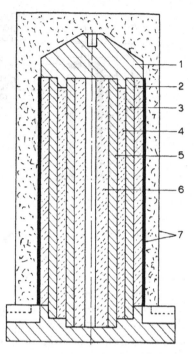

FIG. 4.8   Welding of multiplex cylinders. 1, Cap; 2, foil and reinforcement; 3, base cylinder; 4, soft metal; 5, supporting tube; 6, core; 7, container.

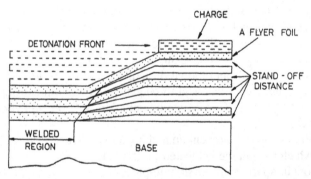

FIG. 4.9   Welding of multilayered foil cylinders.

A special application of the explosive system is that of bonding tube-to-tube plate in power station heat exchangers. Again, both parallel and inclined systems can be used here, but the latter appears to be more popular (Fig. 4.10). In its basic form, the prepared explosive/detonator package, enclosed in a polythene buffer container, is inserted into the

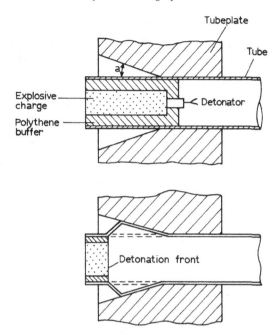

FIG. 4.10   Schematic representation of tube-to-tube plate welding.

appropriate — as yet unbonded heat exchanger tube (Fig. 4.11) — and is detonated. Figure 4.10 shows schematically the deformation of the tube (in the YIMpact system).[5]

Another facility afforded by this type of approach is the elimination of fractured or otherwise faulty heat exchanger tubing, without switching off the whole unit. A metal plug, provided with a suitable charge, is inserted into the tube and the charge is detonated *in situ*, expanding and bonding the plug to the tube (Fig. 4.12). Parallel welding systems are employed for this application.[6, 7]

Again, a simplification of eqn (3.21) can be introduced when considering cylindrical configurations, and the modified equation takes the form of:

$$V_p = (2E)^{0.5}[2R/(2 + R)]^{0.5} \tag{4.4}$$

4.1.4 Energy Requirements

The quoted Gurney-type equations provide means of estimating flyer velocities related to the ratio R and, sometimes, the detonation velocity,

FIG. 4.11   Ultrasonic inspection of a desuperheater — $\frac{1}{2}$ in o.d. × 20 swg stainless steel welded to a low carbon steel tubeplate (courtesy YIMPact, IMI Yorkshire, Leeds).

but, in their usual form, they do not necessarily give a clear, practical guidance to the energy required to produce a satisfactory bond.

Approximate, empirical approaches to this problem have been employed by a number of investigators and have led to the establishment of very useful experimental expressions.

The relevant relationships thus obtained reflect the geometry of the system and therefore fall naturally into the planar and axisymmetrical

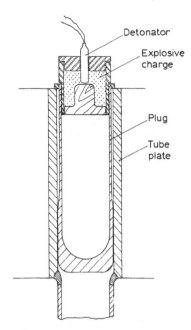

Detonator

Explosive charge

Plug

Tube plate

FIG. 4.12 Explosive plugging of faulty heat-exchanger tubing.

groups. In plate cladding, the estimates of Kowalick and Hay,[8] and Carpenter *et al.*[9] are representative of the main lines of thought.

Making use of dimensional analysis, the following functional relationships have been established[7]

$$R = K_1 \beta^A (V_P^2/E)^B (h/s)^C \qquad (4.5)$$

where β is the dynamic angle, E is the available energy of the charge, s is the stand-off distance, and h is the flyer thickness.

Since the constants A, B and C cannot be established easily, the equation is reduced to an approximate one in the form of:

$$R = K_2 \rho h C_S^2 \beta^2 \qquad (4.5a)$$

where K_2 is a constant, C_S is the velocity of sound in the flyer material, and ρ is the flyer density.

Carpenter *et al.*, on the other hand, considering energy balance, i.e. the kinetic energy of the flyer and the energy of plastic hinge deformation, proposed the following relationship:

$$L_E \simeq (\beta^2 h \sigma_o)/s \qquad (4.6)$$

where L_E is the explosive weight/unit area, and σ_0 is the yield stress of the flyer material.

Using the concept of vectorial length, to account for the axisymmetrical geometries, Blazynski and Dara[1,10] assessed the energy E of the explosive necessary for a good bond as:

$$E = \rho^{1\cdot6}\sigma_0^{-0\cdot6} V_D^{3\cdot2} h^3 f[(D/h)^a (h'/h)^\beta (s/L)^\gamma (\sigma_0^{0\cdot6} L/\rho^{0\cdot6} V_D^{1\cdot2} h)^\delta] \quad (4.7)$$

For the usual range of engineering composites, the above equation reduces to the two following forms:

For explosive systems and short cylinders (in SI units):

$$E = 0\cdot226\, \rho V_D^2 L h^2 (D/h)^{1\cdot5} \quad (4.7a)$$

For implosive systems and long cylinders

$$E = 2\cdot25\, \rho V_D^2 L h (D/h)^{1\cdot5} \quad (4.7b)$$

and for short cylinders $(L < 150 \text{ mm})$

$$E = 0\cdot1\, \rho^{1\cdot1}\sigma_0^{-0\cdot1} V_D^{2\cdot2} h^3 (D/h)^2 (L/h)^{0\cdot83} \quad (4.7c)$$

where D is the outer diameter of the cylinder, and 'dashed' quantities refer to the base.

4.1.5 Mechanical Properties Tests

In spite of continued and dedicated efforts of explosive welding fabricators, no agreement has yet been reached on the possible standardisation of testing and assessing mechanical properties of composites manufactured by this technique.

Apart from the obvious visual examination of, say, transition joints, the basic testing techniques are either ultrasonic or destructive. The choice depends entirely on the final destination of the product and the current requirements of the consumer.

Ultrasonic testing is used routinely to evaluate the continuity of the bond, but although it will distinguish between the bonded and non-bonded areas, it cannot provide any indication of whether the specimen is characterised by a bond of full strength or a continuous, but a weak one.[11]

This particular testing is carried out at frequencies ranging from 2 to 5 MHz and in accordance with either ASTM A578 standard or that of the particular fabricator.

If this type of test is regarded as insufficient, destructive testing is brought into operation. A full set of such tests is likely to consist of one shear, one tensile and two bend tests, although in most cases the shear test will provide all of the required information.

A number of shear testing techniques are in existence. Of these the ASTM A264-44T (Fig. 4.13(a)), and the 'tensile' shear (Fig. 4.13(b, c)) tests are commonly used.

In the former, an axial compressive load is applied to the specimen until failure occurs. In the latter, a series of coupons is machined in the specimen, down to the weld interface, and shearing is generated by the application of a tensile axial load (Fig. 4.13(c)). The coupons are sheared off in succession and so the consistency of weld quality can be established.

Ram tensile tests are also often used to evaluate the bond strength of composites. A test specimen, used by Du Pont, is shown in Fig. 4.13(d).

If the size of the composite allows it, simple tensile tests can be carried out by machining ordinary specimens out of the body of the welded part. An example of this is provided in Fig. 4.14, which shows a strip of a trimetallic composite (on the left of the figure) and a machined out tensile specimen (on the right).

Bend tests are made to assess the response of the weld to tensile and compressive stressing. The three basic tests are shown diagrammatically in Fig. 4.15.

Mechanical non-destructive, micro-indentation hardness tests are also performed, particularly so in thin multilaminates, because the deformation gradients can be illustrated by a microhardness traverse across the weld interface. It is in this region that large plastic deformation takes place and produces strain-hardening. Further, since new metallic phases or intermetallic compounds are likely to form at the interface, their effect on the strength of the bond can be detected without actually damaging the composite.

4.2 PLANAR GEOMETRIES

4.2.1 Clad Plate

4.2.1.1 Applications
Clad plate is used either for further direct conventional forming, e.g. rolling and extrusion, providing an already homogeneously joined material, or for machining out transition joints, distillation columns, flow valves, shafts and fluidised beds.

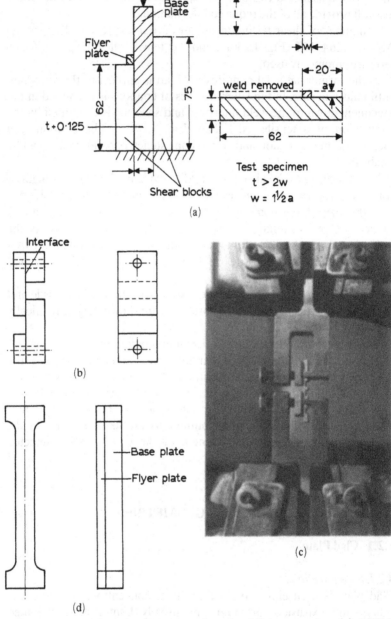

FIG. 4.13   Tests of mechanical properties of welded composites. (a) ASTM shear test; (b) and (c) tensile shear test; (d) tensile ram test specimen.

FIG. 4.14   Standard tensile test specimen (right) machined out of a trimetallic strip (left).

In the area of special applications, aluminium/copper, and aluminium/steel joints — and these material combinations, generally — are of particular commercial importance in the electrical, electronic, chemical and structural industries. Aluminium/steel combinations — often improved by the addition of a layer of silver, nickel, titanium and, possibly, pure aluminium — are used in cryogenic situations.

Each of these applications calls for a specific welding system to be developed — on the lines of those discussed in Section 4.1.2 — but, in turn, will offer a number of possible solutions. One such example (Fig. 4.16) is provided by the arrangements used in the lap-joining rectangular bars[1] intended for the manufacture of copper/aluminium bus-bars (Fig. 4.16(d)) and earthing strip connections, or rail-to-rail copper/steel current carrying joints (Fig. 4.16(c)).

A full discussion of the many industrial, specialised uses of clad plate will be found in Ref. 1.

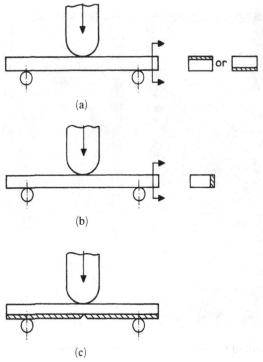

FIG. 4.15   Bend tests. (a) Normal bend test; (b) side bend test; (c) notched bend
test.

4.2.1.2 Interface welds

Although the metallurgy of the welded interface depends on the properties of the bonded materials and, in particular, on their respective acoustic impedances, a suitable manipulation of the welding parameters will afford a considerable degree of control over the type of weld produced.

In most cases, the basic possibilities indicated in Fig. 3.25 can be reproduced in composite plates. For instance, Fig. 4.17, referring to a copper/low carbon steel (M, S) clad, shows the effects of the stand-off distance and of the level of energy delivered. A relatively small stand-off distance, resulting in a lower impact velocity, will produce a very low amplitude, large frequency welding wave of a basically clean, metal-to-metal contact (Fig. 4.17(a), N.B. the black interface outline is due to etching). On increasing the stand-off distance to its optimal value for the

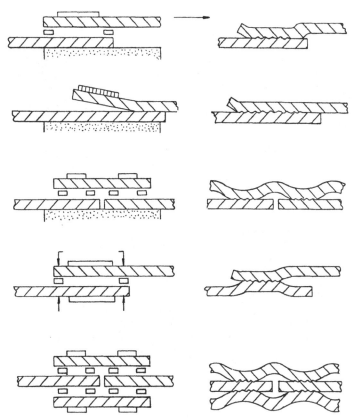

FIG. 4.16   Examples of lap bar welding arrangements (after Ref. 1). ⊏⊐, subsonic detonation front; ⬚⬚⬚, supersonic detonation front; ▫, interplate spacers.

energy supplied (Fig. 4.17(b)), a typical crested wave, with flattened humps, and direct metal-to-metal contact will be formed. When excessively high energy levels are generated by the charge, an irregular wave is created with vortices containing intermetallics of the partially trapped jet. Molten and resolidified material is also present and is distributed intermittently on the interface (Fig. 4.17(c)).

Further examples of the weld patterns are given in Fig. 4.18. The presence of intermetallics in the wave troughs of a stainless steel/silver clad is obvious in Fig. 4.18(a), whereas the acoustic impedance difference between, say, aluminium and silver (Fig. 4.18(b)), combined

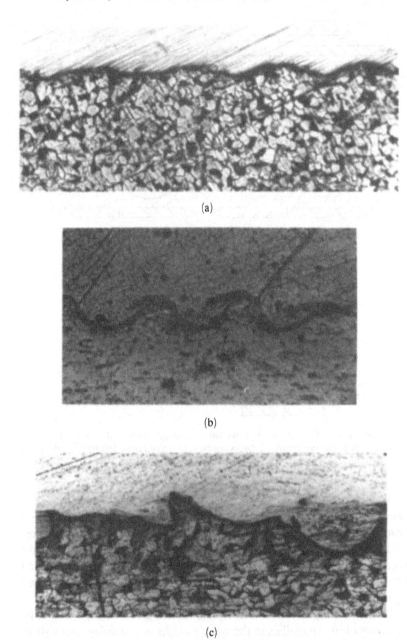

(a)

(b)

(c)

FIG. 4.17   Cu/low carbon steel interface waves (× 115). (a) Small stand-off; (b) sufficient stand-off; (c) excessive energy level.

with the low strength of the former, results in a resolidified intermetallic layer (black) containing also particles of broken away silver.

In many cases in which two metal plates to be bonded are 'incompatible', in the acoustic impedance sense, a thin layer or film of a third metal of properties lying in between those of either of the 'principals', is inserted to reduce the degree of incompatibility. This can be illustrated on the example of, for instance, an aluminium/tantalum/ stainless steel composite (Fig. 4.18(c)) in which tantalum is the 'bridging' material. Again, because of its low strength, aluminium, in this case, breaks away on the interface with tantalum and 'smears' over the latter. The wavy weld form is, however, present on the tantalum/stainless steel interface.

In the trimetallic clads, in which the constituent materials are more or less of comparable thickness, the degree of acceleration required between the individual plates is quite high and can be attained mainly through the medium of increased energy. If the system is not correctly designed, excessive interface melting and resolidification may be present, as in Fig. 4.18(d), and may result in the appearance of a practically continuous layer of brittle intermetallics separating the two materials. The figure is of further practical interest in that the type of weld changes from wavy to linear form across the middle plate.

The presence of intermetallics may well introduce an element of brittleness or reduced ductility, but may not in itself be unacceptable. This will depend on the properties of any given alloy and, naturally, on the final use to which the composite will be put. It should be noted that even with a high volume of melt of some 70%, static tests still produce good results.[12]

However, if the welding parameters are correctly selected, metal-to-metal, intermetallic-free interfaces are obtained both in similar (Fig. 4.19(a)), and in disparate materials such as, say, aluminium and low carbon steel (Fig. 4.19(b)).

Although, generally, wavy form welds are preferable, the straight- or almost straight-line configurations are required in certain circumstances. This is particularly so when the clad is to be conventionally rolled after the welding,[13] and in the composites consisting of aluminium, its alloys, and a range of steels. The main reasons for opting for flat interfaces are the reduction in the ductility of the shocked zone in wavy welds, caused by the adiabatic rise in temperature in the vortices and the consequent formation of molten pockets, high stress concentration factors associated with the presence of sharp peaks, and

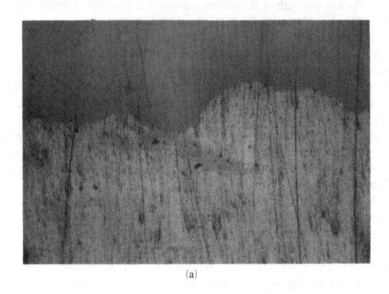

(a)

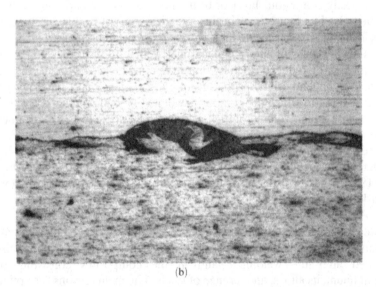

(b)

FIG. 4.18   Examples of interfacial waves. (a) Stainless steel flyer/silver base
(×150); (b) aluminium flyer/silver base (×100); (c) SEM micrograph of
aluminium/tantalum/stainless steel composite plate; (d) copper/brass/low
carbon steel composite (×150).

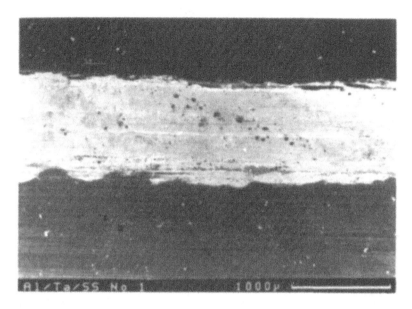

(c)

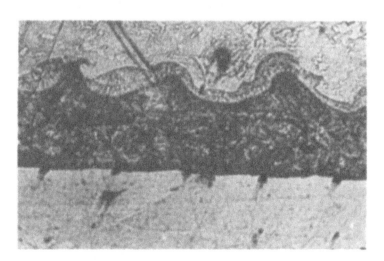

(d)

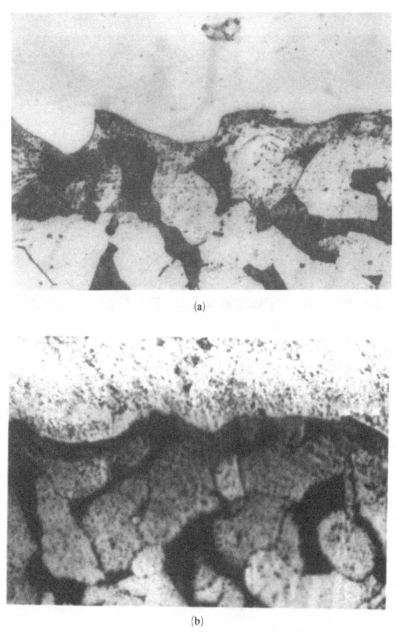

(a)

(b)

FIG. 4.19   Metal-to-metal wavy welds (×120). (a) Stainless steel flyer/low
carbon steel base; (b) aluminium flyer/low carbon steel base.

the fact that adiabatic shear bands, melted zones, and the interfaces are unfavourably orientated with respect to the direction of forming.

Some of these difficulties can be obviated or, at least, reduced when flat-interface welding. In rolling, cracking of the wave crests will be avoided, and in bimetallic combinations metallurgical properties may be improved. For instance, a straight-face stainless/low carbon steel clad, annealed at 850°C for 30 min, proved stronger than a wavy system.[14] This is probably due to decarburisation of the wavy interface around the peaks where the shock induced strain and diffusion rates are high.[15]

The more sensitive to diffusion combination of aluminium and steel is also, in general, better served by a straight faced bond, whereas in commercially clad rippled aluminium/steel interface deterioration sets in at around[16] 260°C, flat interfaces between aluminium and maraging steel can be raised up to 480°C without nucleation and growth of intermetallics.[17] No diffusion bond and no intermetallics, even after a 2 h anneal at 520°C, and no excessive hardening were detected in welded combinations of aluminium flyers (ranging in thickness from 6·5 to 19 mm) and steel bases (12–38 mm thick).[18]

Lower levels of trapped intermetallics are thus possible, while relatively high weld strengths are still available.

4.2.1.3 Mechanical properties

Although well over 300 metallic composites have been produced by explosive welding techniques,[19] most of the directly applicable combinations are listed in Table 4.2. More specialised and sophisticated composite materials, produced in relatively small quantities are discussed briefly in, among others, Refs 1 and 26.

The mechanical properties of the industrially utilised bi- and trimetallic clads are quoted in Table 4.2. All of these strength data were obtained by testing techniques described earlier, and are extracted from a number of specified sources.

Satisfactory properties can only be obtained if optimal or near optimal welding conditions are attained and this, in turn, requires the use of properly constructed 'welding windows'. Although, as indicated in Section 3.3.6, the usual procedure is to superimpose the individual material windows onto each other, some experimental work has already been carried out in this area and some information, albeit limited, is available.

In his work on explosive welding of molybdenum and TZM with Inconel 601 and Kanthal A 1, Pruemmer[24] has produced 'welding

TABLE 4.2
Typical Mechanical Properties of Industrially Utilised Bi- and Trimetallic Clads

Flyer	Base	Average-as clad (MPa)		Reference
		Shear	Tensile	
TM CA35ATi	MS	241	350	19, 20, 22
304 SS	MS	448	611	19, 20
Hastelloy C	MS	391	545	19, 20
11 00H 14Al	MS	96	505	19, 20
DHP copper	MS	152	511	19, 20
Ni alloys	MS	379	350	19, 22
Zr	MS	269		19
Cupronickel	MS	230	325	19, 22, 25
5083Al-3003Al	316L SS		130	21
50 52Al-OFHC Cu	304L SS		175	21
Brass/bronze		280		22
Ta		140		22
AlBr (B171 Alloy 614)			770	25

windows' for these composites. Similarly, Loyer *et al.*[25] have produced the empirical data needed to construct 'windows' for copper/ aluminium, and aluminium/steel systems. Specific 'windows' for structural and soft aluminium alloys were also determined.[27]

4.2.2 Multilaminates

4.2.2.1 Types and systems
Although clad plates may, in certain circumstances, be classified as laminates, the name is usually associated with either stacks of thin-layered materials such as foil, or with relatively small — in overall thickness — multimetallic composites used for the manufacture of transition joints, valves, etc.

Basic welding systems employed in these operations are of the kind shown in Fig. 4.6. Practical problems arising here are those of the difficulty of maintaining a constant stand-off distance between the adjacent thin layers, and of the variation, through the stack, in the mechanical aspects of the process. The effect of the latter is discussed in Section 3.3.5. The combination of changing geometry and the consequent variation in shock effects can lead easily to the generation of a reflected wave of sufficient strength to break the already formed bonds and to separate the layers of the laminate. An example of this is provided by Fig. 4.20.

In many applications, the strength of a laminate, consisting of a stack of thin foils, may not be sufficient and therefore reinforced multi-laminates are manufactured. The reinforcement, in the form of fibres, wires or meshes, can be either metallic or non-metallic, e.g. carbon or boron fibre, but for reasons of both economy and ease of operation, metallic inserts are more often used.

Four basic and self-explanatory systems are employed (Fig. 4.21); the welding proceeding either in a single operation or in a sequence involving pre-welded assemblies.

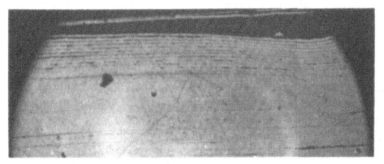

FIG. 4.20   Separation of foil layers in a welded multilaminate.

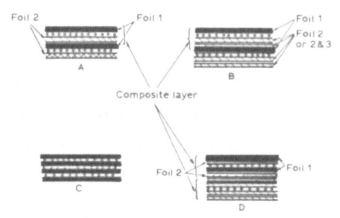

FIG. 4.21   Basic foil/mesh reinforced welding systems.

4.2.2.2 Foil laminates

Utilisation of the foil-type material results in three possible, practical arrangements. These range from a simple single foil sheet–substrate system, through multilayered foil stack (possibly up to a hundred sheets), to the wire or wire mesh reinforced aggregate.

Single foil–substrate systems can range from some 1600 m^2 of 0·2 mm aluminium foil cover[28] to a wide variety of metallic foils welded to alumina.[29] The latter application is taken from the opposite extreme of size of a 4·83 × 0·76 mm platelet of alumina base welded, in turn, to aluminium, copper, molybdenum, nickel, niobium, silver, tantalum, titanium, vanadium and zirconium foils.

To assess the quality of the bond, the authors proposed the concept of a Bond No. defined as the sum of the metal electronegativity and weight of the foil. On this basis, for a constant Bond No., any increase in the charge energy will improve the quality of the bond, but, equally, an increase in the Bond No. for a given level of energy available, will produce poor results. A classification of poor (P), and good (G), bonding, associated with the individual flyers, indicates that for the latter eventuality a value of Bond No. less than 1·7 is required. Considering the varying foil thickness, the results for aluminium and silver (7·62 × 10^{-3} mm) are respectively G (1·6) and P (2·24), for copper, nickel, niobium, tantalum, titanium, vanadium and zirconium (2·54 × 10^{-3} mm) the respective data are, P to G (1·98), P to G (1·91), G (1·68), G (1·67), G (1·45), P to G (1·78), and G (1·54). Molybdenum (3·81 × 10^{-3}) gives P (1·95).

The reinforced systems are well represented by a selection comprising aluminium and aluminium-alloy matrices with Kanthal and chromium/nickel meshes,[30] TZM wires strengthening a columbium alloy matrix,[31] steel wires in aluminium matrix,[32] and copper and aluminium matrices with tungsten wire in the former, and beryllium wire in the latter.[33]

Depending on the charge used, 0·3–1·0 mm thick aluminium and aluminium-alloy matrices reinforced with 0·3–0·8 mm diameter chromium/nickel meshes, gave tensile strengths of the composites ranging from 430 to 830 MPa, and A_{50} elongations of 4–9%. The relatively low elongations were explained in terms of matrix strain-hardening. The relationship between the composite (C), matrix (M), and reinforcement (R), as defined by the stress to fracture (σ_F) and the density ρ, was proposed in the form[30]

$$(\sigma_{CF} - \sigma_M)/(\sigma_{RF} - \sigma_M) \leqslant (\rho_C - \rho_M)/(\rho_R - \rho_M) \qquad (4.8)$$

The reinforcing of two 0·34 mm thick C 1294 columbium alloys with 0·25 mm diameter TZM wires, gives a tensile ultimate strength of the composite of 900 MPa, as compared with 690 MPa for the two plates welded to each other.[31]

Non-reinforced composite of a 3·18 mm aluminium flyer welded to a 6·4 mm aluminium base, fractures in tension at 127 MPa, but fails at 160 MPa when reinforced with 0·2 mm diameter steel wires.[32]

A special and highly sophisticated example of foil stack welding is that of honeycomb core manufacture. Honeycomb panels are widely used in the aircraft and motor industries where dimensionally and materially high quality components are required.

Explosively welded panels are produced by cutting and stacking, in parallel, up to 600 foil sheets with 'stand-offs' provided by printing the required strips on the respective foils. Water soluble gels, sugars and alginates are used and serve the double purpose of the initial layer separators and cell positioning.[34–36] After the welding, the multilayer sandwich is expanded to form a honeycomb panel.

Aluminium alloys, PH 15-7 Mo and 304 stainless steels, Hasteloy X, Rene 41, Inconel 718, Ti 75 A, TD Ni and copper foils are commercially welded.[34–36] Core thicknesses varying from 3·15 to 300 mm, with cell sizes in the range 1·6–6·4 mm are available in foil thicknesses between 0·0128 and 0·0152 mm. If required, multimaterial structures can be produced.

4.2.2.3 Transition joints

Bi- and trimetallic transition joints ranging from aluminium/copper bus-bars, through aluminium/steel with an antidiffusion, 2 mm thick titanium interlayer, to possibly multimetallic tubular, cryogenic connections, are currently produced.

Aluminium/stainless steel cryogenic joints are used in, for instance, liquefied gas storage vessels and are made with an interlayer of either silver or tantalum, depending on the use to which the joint will be put (Fig. 4.22), by welding plate sandwiches and machining the parts out of the laminates. A minimum tensile strength of some 280 MPa at − 196°C is usually achieved.[24]

However, five-layer cryocoup joints, consisting of layers of A 5083 aluminium alloy, pure aluminium, titanium, nickel and 304 stainless steel, and manufactured in sizes ranging from 25 to 450 mm o.d. are also available. At working pressures of about 4 MPa they exhibit no leaking[37] when mass spectrometer helium tested at a sensitivity of 10^{-10} cm^3/s.

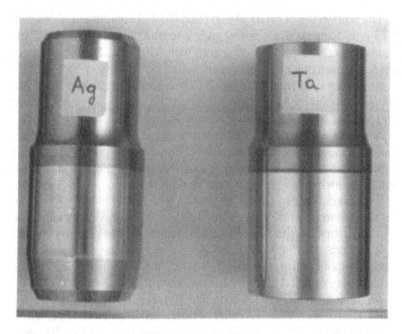

Fig. 4.22   Aluminium/silver/stainless steel, and aluminium/tantalum/stainless steel transition joints machined out of multilaminate plates. (Courtesy, ICI Nobel's Explosives Co. Ltd, Stevenston, Scotland.)

Further applications of explosively welded transition joints arise generally whenever the transition from copper to steel and aluminium, and aluminium to steel is required. For instance, in the aluminium smelting industry Kelomet anodes and cathodes of aluminium/carbon steel are satisfactorily used[38] at a temperature of 260°C, or, with a thin interlayer of ferritic stainless steel at a temperature of 340°C, showing an average voltage drop of 9 mV over a 2 or 3 year period.[38]

4.2.3 Structural Elements

Structural transition elements find their use in marine superstructure construction by providing aluminium/steel connections that can be welded conventionally to steel hulls. High corrosion resistance is usual and crevice-free interfaces are easily obtainable, typical trimetallic clads consist of a 6 mm aluminium alloy element welded to a 9·5 mm interlayer of aluminium bonded, in turn, to a steel backing part 19–25 mm thick.[21] The steel backer can then be welded directly to the ship superstructure. Clads of this kind are manufactured in strips of up to 3·5 mm in length.

Bimetallic aluminium/steel prefabricates are also used for structural purposes[15] in, for instance, the welding of vertical elements (Fig. 4.23(a)), joining of rolled sections (Fig. 4.23(b)), and welding of more complex parts to superstructures (Fig. 4.23(c)).

Single and double explosive clads (2) in Fig. 4.23(a), provide basic joints between aluminium and steel. The joints between two over-lapping aluminium plates and a vertical aluminium component are further strengthened by fillet welds (3), whereas the steel part of the system is fillet welded (1) to a horizontal steel plate. This type of explosively welded interface is shown in the micrograph.

A possibility of using welded rolled sections in structural composites is explored in Fig. 4.23(b), where a T-section is explosively welded to a flat aluminium element (2), and the steel element is further extended by butt welding it to a steel plate, while, at the same time, the aluminium part of the composite is fillet welded (3) to a vertical aluminium element.

Similarly, a double-channel explosively welded (2) to two aluminium flat plates can be fillet welded (1, 3) to any desired aluminium components.

Further possibilities are indicated in Fig. 4.23(c), where a structure consisting of an explosively clad aluminium/steel plate (2) is, in turn, welded to an upright aluminium plate to form a stanchion-like component, and is then joined to large area steel plates, or, for instance, where a T-shaped aluminium structure is welded, again, to a large steel plate through the medium of a composite (2).

4.2.4 Explosive Hardening

The early work of Rinehart on mild steel,[40] and MacLeod[39] on austenitic manganese steel, has established the fact that the passage of a high-pressure shock wave, causing severe lattice deformation and appearance of additional dislocations, will alter the hardness of the material. Depending on the geometry of the system and the magnitude of the charge — in direct contact with the metal — either the bulk of the material or part of it will be affected. In the latter case, metallurgical changes taking place will effectively produce a two-layer composite, the respective elements of which will differ in their properties.

Providing that the magnitude and configuration of the high-pressure wave is carefully controlled and that, therefore, no spalling, cracking or

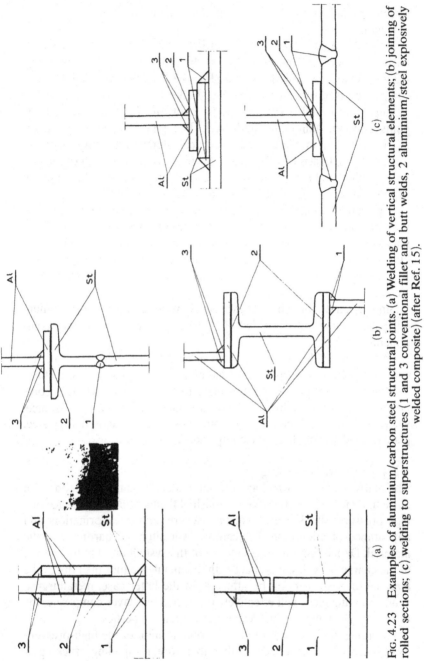

Fig. 4.23   Examples of aluminium/carbon steel structural joints. (a) Welding of vertical structural elements; (b) joining of rolled sections; (c) welding to superstructures (1 and 3 conventional fillet and butt welds, 2 aluminium/steel explosively welded composite) (after Ref. 15).

other gross damage to the metal takes place, two direct practical applications of the effects of these phenomena become clear.

First, there exists a possibility of hardening large metal components, *in situ*, in a manner which offers simplicity of operation combined with a high degree of qualitative uniformity. Second, there is the possibility of autofrettaging or pre-stressing of metal components, resulting from the imposition of plastic deformation. This becomes even more attractive when it is realised that it can be achieved without the use of machine tools of any description.

So far, most of the experimental work has been concerned with hardening and from the practical point of view, the interest has been centred mostly on the austenitic manganese or Hadfield type steels. These, on the one hand, respond more readily to the treatment, and, on the other, have been used for heavy duties in a wide range of applications. The beginning of the use of the explosive welding technique for industrial purposes dates to the mid-1950s, when railway points were first strengthened in this way.[39]

In addition to the railway points, the most usual application of this process is the hardening of blow bars (Fig. 4.24) which constitute the active elements of mineral crushers. In the hardening process, either or both of the leading and trailing faces of the bar carry a single, double or treble strip of explosive charge, extending from the apex of the section. In this way, six distinct test conditions can be arranged and the hardness distribution assessed.

The application of the technique to low carbon steels has been given less consideration, mostly because of the comparatively modest rise in the level of hardness, but also because of the possible problems arising out of the influence of strain rate on the transition temperature between

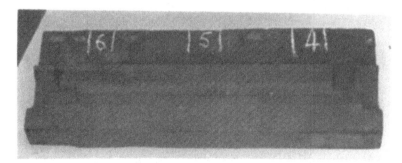

FIG. 4.24   An austenitic manganese steel blow bar in 'as cast' condition.

ductile and brittle fracture. The response of mild steel to hardening was investigated on plates 6 mm thick, the hardening being carried out by exploding single and double charges on both sides of the plates.[40] Figure 4.25 indicates that there exists a well defined relationship between the surface hardness, promoted by a shock wave, and the weight of explosive charge per unit area of contact.

The curves for the austenitic manganese and low carbon steels are almost parallel, and the ratio of surface hardness between the two materials for a given level of energy varies between 1·5 and 1·7. The curve for low carbon steel is interesting also in the sense that it shows that small variations in the carbon content in the steel do not influence the behaviour of the metal to any great extent. The curve was obtained in a series of experiments involving two different steels.

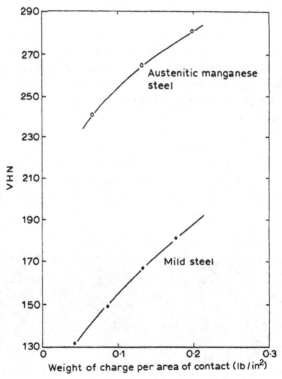

FIG. 4.25   Variation of metal hardness with the weight of charge per unit area of contact (after Ref. 40).

The distribution in depth of hardness can be determined from Fig. 4.26. This deals, again, with the two materials, but the thicknesses involved are of different order.

In the case of blow bars (austenitic steel) it is seen that, in general, the increase in the explosive charge is equivalent to the increase in the surface hardness and also to the increase in the level of hardness throughout the faces of the bar where explosively loaded. It was found that the ratio of the hardness on the surface of the metal to that in the centre of the mass of the component varied between 1·02 and 1·05 for the largest and lowest charges respectively. The corresponding variation in the values of the surface hardness was from 280 to 240 VHN.

When the leading face only of the bar was subjected to the explosive treatment, the surface hardness, for a given charge, was the same as that in the previous case, but the hardness then decreased to a lower value and this was maintained over a certain depth. The hardness plateau thus formed was then followed by a further reduction in hardness, progressing at a higher rate than that nearer the surface of contact with the

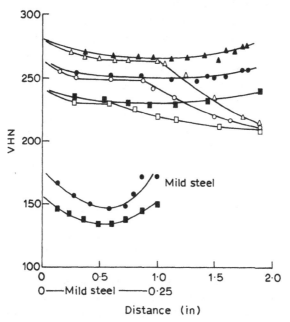

FIG. 4.26 Distribution in depth of metal hardness (after Ref. 40). Single charge: □, leading face, ■, both faces; double charge: ○, leading face, ●, both faces; treble charge: △, leading face, ▲, both faces.

explosive, and leading to the original condition of the metal. In the case considered, the original hardness was about 215 VHN, and this would be reached near the other face of the wedge.

It is clear, therefore, that, unlike the case of shot-peening or case-hardening, where a hard thin layer of metal rests on a relatively soft cushion — with consequent tendency to surface cracking and fracture — the hardness produced by a high-pressure shock wave is reduced gradually throughout the thickness of the metal. The harder layer of metal near the surface is supported by a hardness plateau of varying length which acts as a protecting barrier between it and the softer metal near the other face.

The occurrence of the plateau was first reported by Rinehart[40] who ascribed its existence in low carbon steels to the formation of Neumann bands or shock twinning. Further, Rinehart's results seemed to indicate that the difference between the hardness of the plateau and the original hardness of the metal was constant irrespective of the magnitude of the explosive charge. The only influence that the magnitude of the charge might have on the formation of the plateau was the bringing of it nearer the surface and possibly extending its length.

Some of these effects can be extended to the austenitic manganese steels, but there are notable exceptions. It is clear that the increase in the charge will extend the length of the plateau, although it may not necessarily bring it nearer the surface, but it is also clear that the difference between the hardness of the plateau and that of the original material depends on the magnitude of the charge.

The possible explanation of this effect lies in the fact that, although the increase in the charge will not materially increase the peak value of the pressure wave, it will prolong the time interval over which a high-level uniform impulse is delivered to the metal. In other words, the time interval during which the metal is subjected to intercrystalline changes will be extended and, consequently, an increase in the hardness level will be expected — at least in those ferrous alloys which respond more readily to such changes.

The explosively treated and untreated, matched pairs of blow bars were used in an operational crusher and the wear, in terms of loss in weight, was determined. The loss in weight in untreated bars was approximately three times that in the treated components. Again, the difference between explosive loading on both faces and on one face only was of the order of 23% in favour of the symmetrical type of loading.

4.3 AXISYMMETRICAL GEOMETRIES

4.3.1 Multiplex Tubular Components

4.3.1.1 Applications and techniques

The high degree of sophistication required in the rapidly developing technology of the manufacture of petrochemicals, electronic devices, particularly cybernetics, cryogenic systems, and of atomic pile and toxic metals remote control handling systems calls for the development of a new range of engineering composites. Some of these can either no longer be produced by conventional techniques alone, or can be made at fairly high cost only. A typical example of the first group is a multi-layer, possibly multimetallic, cylindrical pressure vessel, and of the second a semiconductor system enclosed permanently in a protective metal sheath.

In between these two extremes, there lies a wide range of more commonplace components such as special types of bi- or multimetallic heat exchangers combining structural strength with anticorrosive properties and, most important, ensuring a rate of heat flow as good as that displayed by the individual metals of the combination. It is in this last respect that the conventional co-drawing or co-extruding methods of production are likely to give less satisfactory results since, even with a high degree of process control, it is practically impossible to prevent permanent contamination of the mating surfaces. The lack of cohesion between the original components of such an assembly is likely to result in a very high degree of differential deformation leading to the appearance of inhomogeneous strains and consequently to an increase in the redundancy in the system. Technologically, this will result in poor quality of the product and a high ratio of rejects.

Many of these limitations and difficulties can be avoided if the integrity of the component in question is assured by, for instance, implosive welding prior to conventional processing. For instance, bi- or trimetallic, thick-walled welded cylinders offer a solution to the problem of, conventionally very expensive, the manufacture of multiplex pressure cylinders used in a variety of hydraulic presses. The normal production route is that of shrinking, with interference fit, of the outer onto the inner cylinder. The high degree of dimensional tolerance required calls for the honing of often large diameter cylinder surfaces and for a careful assessment of the range of temperatures.

In a welded, say, bimetallic system, the plastic deformation of the flyer (Fig. 4.27) is sufficient to lead to the development of the residual stress field in the duplex cylinder, which in an implosive operation is compressive. Effectively, therefore, pressurisation of the cylinder during its working life can be increased substantially, because initially the working pressure has to counterbalance the compressive stress system. In addition, since a stand-off distance is necessary, honing is not required and ordinary machine finish of the interfaces is quite sufficient.

The welding of a multiplex tube is carried out in either an explosive or implosive system (Figs 4.7 and 4.8). In an explosive system, the mainly tensile residual stresses are high and therefore the range of applications of the tube, as an internally pressurised cylinder, is limited.[41,42]

Although, basically, no limit on the length of the tube welded in an implosive system need be imposed, in practice the welding of long, 8–10 m, and perhaps heavy-walled tubing will call for relatively high levels of energy and may not necessarily be acceptable in industrial situations. It is for this reason that short lengths, larger diameter tubing is welded and then conventionally processed to give the required

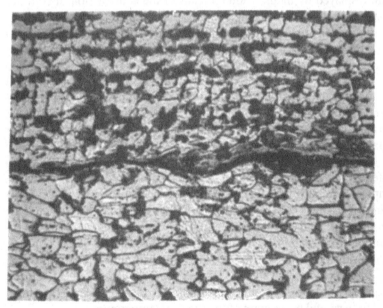

FIG. 4.27   Plastic deformation of the flyer (upper layer) in a duplex cylinder.

lengths and wall thicknesses. In view, however, of the severity of cold-drawing or extrusion operations the quality of the weld obtained at this stage should be as high as possible.

4.3.1.2 Interface morphology

Although the basic features of the explosive welding interface will be present in an axisymmetrical tubular configuration, the practical side of the operation imposes limitations on the geometry of the system. Only parallel configurations are economically viable and the relatively small stand-off distances reduce the value of impact velocities. In consequence, the wave amplitude tends to be smaller than in plate welding, and the frequency is increased. With the limited energy levels available in explosive systems, these differences become even more pronounced (Fig. 4.28) in this case.

Again, as with planar geometries, clean metal-to-metal interfaces can be obtained (Fig. 4.29), or trapped jet situations can arise (Fig. 4.30). A 'composite' made up of the original and a new phase material can also be formed (Fig. 4.31). Although this particular brass/brass combination is not necessarily of practical interest, it does illustrate the point that metallurgical changes can be easily effected in the material(s) and thus composites can be formed in the course of implosive welding.

A trimetallic cylinder exhibits slightly different features (Fig. 4.32) in that the wave amplitude clearly increases radially towards the base, although loss of energy supplied by the detonation front does occur in that direction. The explanation lies probably in the welding conditions approximating, in this situation, to those associated with the stress mechanism of wave formation (Section 3.3.2).

4.3.1.3 Processing and properties

Depending on the ultimate destination of the composite, post-welding processing may or may not be necessary.

Where multiplex pressure cylinders or pressure vessels are concerned, the quality of the 'as-welded' component is sufficiently high to allow its assimilation into the working system as part of a machine or an element of a production cycle. However, if the original composite was manufactured as a short, but large diameter tubing and is required to serve as an element of a pipeline or be part of, for instance, a heat exchanger, it will be necessary to process it further by cold-drawing or extrusion. In view of the special interface properties of such a composite, conventional metal-forming may create certain problems which

(a)

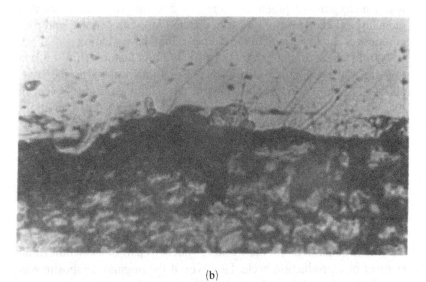

(b)

FIG. 4.28   Duplex cylinder brass (flyer)/low carbon steel (base). (a) Implosive system with more pronounced waves; (b) explosive system.

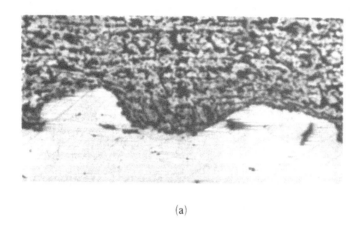

(a)

(b)

FIG. 4.29   Bimetallic cylinder, metal-to-metal welds. (a) Stainless/low carbon steel ($\times 100$); (b) copper/low carbon steel ($\times 120$).

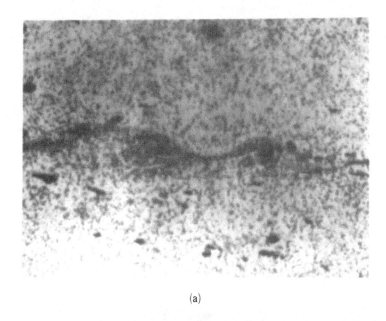

(a)

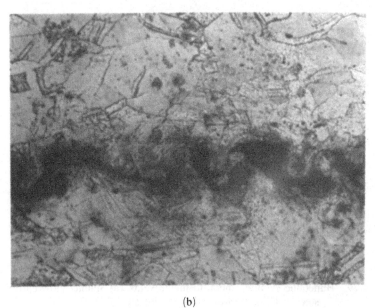

(b)

FIG. 4.30   Bimetallic cylinder, brass/brass interface with new phase in the
wave tails and trunks (× 600).

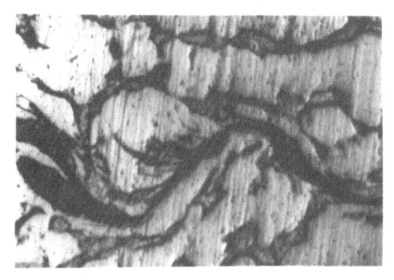

FIG. 4.31   Development of a different phase on the interface of a brass–brass welded duplex cylinder.

FIG. 4.32   A section through a trimetallic brass/copper/low carbon steel welded cylinder (× 15).

will then be reflected in the final mechanical properties of the product. These are specific to particular cases and have to be reviewed accordingly.

(a) Duplex cylinders. The interaction of process parameters is best discussed with reference to actual engineering material combinations, welded in either an explosive or implosive system.

In a particular investigation,[10] aluminium (BS 1471 HT 30 WP), brass (Br) (BS 249), low carbon steel (MS) (BS 3602), stainless steel (SS) (EN 58), and commercially pure copper (Cu) tubing was welded in a variety of combinations. The characteristics of the respective systems were as follows:

Explosive system: Flyer o.d. 45–48 mm, wall thickness (t) 1·2–1·8 mm, base wall thickness (t') 4–10 mm, average length of the cylinder 140 mm.

Implosive system: Flyer o.d. 50–80mm, $t = 1·2–3$ mm, $t' = 3·5–8·5$ mm, average length 400 mm.

Stand-offs (c) varied from 0·23 to 0·75 mm and from 0·04 to 0·60 mm for the two systems respectively.

The actual size ranges used in particular combinations, and the associated amplitudes and wave-lengths are given, for comparison, in Table 4.3.

In general, for a given flyer D/t ratio, the amount of energy required is lower in the explosive than in implosive systems. The loss of available energy when welding implosively is much higher since the rate of dissipation is unimpeded by any external constraint. Pressure waves reflected from the wall of the inner cylinder are re-utilised in elasto-plastic, plastic and hydrodynamic transformations that take place in the flyer and on the weld interface. However, as the bore increases, i.e. the relative D/t ratio is reduced, the loss of energy will be greater and the required energy levels will become similar for both systems. In the set of conditions investigated, the equality would be expected at about $D/t = 100$.

The mechanical properties of the welds can be assessed from the hardness distribution across the cylinder section (Table 4.4), and the tensile shear test (Fig. 4.13, Table 4.5).

The distribution of hardness in the flyer suggests the appearance of a plastic layer of the material near the charged surface of the component, followed by a band of the elastic metal retaining its original hardness

TABLE 4.3
Materials, Sizes and Wave Characteristics

| System | Test No. | Material | | D (mm) | t (mm) | t' (mm) | c (mm) | CLA (μm) | | Wave | |
		Flyer	Base					Flyer	Base	Amplitude (mm)	Length (mm)
Implosive	1	Cu	MS	53·3	2·41	4·66	0·08	0·7	0·25	0·01	0·03
	2	Brass	Brass	53·3	2·50	5·00	0·07	1·9	1·5	0·025	0·09
	3	Al	Al	53·3	2·50	5·00	0·04	2·0	1·4	0·01	0·06
	4	Brass	Cu	53·3	2·50	3·70	0·11	1·3	> 5	0·02	0·06
	5	MS	Cu	60·8	2·20	7·70	0·20	> 5	0·76	0·03	0·24
	6	Al	Al	72·0	2·90	7·70	0·08	4·3	1·0	0·02	0·07
	7	Brass	Brass	51·0	1·30	4·80	0·05	> 5	2·0	0·03	0·09
	8	Brass	MS	60·0	2·10	5·70	0·22	> 5	> 5	0·01	0·07
	9	MS	Al	51·0	1·20	3·60	0·09	2·0	1·0	0·02	0·14
Explosive	10	Cu	MS	42·0	1·80	19·9	0·23	0·25	1·0	0·06	0·19
	11	Cu	MS	44·6	1·40	10·2	0·30	3·8	> 5	0·08	0·20
	12	Brass	MS	48·0	1·25	4·00	0·75	> 5	0·25	0·06	0·17
	13	Al	MS	48·0	1·50	4·00	0·50	4·6	0·76	0·01	0·16

TABLE 4.4
VHN Hardness Measurements

System	Test No.	Material		Flyer			Base		
		Flyer	Base	o.d.	MID	Weld	Weld	MID	i.d.
Implosive	1	Cu	MS	135	64	143	206	170	170
	2	Brass	Brass	160	116	170	170	116	116
	3	Al	Al	99	83	97	97	61	61
	4	Brass	Cu	69	67	122	160	135	135
	5	MS	Cu	206	193	302	133	96	96
	6	Al	Al	140	148	175	175	97	94
	9	MS	Al	181	148	206	97	80	80
Explosive	10	Cu	MS	98	110	151	309	202	202
	11	Cu	MS	98	110	151	302	202	202
	12	Brass	MS	181	176	181	297	206	213
	13	Al	MS	61	83	93	220	181	206

TABLE 4.5
Strength of Representative Welds

System	Test No.	Material		Coupon (shear stress in MPa)				
		Flyer	Base	A	B	C	D	E
Implosive	1	Cu	MS	122·6[a]	167·5[a]	177·4[a]	176·5[a]	—
Implosive	2	Brass	Brass	121·3	152·4	154·4	153·6	164·9
Implosive	3	Al	Al	116·7	119·2	128·4	122·3	118·6
Implosive	9	MS	Al	54·3	72·1	74·4	76·5	72·4
Explosive	10	Cu	MS	253·8[a]	192·5[a]	189·1[a]	196·4[a]	189·1[a]
Explosive	13	Al	MS	38	36·4	42·8	38	—

[a]Copper failure.

(MID), and, finally, a plastically deformed layer in the vicinity of the weld. The effects of the transmitted pressure wave are apparent in the base, but only in the immediate vicinity of the weld interface. The base remains, essentially, rigid.

The properties of the weld interface depend greatly on whether an intermetallic alloy is formed. The presence of an alloy will, generally, increase the hardness level and consequently reduce the degree of ductility.

For geometrically similar systems, the levels of weld hardness reached in a given material are basically independent of whether that particular metal acts as the base or the flyer. In other words, in the absence of an intermetallic alloy, the level of hardness is governed only by the mechanical properties of the materials involved. A comparison between the two welding systems supports this observation in the sense that, for a given D/t, the different energy levels employed do not appear to influence the maximum values of hardness in the respective components.

With regard to the shear strength, it is seen that this is equal to, at least, the shear yield strength of the weaker material. Where the difference in yield strengths between the welded metals is significant, the failure tends to occur within the weaker material and not on the weld interface even though an element of brittleness may be introduced on that surface.

Of particular practical interest and importance is the question of the type and distribution of residual stress fields in multiplex cylinders. As already mentioned, building-in a basically compressive field will increase the pressure bearing capacity of the cylinder and thus improve its efficiency and working life.

A useful comparison between the effects of the respective welding systems is provided by Fig. 4.33 which shows the distributions of radial (σ_R), circumferential (σ_ϕ), and longitudinal (σ_Z) residual stresses in MS/MS and Cu/MS cylinders.[10,41] Essentially, either mechanism of welding operation will give rise to a reasonably well defined residual stress pattern.

Considering, for instance, the circumferential stresses, it is seen that, as a result of the application of pressure, the flyer will experience a tensile stress. As the flyer accelerates towards the base, it retains and possibly increases its tensile residual stress and then passes into the plastic regime. Depending on the level of the available energy, the flyer will either only touch the base or it will impact the component. In either case, the base will acquire a compressive stress which, in the case of impact, may be numerically significant. After the welding has taken place, there will exist a tendency to spring back, which will result in a decrease in the level of tensile stress in the flyer, and in a possible

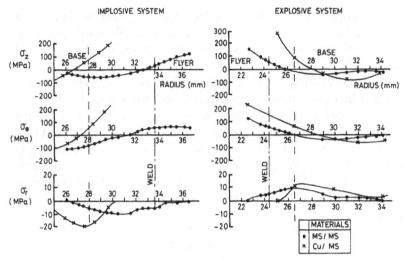

FIG. 4.33   Comparison between residual stress fields in explosively and implosively welded duplex cylinders.

increase in the compressive stress in the base. These tendencies are clearly visible in the figure.

As would be expected, the radial stress systems are, because of the system geometries, opposite in their nature. Consequently, compressive stresses are acquired by both the flyers and bases in the implosive system, and tensile stresses are generated in the explosively welded cylinders.

Longitudinal tensile stresses characterise flyers in both systems, with bases remaining predominantly compressed.

Stress levels are, on the whole, low, especially for like metallic combinations, but are slightly higher for bimetallic systems. These differences reflect the mis-match between the mechanical properties of the alloys involved.

The highest recorded stresses were of the order of 300 MPa and were therefore substantially lower than the levels expected to develop in duplex cylinders manufactured by conventional shrinking technique.

It is clear that the general level of circumferential stress is bound to be affected by the amount of strain imposed, i.e. the magnitude of the annular stand-off, and that consequently the stress sign and level can change.

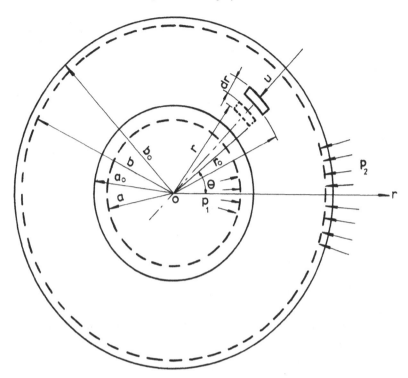

FIG. 4.34   Motion of an element of tube wall during the loading phase (after Refs 41 and 42).

From a practical point of view, it is important to be able to predict stress patterns *a priori* and therefore use is made of a mathematical model proposed by Blazynski and Dara,[41,42] based on a modified (by them) Appleby's approach to the response of a pressurised cylinder.[43] This particular model can be summarised as follows: using the notation of Fig. 4.34, and assuming $\phi(T)$ to be an arbitrary function of time, taking also p_1 as the internal pressure exerted on the bore of the cylinder by the supporting plug, and p_2 as the pressure generated by the charge, the radial and circumferential stresses during the loading (welding) (L) phase of the operation are given by:

$$\sigma_R = -p_1 \ln(r/a)[2K + pd^2\phi/dT^2]$$
$$+0{\cdot}5\,d\phi/dT(1/a^2 - 1/r^2)[\lambda - \rho\,d\phi/dT] \tag{4.9}$$

and

$$\sigma_\theta = 2K + \sigma_R + (\lambda/r^2)\, d\phi/dT \tag{4.10}$$

with the plane strain condition prevailing in implosive systems

$$\sigma_Z = 0{\cdot}5(\sigma_\theta + \sigma_R) \tag{4.11}$$

where λ is the coefficient of viscosity, K is the dynamic yield stress in shear, and

$$a = [a_o + 2\phi(T)]^{0{\cdot}5} \tag{4.12}$$

also

$$b = [b_o + 2\phi(T)]^{0{\cdot}5} \tag{4.13}$$

The relationship between the two pressures is defined by

$$p_1 - p_2 = \rho \ln(K)\, d^2\phi/dT^2$$
$$+ 0{\cdot}5(1/a^2 - 1/b^2)\, d\phi/dT\,[\lambda - \rho\, d\phi/dT] + 2K \ln(K) \tag{4.14}$$

where $K = b/a$.

If it is assumed that the motion of an element within the cylinder body approximates closely to sinusoidal, $\phi(T)$ becomes

$$\phi(T) = A^2 b^2/2[(1/A + 1/\omega)^2 - b_0^2/2 \tag{4.15}$$

where $A = \omega/2[b_1/b_o - 1]$, and ω is the frequency.

The unloading (U) or springback phase is purely elastic and the classical Lamé equations determine the three stresses associated with part of the process.

In consequence,

$$\sigma_r = [p_1(r^2 - b^2)/r^2(k^2 - 1)] + [p_2(r^2 - a^2)/r^2(1/k^2 - 1)] \tag{4.16}$$

$$\sigma_\theta = [p_1(r^2 + b^2)/r^2(k^2 - 1)] + [p_2(r^2 + a^2)/r^2(1/k^2 - 1)] \tag{4.17}$$

$$\sigma_Z = 2\mu[p_1/(k^2 - 1) + p_2/(1/k^2 - 1)] \tag{4.18}$$

where μ is Poisson's ratio.

The residual stresses (σ_R) represent the respective stress differences and therefore

$$\sigma_{RR} = \sigma_{LR} - \sigma_{UR} \tag{4.19}$$

$$\sigma_{R\theta} = \sigma_{L\theta} - \sigma_{U\theta} \tag{4.20}$$

$$\sigma_{RZ} = \sigma_{LZ} - \sigma_{UZ} \tag{4.21}$$

The validity of this approximate approach was tested on two sets of duplex cylinders of MS/MS and MS/Cu composition respectively.[42] The relevant dimensional and material data are given in Table 4.6, and the calculated and experimentally determined residual stress distributions are given in Fig. 4.35.

In the theoretical assessment, the time of deformation T_1 was assumed to be approximately equal to the total time of detonation; the assumption is reasonable since the radial stand-off distances are small and therefore, with a high energy loss inherent in an implosive system, the detonation pressure pulse would approximate closely to the total time of detonation of the charge. On this basis, the total time for specimen No. 1 was 120 μs, and for specimen No. 2, 117 μs.

The value of the dynamic yield stress in shear presents some difficulty. Since this quantity depends on the strain rate, it is necessary to evaluate the latter first. In the case considered, the respective values were obtained as 100/s and 64/s, and so the values of K (from the appropriate stress–strain curve) were taken as 183 MPa for steel and 130 MPa for copper. The transient effect of temperature is not accounted for in these values.

It will be noticed that because at the end of loading $\dot{\phi}(T_1) = 0$, the viscosity terms do not enter into the calculations. Basically therefore, the assessment of the stress field is confined to the plastic and inertial terms of the respective equations. The viscosity term vanishes in this case as a result of the assumed form of the time function.

The agreement between the calculated and experimentally obtained results is satisfactory, in spite of the simplifications introduced by the approximate K values and by neglecting the effect of viscosity. The inclusion of the latter, particularly at higher velocities of deformation, would be desirable, but this would, of course, call for a different type of time function.

TABLE 4.6
Dimensional and Material Data (all dimensions in mm)

a_0	b_0	a_1	b_1	a_2	b_2	c	t	t'	l
Specimen No. 1 MS flyer — MS base									
26·2	37·0	25·9	36·61	25·8	36·56	0·13	2·95	7·75	357
Specimen No. 2 MS flyer — copper base									
20·3	30·4	20·2	30·22	20·1	30·18	0·20	2·20	7·70	348

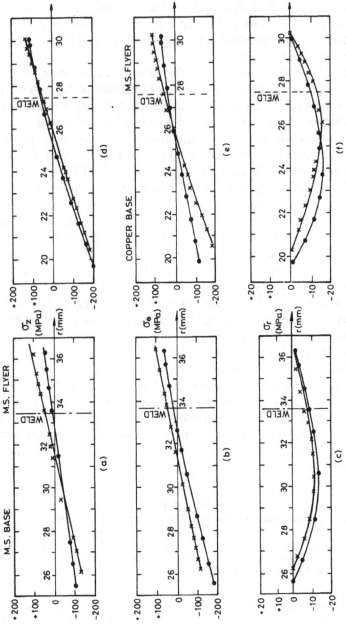

Fig. 4.35 · Comparison between the calculated and experimental values of residual stresses in implosively welded duplex cylinders: ×, experimental; ●, —, theoretical (after Refs 41 and 42).

(b) Triplex cylinders. The use of triplex tubing has been established in a variety of situations ranging from pressure vessels for the petrochemical industry to the provision of a base for semiconductor systems. Further, a feasibility study[44] of the explosive welding of triplex tubing incorporating axial leak detection channels indicates yet a different application for this kind of composite. The mechanics of the operation, involving first the welding of two components, followed by the impact of the new, double flyer onto the base, bring the process nearer the situation that obtains in multilaminates. In consequence, the interface morphologies are slightly more complex than in duplex cylinders (Fig. 4.32).

Again, the basic tendencies and properties of this type of composite are best discussed with reference to an investigation relating to a combination of three engineering materials.[45] BS 3602 low carbon steel (MS), BS 2871/CZ 105-885, 70/30 brass (Br), and BS 61 1947 copper (Cu) tubes, in Br/Cu/MS and MS/Br/Cu combinations were implosively welded. The brass and copper tubing was annealed at 650°C prior to welding. The dimensions of Flyer 1 (outer) varied from 50 to 60 mm o.d., and from 0·3 to 14 mm in thickness, those of Flyer 2, from 49 to 53 mm o.d., and 2·6 to 4 mm in thickness. The base dimensions ranged from 44·5 to 49·7 mm o.d., and from 3·7 to 9·5 mm in thickness.

Depending on the available levels of energy and also the annular stand-off distances, the three basic types of interfaces are identified. With just a sufficient rate of energy dissipation, the presence of either the wavy or line metal-to-metal bonds is noted, whereas an excess of energy results in the appearance of either wave or line welds with pockets or layers of intermetallics or solidified melt. The amplitude of the wave increases with the stand-off distance (s) for any given material combination. For equal, but relatively small radial spacings, the welds are practically linear and the amount of plastic deformation — indicated by crystal orientation — is also small.

The importance of the quality of the welds depends here, as in a duplex tubing, on the final objective of the operation. Although an element of brittleness introduced by the presence of trapped jet or solidified melt is responsible for the weakening of the weld, the latter will normally remain strong enough to allow further processing, e.g. cold drawing, of the composite. The heat transfer conductivity and other physical properties of the component are not likely to be strongly affected.

Tensile shear tests of the type shown in Fig. 4.36 indicate that the shear strength of the Br/Cu combination varied from 200 to 240 MPa

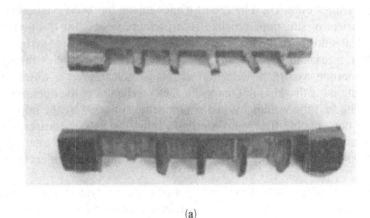

(a)

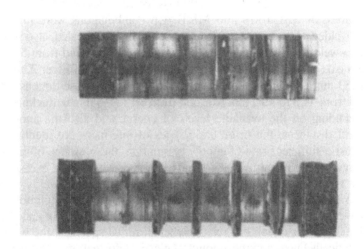

(b)

FIG. 4.36   Tensile shear test specimens machined out of multiplex cylinders. (a) Side view showing shearing and bending of flyer coupons; (b) front view.

and the failure of Flyer 1 was in shear. The Cu/MS base strength ranged from 100 to 240 MPa and failure by necking of the base would occur.

In the MS/Br systems, the strength varied from 40 to 185 MPa with the failure of the weld taking place, and in the MS/Cu base combinations the strength ranged from 145 to 210 MPa with shearing being the mode of copper failure.

The brass/copper assembly tended to produce hard layers of solidified melt on the interfaces. Microhardness tests on a MS/Br/Cu composite showed an increase in hardness on the surface of Flyer 1 that
reflects small plastic deformations of the material in contact with the
charge. Little or no increase in hardness was noticed in the tube wall
until the interface was reached. A gradual increase in hardness —
reaching a maximum either near or on the surface — would be noted,
the increase is partly due to the local plastic deformation and partly,
where applicable, to the presence of the solidified, trapped jet. On those
surfaces which consist of continuous intermetallic layers, the hardness is
always much higher than that of the metal substratum in the vicinity of
the weld.

(c) Special applications. A special case of multiplex cylinders arises
often in those industrial applications in which high temperatures,
required to improve energy conversion efficiency, are generated in
systems that necessitate the use of highly anticorrosive, but mechanically strong materials. This is of particular importance at temperatures exceeding 1000°C in solar heat absorbers, thermo-ionic energy
converters, etc. The necessary working properties of suitable composites can be successfully, and relatively simply, obtained by implosive
welding of refractory materials to the corrosion resistant metals.

Typical examples of compound tubing, and of discs machined out of
welded composites is provided by the work of Prümmer and Henne,[46]
who manufactured Kanthal Al/molybdenum tubes, and Inconel 601/
molybdenum cups. The results were generally satisfactory in that gas
penetration of hydrogen only was detected. The hot shells, produced by
implosive welding, were UHV-tight.

(d) Plug drawing. Conventional, cold plug-drawing operations[47-49]
are used to reduce the wall thickness and outer diameters of composite
tubing, and to increase it to a useful length.

Although the discussion of the operation and its general outcome is
limited here to an anticorrosive MS/Cu duplex, and the already
mentioned MS/Br/Cu triplex systems, the existing experimental data
indicate that the conclusions reached in those cases are applicable to
other material combinations.

Two basic problems merit consideration in the drawing operation of
this type. One is the mechanics of the process itself, associated,
naturally, with the performance of the plant, and the other is the quality

of the final product. In the context of the former, the effect on the drawing stress of the volume fraction (f) of the flyer, in the interchangeable material combinations, is of interest, as is also that of the lubricant. In the context of the latter, the effect of the deformation on the quality of the weld must be considered. The relatively small changes in the patterns of hardness distribution are of minimal importance, at this stage, and in the interest of clarity can be neglected in the present discussion.

Figure 4.37, giving the variation of the drawing stress with the total homogeneous strain, indicates that for the medium valued fractions (f), interchangeability of the component materials does not affect the drawing stress to any noticeable degree. For instance, in the case of interchangeable low carbon steel/copper systems, a nominal average volume fraction of 0·55 produces the same levels of drawing stress for both copper and steel flyers. An increase in the value of f to a nominal average of 0·65 raises the drawing stress, for a given strain, by about 4%, but, again, the order in which the materials are arranged does not seem to be important. It is clear, of course, that for very low fractions, in other words for very thin sheath flyers and thick bases, and vice versa, the effect of the volume fraction and that of the flyer material will be pronounced. The levels of drawing stresses do not however differ from those expected in the processing of similar standard monometallic, seamless tubing.

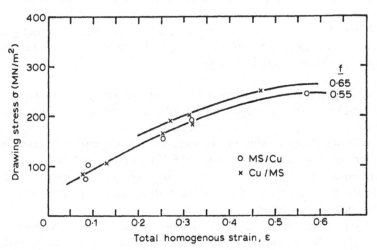

FIG. 4.37   Variation of the drawing stress with total homogeneous strain and flyer volume fraction (f) in cold plug drawing of duplex tubing (after Ref. 46).

With regard to lubrication, it was found that Rocol RL5 gave a satisfactory surface finish irrespective of the material system. Variations in the drawing stresses observed in identical material and deformation conditions were found to have resulted more from the distortions of the duplex tubes, produced during the welding, than from the failure of the lubricating film.

The effect of drawing on the quality of the product can be deduced from Table 4.7. Bearing in mind that the 'before' and 'after' drawing measurements cannot, of course, correspond to the same physical sections of the tube, it is noted that, in general, a deterioration in the weld strength takes place. The extent to which the strength falls depends on the type of the weld and, for the wave profiled welds, on their amplitudes.

Failure of the welds during machining of the specimens, e.g. Tests No. 17 and 18, is indicative of the inherent local weakness produced during the welding stage and points, again, to the necessity for careful control of the welding operation. It is clear that the drawing operation, in addition to weakening the weld by reducing its amplitude, changes also the mode of failure. Whereas failure of the base by necking appeared to be the dominant feature of the strength tests after the welding, failure of the weld itself is observed in these tests after the drawing. Since, however, the drawing takes place mainly in the conditions of plane compressive strain, even the complete failure of the weld along its whole length would not be catastrophic and the tube could still be drawn, although the amount of relative differential deformation could be high. The advantages of retaining clean interfaces between the respective tube components would still outweigh the effect of weld failure.

The drawing of triplex tubing[45] (Fig. 4.38) leads to similar conclusions. Again, general deterioration is noticed in the weld quality, accompanied, for the Br/Cu part of the composite, by a change in the mode of failure. In spite, however, of the possible disappearance of the weld on parts of the surfaces, the drawn tubing retains its enhanced heat transfer properties since these, clearly, do not depend on the weld strength. The strength, for an MS volume fraction of 0·6 and a brass fraction of 0·3, varies between 0 and 110 MPa along the tube for the MS/Br interface, and between 80 and 180 MPa for the Br/Cu interface.

From a practical point of view, mathematical modelling of the drawing process is necessary to establish the optimal working conditions. A suitable analysis, based on Shield's approach, was proposed by Townley and Blazynski.[50,51] The analysis is based on the assumption

TABLE 4.7
Weld Strength in Shear (all data in MPa)

No.	Before drawing Coupons								After drawing Coupons								Type of weld
	1	2	3	4	5	6	7	8	1	2	3	4	5	6	7	8	
1	135b	140b	165b	167b	170b	172b	180b	175b	148b	193b	183b	167b	188b	185b	172b	178b	Wavy
7	185s	125s	148s	176s	200s	154s	160s	140s	180w	121w	174w	131s	154s	109w	87w	126s	Wavy
16	58b	71b	88b	86b	m	64b	77b	m	40w	60w	67w	65w	50w	47w	49w	52w	Wavy
17	185w	189w	210w	248w	220w	237w	195w	212w	199w	210w	157w	104w	98w	159w	m	m	Wavy
18	191b	173b	160b	168b	123b	106b	121b	166w	91w	73w	m	m	123w	106w	121w	96w	Wavy
19	222b	237b	257b	268b	265b	269b	212b	243w	161w	164w	173w	131w	197w	188w	194w	186w	Wavy

b, failure of base by necking; m, failure in machining; s, failure of flyer in shear; w, failure of the weld.

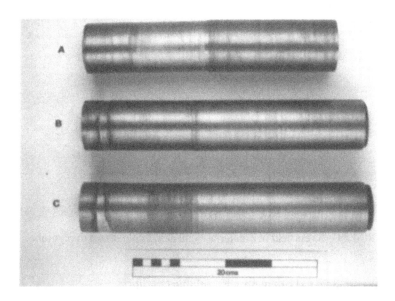

FIG. 4.38   Low carbon steel/brass/copper triplex tubing cold plug drawn.
A- 17%, B- 21%, C- 23% (after Ref. 45).

that the drawing is effectively a plane strain process, and friction is
defined by means of a factor, m, such that the limiting shearing stress
$\tau = mY/V^3 = mk$ (Y is the yield stress in simple tension, and k that in
shear).

A consideration of the flow at the interface between materials 1 and
2 (Fig. 4.39), i.e. when $\theta = \beta$, leads to the conclusion that shear stresses
on that boundary must be equal. Hence,

$$k_1 c_1 = k_2 c_2 \qquad (4.22)$$

where c reflects the conditions of flow and yielding.[50] The mean
drawing stress in the general case is given by

$$\sigma_M = k_1 c_1 \ln(r_0/r_1) \qquad (4.23)$$

where r_0 and r_1 are the pass entry and exist radii. For frictionless
drawing, the mean drawing stress becomes

$$\sigma_M = 2k^* \ln(r_0/r_1) \qquad (4.24)$$

with k^* representing an equivalent yield stress in shear that corresponds

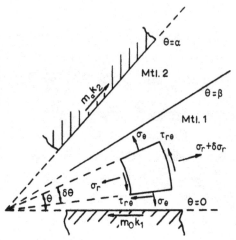

FIG. 4.39   Stress field of a material element in a cold drawn duplex tube.

to the common, to both materials, yielding condition.[52]

$$k^* = [(1 - f) k_1 + f k_2]$$ (4.25)

The angular distribution of shearing stress, at any radius r within the die, is given by the following expression:

Material 1 $(c_1 > 2)$

$$\theta - 0 = 0 \cdot 5 [2c_1 \tan^{-1}(A^{0.5} \tau)/(c_1^2 - 4) - 2 \tan^{-1} \tau]$$ (4.26)

where

$$A_1 = (c_1 + 2)/(c_1 - 2),$$

$$B_0 = [1 - (1 - m_0^2)^{0.5}]/m_0,$$

$$B_1 = [1 - (1 - m_1^2)^{0.5}]/m_0, \text{ and } m_1 \text{ is the friction factor at } \theta.$$

Material 2

$$\alpha - \theta = [2c_2 \tan^{-1}(A_2^{0.5} \tau)/(c_2^2 - 4) - 2 \tan^{-1} \tau]$$ (4.27)

where

$$A_2 = (c_2 + 2)/(c_2 - 2),$$

$$B_2 = [1 - (1 - m_2^2)^{0.5}]/m_2,$$

$$B_\alpha = [1 - (1 - m_a^2)^{0.5}]/m_a, \text{ and } m_2 \text{ corresponds to } \theta.$$

The limits of homogeneity of flow are imposed by the magnitude of either the interfacial shearing stress or the drawing stress.

In the first instance, assuming a good bond between the constituent materials, the flow will cease to be uniform when the interfacial shear stress reaches the value of the yield stress in shear of the weaker material. In the latter situation, the tube is likely to fail in tension if the drawing stress at the exit from the die reaches the value of the mean tensile yield stress of the composite. With regard to the effect of the interfacial shear stress, it is noted that even with weak or non-existent intermetallic bonds, the stress can be supported by frictional means because the deformation takes place while normal pressure is acting on the interface. The frictional effect will be at a minimum at the exit from the die and consequently this mode of failure would be likely to originate in that section of the pass.

An ideal case of perfect bonding can be discussed with reference to Fig. 4.40. Since the shear stress is a function of θ only, its distribution across the wall is representative of the situation obtaining in any other section of the pass. The distribution is defined by two straight lines (Fig. 4.40(a)) whose point of intersection coincides, in the particular case used here as an example, with the interfacial position $\theta = 0.6\alpha$ ($f = 0.4$). The maximum shear stress in the weaker material occurs at the interface where $\tau = 0.3k$ and therefore failure due to shear is unlikely.

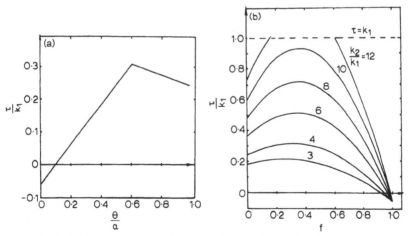

FIG. 4.40   (a) Shear stress profile across the wall ($f = 0.4$); (b) interfacial shear stress and volume fraction for die angle $\alpha = 7.5°$, $m_\alpha = m_o = 0.06$.

Since the influence of the friction factors is small, the interfacial shear stress can be regarded as being a function of mainly k_2/k_1 and f. It is thus seen (Fig. 4.40(b)) that the interfacial shear is particularly high for $0.2 < f < 0.5$, and that for $k_2/k_1 = 11$ the stress will be equal to the shear yield stress when $f = 0.4$. As the ratio is increased above this critical value, the interfacial shear will be less than the shear yield stress for a limited range of f values only.

Figure 4.41 (representing here only one die angle) is used to clarify the situation further. The figure indicates that, for instance, for $k_2/k_1 < 10.6$, the interfacial shear stress will not reach the shear yield stress of the weaker material. This minimum k-ratio value of the regime boundary occurs for $0.3 < f < 0.4$. For high f values, lying outside this range, the permissible k-ratio values can be higher.

The other limiting factor, the value of the mean drawing stress, affects the maximum possible amount of deformation. For any set of friction factors m, the mean drawing stress depends on the die angle. The stress decreases with increasing die angle because for a specified reduction any decrease in the die angle is equivalent to an increase in the area of contact and therefore in the frictional effect. The permissible k-ratio range also decreases with the increasing die angle. This tendency is associated with a rise in the interfacial shear stress. The total strain that can be imposed without causing tensile failure at exit from the die also increases with any increase in the die angle.

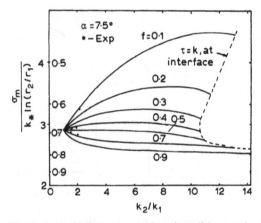

FIG. 4.41   The regime of homogeneous flow (conditions as for Fig. 4.40).

4.3.2 Foil and Mesh Reinforced Cylinders

4.3.2.1 Types and applications

Although they differ in their geometries, the principles behind the manufacture of axisymmetrical foil cylinders are the same as those discussed for the corresponding planar configurations.[53]

The two basic systems are those of multilayer foil, and multilayer foil–metallic mesh reinforced assemblies. Both are manufactured by preparing flat sandwiches[54,55] — using one of the arrangements of Fig. 4.21 for mesh reinforced cylinders — and then wrapping them onto rigid bore-supporting mandrels. Thus prepared assemblies are, subsequently, welded in a manner shown in Fig. 4.8.

In the foil/foil systems, the welding takes place directly on the adjacent interfaces, but when foil layers are interleaved by metallic mesh, the levels of energy supplied need to be adjusted so that the impacting foils can actually fill-in the mesh and therefore the welding between the foils and strands of the mesh can be effected. It is only in this way that an acceptably 'homogenised' structure of the component can be obtained, i.e. a structure which will not contain randomly spaced voids and unwelded areas.

The main area of applications of either type of cylinder lies in the storage and transportation of pressurised fluids by offering a low mass, but structurally strong container.

4.3.2.2 Metallurgical aspects

The combination of strength and relatively low mass ratio of a cylinder can be achieved by making extensive use of aluminium foil, strengthened either by a foil of a 'stronger' material, or by treating aluminium as the matrix and reinforcing the structure with a strong metallic mesh.

In a foil/foil system, the quality of the weld between any two surfaces depends, in terms of the frequency and amplitude of the interface wave, not only on the material characteristics of the respective foils (as reflected by their acoustic impedances) but also on the effect of the stress wave reflection from the successive layers and, eventually, the bore supporting mandrel. The material properties of the mandrel are therefore of considerable importance in this case. Typical material combinations used[53,54] are indicated in Table 4.8.

Since the impacting, originally compressive, pressure/stress wave will be partly transmitted and partly reflected through and from the successive interfaces, the reflected wave will become tensile on alternate

TABLE 4.8
Materials and Sizes of Welded Cylinders[54]

Test No.	Base tube Material	o.d. (mm)	i.d. (mm)	Foil 1 Material	t (mm)	Foil 2 Material	t (mm)	Total o.d. before welding (mm)
1	Cu	58·4	50·8	Cu	0·1016	Brass	0·1016	61·0
2	Cu	58·4	50·8	Cu	0·1016	Al	0·1016	61·0
3	MS	58·4	50·8	MS	0·0762	Al	0·1016	61·0
4	MS	58·4	50·8	MS	0·0762	Cu	0·1016	61·0
5	MS	58·4	50·8	MS	0·0254	Brass	0·1016	60·0
6	MS	58·4	50·8	MS	0·0762	MS	0·0762	60·4
7	MS	58·4	50·8	Brass	0·1524	Al	0·1016	62·2
8	MS	58·4	50·8	Al	0·1016	Al	0·1016	61·0
9	Al	58·4	50·8	Al	0·0127	Al	0·0127	60·0

surfaces. An increasing number of stress disturbances will propagate back towards the free cylinder surface and this, in turn, will modify the amplitude and frequency of the wave at any given interface. If in two materials of differing acoustic impedances the prevailing system conditions allow significant waves to be formed, these will appear on the surfaces on which the incident compressive stress wave passes from the medium of a lower acoustic impedance to that of a higher one. The propagation of stress disturbances towards the free surface increases the frequency of the welding wave in that direction, but reduces its amplitude. The transition from a higher to a lower acoustic impedance results in a line-type weld.

It is clear therefore that the order in which the successive material layers are initially arranged can be of importance insofar as the mechanical properties of the system are concerned.

These points are illustrated in a selection of microphotographs in Fig. 4.42. For instance, in a MS/Al/MS mandrel system of Fig. 4.42(a), line-type interfaces are observed with the ratio of acoustic impedances being 2·71. The mandrel is welded to the aluminium foil, but pockets of solidified melt are present. On the other hand, the same material combination, welded in a system with considerably enlarged radial stand-off distances (Fig. 4.42(b)), shows the presence of wavy welds which increase in amplitude radially inwards. At the same time, however, the high intensity reflected tensile stress waves produce separation of the already welded layers (dark straight interlayers).

An Al/Cu system (ratio of impedances of 0·42) (Fig. 4.42(c)) is characterised by the presence of both wavy and line welded interfaces. The interplay between the stand-off and stress-wave effects is such that the wavy welds occur on alternative surfaces. Again, the wave amplitudes increase towards the bore supporting mandrel.

With well matched ratio of impedances, as for instance in a Cu/Br (ratio 1·14) system of Fig. 4.42(d), good wavy welds are obtained, with only small quantities of jet material being trapped, intermittently, in wave troughs.

The foil/mesh/foil systems are manufactured in the same way as the foil/foil assemblies, but some predetermined layers of foil are replaced by closely woven mesh with initially unattached warp and weft wires. The thickness of the foil matrix, the number of foil sheets for each layer of the mesh, and the geometry of it — as defined by the mesh number and wire gauge — determine the volume fraction, f, the required average radial stand-off, the total thickness of the composite layer, and

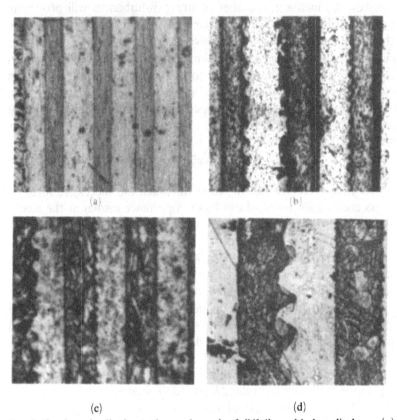

(c)   (d)

FIG. 4.42   Longitudinal sections through foil/foil welded cylinders. (a) Aluminium/low carbon steel ($\times$ 100); (b) aluminium/low carbon steel with layer separation ($\times$ 100); (c) aluminium/copper ($\times$ 160); (d) copper/brass ($\times$ 160).

the number of weld interfaces. The resulting composite is a multi-layered, bidirectionally reinforced cylindrical shell.

An insight into the metallurgical properties of these composites is provided by, for instance, the investigations of Blazynski and El-Sobky.[54,55] A range of materials, shown in Table 4.9, was employed. The two flat mesh angles θ_1 and θ_2, itemised in the table, are defined as:

$$\theta_1 = \tan^{-1}[(t-d)/(a+d)]$$

and

$$\theta_2 = \tan^{-1}[(3d-t)/(a+t)] \qquad (4.28)$$

TABLE 4.9
Reinforced Foil–Composite Cylinder Geometry and Dimensions of Foil Matrix and Wire Mesh

Test No.	Matrix material	Welding system	Mesh	Wire dia. (mm)	Mesh thickness (mm)	θ_1/θ_2 (°)	Foil thickness (mm)		% Vol. fraction f_w	Base tube material
							1	2		
1	Al	D	20 × 20 × 33	0·254	0·55	13·13/9·48	0·1524	0·1524	11·77	Al
2	Al	A	20 × 20 × 35	0·213	0·41	10·28/8·80	0·1016	0·1016	21·94	Al
3	Al	A	14 × 14 × 34	0·234	0·55	10·81/6·90	0·1016	0·1016	19·11	Al
4	Cu	A	20 × 20 × 35	0·213	0·41	10·28/8·80	0·1524	0·1524	15·78	Cu
5	Cu	A	20 × 20 × 35	0·213	0·41	10·28/8·80	0·0762	0·0762	27·26	Cu
6	Cu	A	20 × 20 × 33	0·254	0·55	13·13/9·48	0·1016	0·1016	28·59	Cu
7	Cu	A	20 × 20 × 33	0·254	0·55	13·13/9·48	0·0762	0·1270	28·59	Cu
8	Cu	A	20 × 20 × 27	0·416	1·00	24·71/11·06	0·1778	0·2032	37·24	Cu
9	Cu	B	20 × 20 × 33	0·254	0·55	13·13/9·48	0·1524	0·0762	21·97	Cu
10	Cu	C	20 × 20 × 35	0·2134	0·41	10·28/8·80	0·254	—	18·35	Cu
11	Brass	A	20 × 20 × 33	0·254	0·55	13·13/9·48	0·1016	0·1016	28·59	Cu
12	Brass	A	20 × 20 × 33		As above		0·1524	0·1524	21·07	Cu
13	Brass	A	20 × 20 × 33		As above		0·254	0·254	13·80	Cu
14	Brass	A	20 × 20 × 30	0·2946	0·66	11·51/7·11	0·254	0·254	12·97	Cu
15	Steel	A	20 × 20 × 30		As above		0·127	0·127	24·26	Steel

where a is the distance between two parallel mesh wires, d is the wire diameter, and t is the mesh thickness. Where appropriate, subscripts m and w refer to matrix and wire respectively and the welding system classification is that of Fig. 4.21.

The two angles define the stiffness of the mesh, whereas the volume fraction f_w indicates the degree of strengthening of the composite by the reinforcement.

Because of the foil size effect some difficulties can be experienced when welding thin steel foils. Since foils are produced by cold rolling, their mechanical properties, as reflected by the hardness, correspond to an increase in the yield stress and, consequently, a tendency to separation of the layers during the welding operation is likely. This alone, irrespective of the mesh size effects, will produce a faulty composite.

In most engineering alloys, however, this problem is of relatively lesser importance in comparison with the effects of a possible lack of compatibility of the volume fractions of the mesh and matrix. In a correctly assembled system, welding of the successive layers of foil to each other and to the wires of the mesh takes place satisfactorily (Fig. 4.43(a)), with some deformation of the wires occurring in this particular Cu-matrix/SS-mesh combination. When more ductile materials are welded, such as for instance Al-matrix/Cu-mesh (Fig. 4.43(b)), considerable deformation and distortion of the mesh is apparent, with warp and weft wires becoming welded to each other, as well as to the adjoining layers of the foil. In a slightly volume fraction mismatched system (Fig. 4.44(a)) of a Cu/SS combination, the mesh areas are not fully filled with the foil and small voids remain in the structure. The integration of the mesh, through the medium of a wavy welding interface, is noted in two different material combinations shown in Fig. 4.44(a, b).

The three main modes of structural distortion, normally observed, are shown in Fig. 4.45. In Fig. 4.45(a), showing a transverse section through a composite, the layers of foil change from the originally parallel to a geometry conforming to that of the mesh. The foil interfaces are welded, but they contain small areas of resolidified melt. The mesh wires are distorted randomly to, basically, elliptical shapes and are welded to the neighbouring foil layers.

In a ductile material combination, the weft mesh component can be distorted by the welding operation to such an extent that it assumes a large wavy form (Fig. 4.45(b)). A clearer example of this is afforded by Fig. 4.45(c), in which the foils assume the geometry of the mesh, while the latter welds to the matrix in either a large amplitude wave or in line interface.

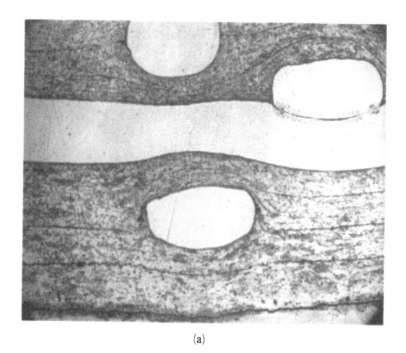

(a)

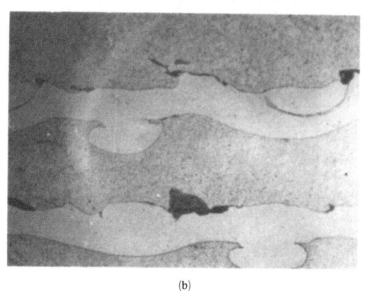

(b)

FIG. 4.43   Foil/mesh/foil explosively welded shells. (a) Copper matrix/stainless steel mesh (× 100); (b) aluminium matrix/copper mesh (× 120).

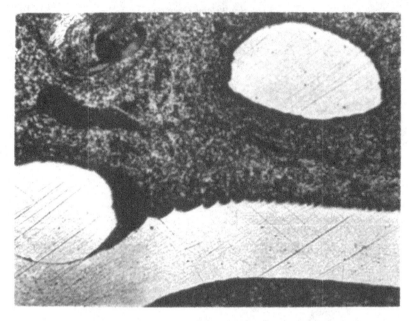

(a)

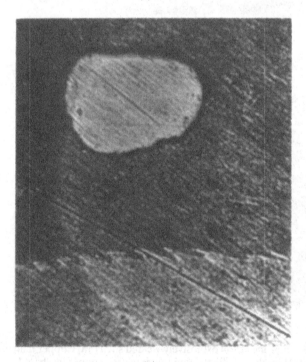

(b)

FIG. 4.44   Welding of wire mesh to foil matrix in foil/mesh/foil composites.
(a) Copper matrix/stainless steel mesh (×100); (b) aluminium matrix/stainless
steel mesh (×150).

The already discussed stress mechanism of wave formation assumes a special significance in the foil reinforced systems. Rotational flow (Fig. 4.46) takes place with the amplitude of the wave often in excess of the foil thickness and mesh wire diameter. This type of flow can be damaging to the structure (Fig. 4.46(b)) by dislocating and, sometimes, breaking the wires of the mesh.

4.3.2.3 Mechanical properties

Whereas the quality of the weld, and therefore the likely behaviour of pure foil/foil systems, can be judged reasonably well from the photomicrographs and tensile tests, the mechanical properties of the foil/mesh/foil systems are more difficult to assess. Because of the mesh geometry, the actual length of a wire element in a unit length of the welded system exceeds unity by a factor of $1/\cos \theta_1$ in the longitudinal, and by $1/\cos \theta_2$ in transverse directions respectively. The wire reinforcement may therefore, in any given loading situation, undergo an initial displacement before it can start to contribute to the load bearing capacity of the composite. The tests carried out on samples of wire mesh indicate dependence, in this respect, on the number of mesh filaments. The lower the number the greater the slack and the free displacement, before interlocking occurs at the cross-over points.

These effects are displayed by basic tensile load-extension diagrams of the welded composites. A typical Cu-matrix/SS-mesh composite (Fig. 4.47), for instance, produces five distinct zones. Zone I is associated with the elastic deformation of both the matrix and the mesh. It is in this zone that the effect of slackness manifests itself when the apparent strain appears to be greater than the actual. In Zone II, the mesh is still in the elastic stage of deformation, but the interlocking of the wires has been completed. The matrix begins to deform plastically. A discontinuity in the curve defines the boundary between the two zones. Zone III represents plastic flow in both elements caused by the failure of the interfacial bond. The wires begin to break in Zone IV which is characterised by a considerable drop in the load bearing capacity of the composite. The load is then taken entirely by the plastically deforming matrix until the latter fractures within Zone V.

Tensile tests were carried out, on a representative range of welded composites, to provide a guide to the likely values E of Young's modulus, and Poisson's ratio. Strains in longitudinal and transverse directions were measured (Fig. 4.48) by means of electrical resistance strain gauges. A simple method of calculating E, based on the rule of mixtures, was also used and a comparison between the calculated and

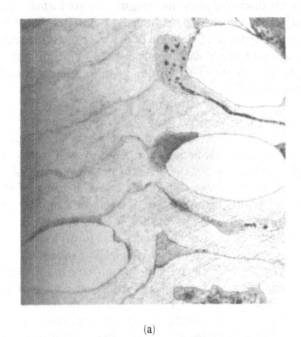

(a)

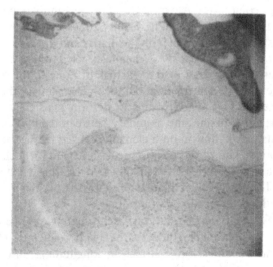

(b)

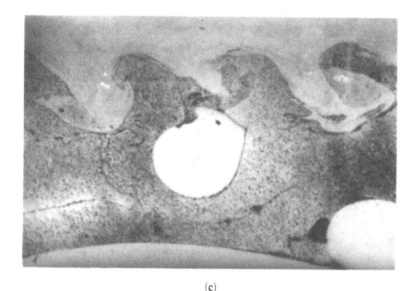

(c)

Fɪɢ. 4.45   Types of structural distortion in the foil/mesh/foil composites. (a) Displacement of wires; (b) heavy deformation of a wire; (c) wave formation in a wire with jet entrapment in the wave troughs.

measured quantities is provided in Table 4.10. Material combinations of Fig. 4.48 can be identified by their test numbers referred to in the table.

Since the system in question is bidirectionally reinforced, a typical layer of the composite is modelled, in the calculation of E, by two unidirectionally reinforced layers of the same volume fraction, but superimposed onto each other at right angles. The effective, equivalent modulus is therefore given by:

$$E = 0 \cdot 5\, E_L + 0 \cdot 5\, E_T \qquad (4.29)$$

where E_L and E_T are the longitudinal and transverse moduli respectively.

$$E_L = f_w + f_f E_f \qquad (4.30)$$

and

$$1/E_T = f_w/E_w + f_f/E_f \qquad (4.31)$$

Table 4.10 indicates that the Poisson's ratio of the composite is usually lower than that of any of its constituent materials. The calculated values

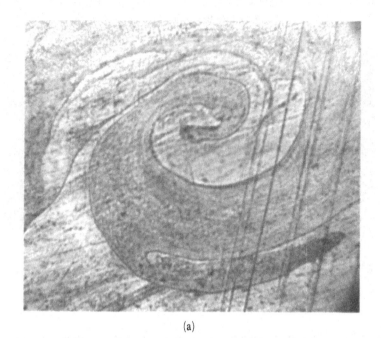

(a)

(b)

FIG. 4.46 Examples of rotational flow in foil/mesh/foil systems. (a) Large mechanical foil wave (×150); (b) arrested base rotational flow showing structural grain distortion and deformation (×160).

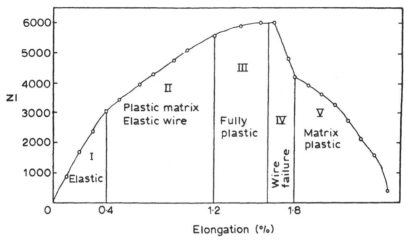

FIG. 4.47 An experimental load/extension curve for a copper foil matrix/ stainless steel mesh reinforced composite (after Ref. 41).

of E are generally in excess of the experimentally established, on average by about 12%, although, in some cases, the discrepancy can be as high as 23 or 34%. This is attributed partly to the fact that the modelling procedure (eqns (4.29)–(4.31)) does not take into consideration the geometry of the wires, and partly to the free displacement of the mesh wires and the resulting apparent high strain values.

Wire, but not mesh, reinforced composites are used for special applications and display better mechanical properties than those of their matrices. An early example[31] of such a material (columbium matrix/molybdenum wire) is provided by Fig. 4.49.

4.3.3 Multimetallic Remote Control Systems

4.3.3.1 Applications and manufacture
Special industrial needs, particularly those of the atomic energy and chemical industries, require, occasionally, the production of 2–3 m long and geometrically intricate components that are capable of combining strength with good protective and/or heat transfer characteristics. In the former case, refractory metals are likely to be involved, whereas in the latter situation, extended surface areas of specified shapes and anti-corrosive properties may be required. These, at times diverse requirements can be fulfilled if a satisfactory technique for manufacturing

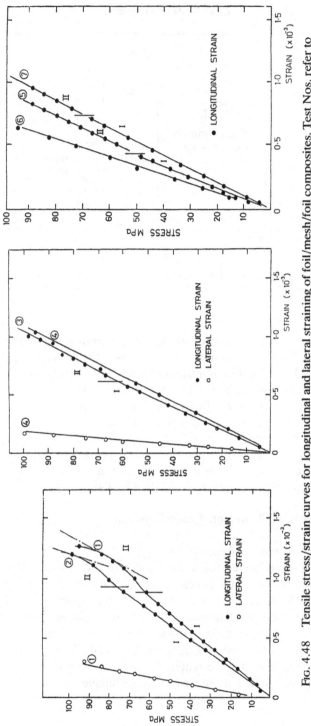

FIG. 4.48   Tensile stress/strain curves for longitudinal and lateral straining of foil/mesh/foil composites. Test Nos. refer to Table 4.10, and the respective zones are denoted in Roman numerals.

TABLE 4.10
Tensile Tests of Specimens Machined from Reinforced Composite Cylinders

Test No.	Matrix material	Specimen No.	f_f	Modulus E (GPa)		Poisson's ratio (measured)	Area (mm²)
				Measured	Calculated		
1	Al	1	11·77	67·7	79·1	0·30	16
2	Al	2	21·94	80·0	88·245	0·22	16
3	Br	12	21·07	96·8	118·638	0·17	20
4	Br	13	13·8	101·0	113·324	0·20	12
5	Cu	4	15·78	106·0	120·845	0·30	20
6	Cu	7	28·59	148·0	129·84	0·28	12
7	Cu	9	21·97	93·3	125·13	0·28	20

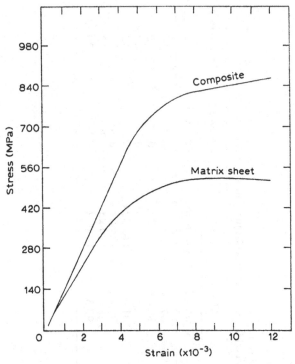

FIG. 4.49   Stress/strain curve for a columbium matrix/molybdenum wire reinforced composite (after Ref. 31).

assemblies of mono- or bimetallic multilayered, coaxial rods is developed.

Conventional metal-forming processes which would normally provide means of elongating the initially short refractory metal combinations are often difficult to operate. Drawing or extrusion of loose rod arrays, held in position in locating outer sheaths, may lead to unsatisfactory results associated with the flow characteristics of, perhaps, widely different constituent materials. Differential elongation will be aggravated further by relative slipping of the individual rods, or of the whole layers, with respect to each other. A solution to these problems is, however, provided by initial attachment of the constituent elements of the assembly to each other, followed by a conventional forming operation. In this type of approach, the welding operation ensures the integrity and rigidity of the composite and therefore facilitates further processing.[56-59] A convenient operational method is that of implosive welding of individual rods, and of the successive mono- or bimetallic layers to the neighbouring elements. The technique derives directly from that used in the implosive welding of solid multiplex and multilayered metal foil cylinders.

The process of manufacturing integral assemblies of bimetallic rods consists of two distinct steps, i.e. implosive welding and subsequent hydrostatic extrusion.[59,60] The primary objectives of the former are to give, naturally enough, a good quality weld between the respective elements of the array, to ensure that the basic shape of the assembly is affected as little as possible, but to provide, at the same time, a billet suitable for extrusion.

Two variants of the welding techniques are normally applicable. Depending on the final objective of the exercise, loose rods can either be contained within a tubular metallic sleeve or, alternatively, within a plastic one. In either case, the sleeve acts as a gas shield, since it prevents scattering of the rods by the detonation gases, and, also, as a means of ensuring uniform pressure distribution on the outer, or flyer, elements of the assembly. Since the rods are closely, but loosely, held together the stand-off distance between the individual elements cannot be uniform throughout. The non-uniformity in the magnitude of individual clearances may slightly affect the quality of welds produced between the respective parts of the assembly.

The presence or absence of a metallic sleeve has, however, a major influence on both the pattern of behaviour of the array during the welding, and on the final properties of the welded system. A more complicated situation arises when a metallic outer sleeve is employed. The modes of collision and of subsequent welding that develop in this case can be examined with reference to Figs 4.50 and 4.51. Figure 4.50 illustrates a basic prewelding layout of an array of rods contained in a sleeve.

When an explosive charge is detonated, the impact between the outer flyer rods and the sleeve takes place along straight lines. Almost instantaneously the rods collide with one another and with the central rod. Figure 4.51(a) shows the expected mode of deformation of a typical element. Rod 2 deforms under the impact at points B, B', D and D' which are symmetrical about the line BB'. The points of contact grow into straight lines towards the centres of cavities. The final shape of the central rod (core) takes the form of a uniform hexagon, as indicated in Fig. 4.51(b), and the portions of the flyer rods below the initial impact points D, D', etc., do likewise.

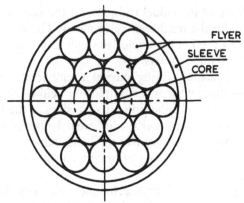

FIG. 4.50   A pre-welding, basic arrangement of an array of rods.

The arcs A′D, DB′ and B′A′ bounding the cavity A′DB′ (Fig. 4.51(c)) become smaller in length and greater in curvature as the three impact lines propagate towards the centre of the cavities. Collision between the three arcs, approaching each other at high velocity, takes place at the centre of the cavity.

It is at this stage that heat effects become more pronounced. The heat generated as a result of the loss of kinetic energy is high enough to cause a considerable amount of localised surface melting to occur. At the same time however, the presence of the relatively cold surrounding rods will be responsible for fast cooling of the region and may lead to phase transformation and to a change in the crystalline structure.

The portion of the metallic sleeve facing the cavity between any two adjacent rods acts as a liner to a circular cavity charge (Fig. 4.51(a,c)). A high speed jet is formed which fills the gap and penetrates, between the rods, into inner gaps. The collision between the rods follows this action and the jet material is squeezed in front of point D. Some of the jet is thus trapped in the inner cavity. It is clear that this action immediately precedes that of the collision and of the subsequent welding of the individual rods. The sequence of events is as follows:

(i)   Formation of jet made of the sleeve material.
(ii) Momentary separation of loose rods.
(iii) Penetration of sleeve jet into the inner cavities.
(iv) Collapse of rods onto each other.
(v)   Welding of rods and layers.

Although the inner core rod assumes, generally, a hexagonal shape, the precise outline of its boundary depends on the properties of the

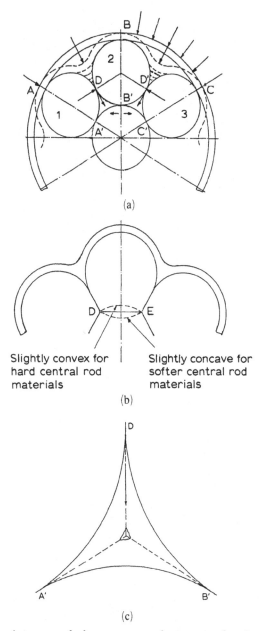

(a)

(b)

Slightly convex for
hard central rod
materials

Slightly concave for
softer central rod
materials

(c)

FIG. 4.51 Development of the geometry of an array of rods welded in a metal gas sheath. (a) Expected mode of deformation of a typical element; (b) convex or concave inter-rod boundaries; (c) incidence of melting in the inter-rod space.

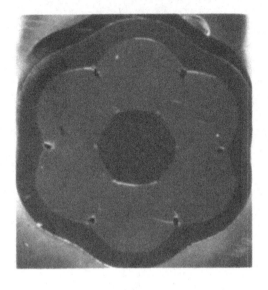

(a)

(b)

FIG. 4.52   Examples of implosively welded arrays of rods. (a) Copper sleeve/ brass rods/copper core; (b) titanium assembly — plastic sleeve; (c) low carbon steel rods/copper tube/stainless steel tube — plastic sleeve; (d) multirod assembly of low carbon steel rods with silver steel inter-rod fillers — plastic sleeve (the starting rod sizes are indicated at the bottom).

(c)

(d)

respective materials. For a core material that is harder than that of the flyer rods, the boundary is slightly convex (Fig. 4.51(b)), whereas for cores softer than the flyer elements it becomes faintly concave. For non-metallic gas shields and monometallic arrays, the boundary remains concave. Some penetration of plastic sleeve material into the inner cavities is expected, but very little actual deposit is normally observed.

4.3.3.2 Interface morphology and properties

The basic geometries generated in the welding systems discussed are shown in Fig. 4.52. A copper sleeve/brass rod flyer/copper core assembly (Fig. 4.52(a)) shows the existence of small voids on the outer rod separating surfaces, but a change to a PVC sleeve (Fig. 4.52(b)) removes this particular feature from a titanium rod/core system. The need for and the result of hydrostatic extrusion, following the welding process, is illustrated in Fig. 4.52(c). A welded MS outer rod flyer/inner copper tube flyer/SS tube core assembly (left) is first machined to a circular outer surface geometry and after extrusion (right) assumes a homogenised, highly deformed structure.

To prevent any penetration and deposition of the sleeve material — whether metallic or plastic — in the internal cavities, these can be filled, *a priori*, with a suitable material. Figure 4.52(d) shows such an arrange-ment in which the major cavities had been filled with silver steel rods prior to welding.

The effect of non-uniformity of the annular stand-off distances is illustrated in Fig. 4.53. This refers to a copper sleeve, cold drawn steel flyer core rod system (central element of the figure). Magnified cross-sections of the individual rod interfaces and inter-rod cavities, with their varied metallurgical features, are shown on the radially arranged photomicrographs.

It is clear from the figure, that the combined effect of the excess of energy supplied and of the small annular clearance is to facilitate, to various degrees, the penetration of the molten sleeve-jet towards the inner cavities. This may well result in a permanent, complete or partial, separation of the rods and in a poor quality weld. An increase in the stand-off distance, for the same level of energy, results however in a partial trapping only of the jet in the vicinity of the outer boundary of the system, and in filling of the inner cavity. Intimate contact between the neighbouring rods is regained and a satisfactory weld between them is obtained.

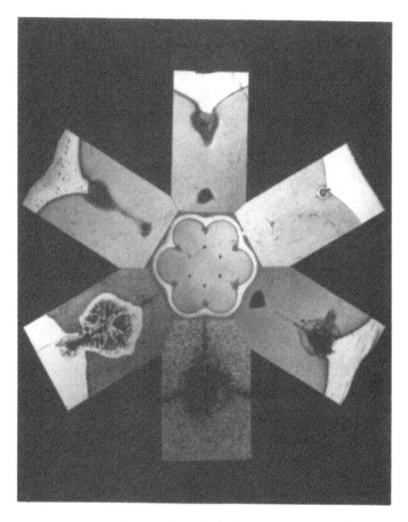

FIG. 4.53   The effect of sleeve-jet penetration into a copper sleeve/low carbon
steel flyer/core rod assembly.

The respective figures show also cracking and formation of voids in
the molten metal zones, which are indicative of a high rate of directional
cooling.

The regained contact between the neighbouring rods in a MS sleeve/
MS flyer/brass core system, is shown clearly in Fig. 4.54(a), where only

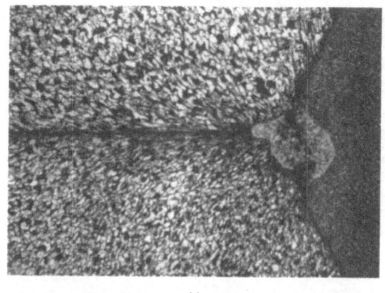

(a)

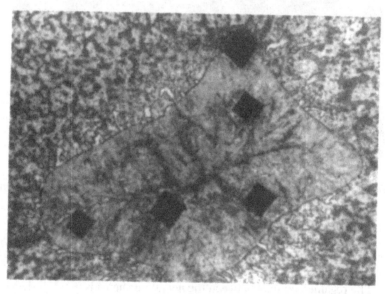
(b)

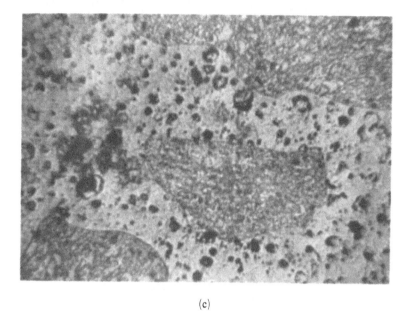

(c)

Fɪɢ. 4.54   Sleeve-jet penetration. (a) Low carbon steel sleeve/low carbon steel outer flyer/brass core (× 100); (b) area of resolidified melt in a system similar to (a) (× 160); (c) entrapment of separated elements of the steel flyer in resolidified copper sleeve melt (× 240).

a small resolidified melt element of the sleeve has penetrated into the flyer/core cavity. An inclusion of this type (Fig. 4.54(b)) is likely to introduce an element of localised brittleness which ought to be assessed (by microhardness testing shown in the figure) routinely before the welded array is incorporated into a working cycle.

On occasions, the situation can be further complicated by the breaking away of the rod elements by the penetrating sleeve-jet. In this case, in addition to the presence of the resolidified jet material, small particles of flyer rod may be entrapped within the melt. This is shown in, for example, Fig. 4.54(c), in which elements of a steel rod flyer are seen embedded in the resolidified copper sleeve melt.

Basically, however, a correct choice of the materials of the assembly, combined with proper alignment of its constitutive elements will produce satisfactory welding conditions. The individual strands of an array will be welded to each other in 'standard' line or wavy type interfaces (Fig. 4.55).

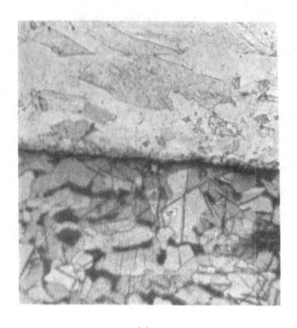

(a)

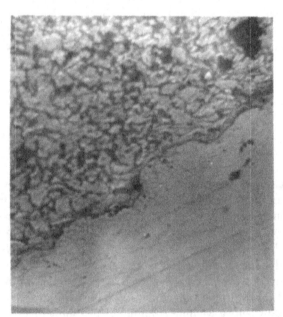

(b)

FIG. 4.55 Satisfactory weld interfaces. (a) Line weld in a copper outer flyer/
low carbon steel core system (×100); (b) wavy weld in a low carbon steel
sleeve/brass rod system (×330).

The mechanical and physical properties of the welded arrays will vary according to the material and process parameters interplay. Some insight into this variation is afforded by, for instance, the investigations of Blazynski, and others.[56-60] A selection of material combinations examined is given in Table 4.11. The weld strengths quoted in the table were assessed by a shear test carried out by means of sets of tools shown in Fig. 4.56.

Hardness levels and distribution are of particular importance in this type of composite. The variation in the hardness profile across a section of the composite is, of course, related to the amount of plastic deformation undergone by the individual elements of the assembly. The presence of solidified sleeve material, if any, in the inter-rod voids will be responsible for more pronounced local variations in hardness, as, naturally, will the presence of a continuous molten layer on the welded interfaces. In the absence, however, of these further complications, the level of hardness is seen to fall away from the surfaces subjected to plastic strain and to increase considerably at the welded interface or in its vicinity. Figure 4.57 illustrates these tendencies for three types of composites.

In the relatively simple case of a metal sleeve/bimetallic rod array (Fig. 4.57(a), Test No. 3), the considerable plastic deformation suffered by the sleeve is reflected in the high level of hardness on the sleeve/flyer interface. The material of the flyer, away from that interface, is less deformed and consequently the hardness is reduced. The flyer/core interface shows an increase in hardness produced in both materials by the local plastic strain. A similar trend is observed in the case of a metallic sleeve/double flyer/core assembly (Fig. 4.57(b), Test No. 8) where high levels of hardness are present only in the vicinity of the welded surfaces. A change in the geometry of the system, introduced by the use of tubular elements (Fig. 4.57(c), Test No. 13), does not affect the basic trend.

Should the pattern of hardness distribution or its magnitude be unacceptable in any given situation, heat treatment of most material combinations is possible and usually produces the desirable result without distorting unnecessarily the composite in question.

When plastic and not metallic gas shields are used, a slightly different hardness pattern is observed. The amount of deformation on the outer surface of the composite is very small and therefore the level of hardness is near that corresponding to the original material. Brittle monometallic phases may be formed occasionally in the inter-rod voids and will be responsible for local spot hardening.

TABLE 4.11
Composite Systems and their Properties (all dimensions in mm)

No.	Sleeve		Outer flyer		Inner flyer		Core		Weld strength (MPa)
	Metal	Wall thickness	Metal	Dia. × No.	Metal	Dia. × No.	Metal	Diameter	
1	MS	1·50	MS	6·35×6			MS	6·35	286
2	MS	1·50	MS	6·35×6			Br	6·35	237
3	MS	0·90	MS	6·35×6			Cu	6·35	228
4	MS	0·90	MS	4·76×9			Cu	6·35	144
5	MS	1·50	Br	6·35×6			Br	6·35	102
6	Cu	1·17	Cu	6·35×6			MS	6·35	244
7	Cu	1·17	Cu	4·76×9			MS	6·35	196
8	Cu	1·58	MS	3·17×12	MS	3·17×6	MS	3·17	209
9	Cu	1·21	MS	6·35×6			MS	6·35	302
10	Cu	1·21	MS	6·35×6			Br	6·35	192
11	Cu	1·21	Br	6·35×6			Br	6·35	108
12	Cu	1·21	Br	6·35×6			MS	6·35	237
13	Cu	1·27	MS	6·35×11	Cu tube	$\phi16 \times 2\cdot75$	SS tube	$\phi10\cdot3 \times 2\cdot5$	366/337
14	PVC	2·75	Ti	6·35×12	Ti	6·35×6	Ti	6·35	331/415
15	PVC	2·75	MS	6·35×11	Cu tube	$\phi16 \times 2\cdot75$	SS tube	$\phi10\cdot3 \times 2\cdot5$	330/300

Fig. 4.56   Sets of tools used for the assessment of shear strength of welds in
rod and tube arrays.

4.3.3.3 Hydrostatic extrusion

To assess the magnitude and relative importance of the various
extrusion parameters, the response of a bimetallic Cu/MS combination
was examined by Blazynski and Matin.[60] The extrusion was carried out
on a press at a constant velocity of 6 mm/s, using commercially clean
castor oil. The direct extrusion technique was employed in which the
pre-shaped leading end of a billet was placed in the die and the billet
was extruded by applying pressure to the punch which does not,
naturally, come into contact with the workpiece. To effect the necessary
sealing of the billet in the die, careful dressing of the latter was necessary. The reason for the adoption of the more complicated and tech-
nically demanding hydrostatic extrusion process is evident from Fig.
4.58 which shows the invariable result of attempting to extrude in a
conventional operation (left), and the satisfactory outcome of the
hydrostatic extrusion process (right). Prior to extruding, the billets are
dressed by removing any fluting that remains after welding. A range of
extruded billets is shown in Fig. 4.59(a), with Fig. 4.59(b) showing the

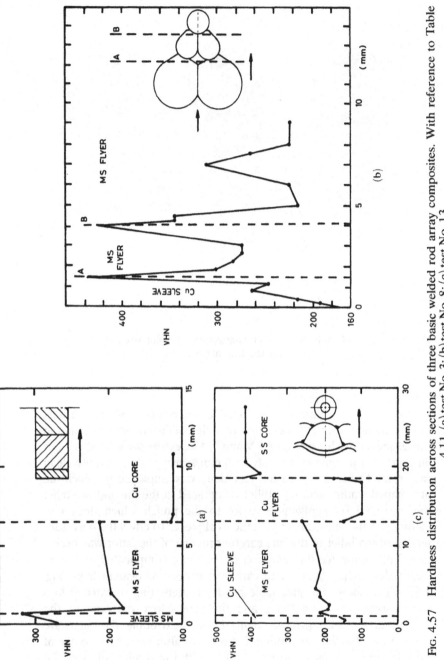

FIG. 4.57 Hardness distribution across sections of three basic welded rod array composites. With reference to Table 4.11, (a) test No. 3; (b) test No. 8; (c) test No. 13.

FIG. 4.58   The effect of conventional (left) and hydrostatic (right) extrusion of copper/low carbon steel metal sleeve implosively prewelded arrays of rods.

effects of differentiality of deformation associated with the variation in the mechanical properties of the constituent elements.

The extrusion was effected in conical dies of three die semi-angles (α) of 10, 15 and 20°. Low carbon steel and commercially pure copper were used as test materials. The average diameters of the welded billets ranged from 21 to 28 mm, with the initial flyer rod diameters ranging from 2·4 to 6·4 mm, and the core rod diameters from 4·8 to 9·1 mm. The steel and copper sleeves varied in their outer diameters from some 21 to 28 mm, and in their wall-thicknesses from 4 to 6·3 mm. All of the billets were originally 70 mm long. The effect of the variation in the volume of the given material was established by varying the copper volume fraction (f) between 0·08 and 0·92.

The billets were extruded to give a range of extrusion ratios (R) lying between 1·2 and 1·8, or of percentage reduction of area (r) ranging from 10 to 45%.

Irrespective of the actual geometry of the pass, the hydrostatic extrusion process results in the consolidation of the structure of the specimen and therefore in 'homogenisation' of the assembly — not so much in terms of the material properties, as in the mechanical pattern of flow. A clear indication of this is the closing, and even elimination of the inter-rod voids present after welding. The core thus becomes more

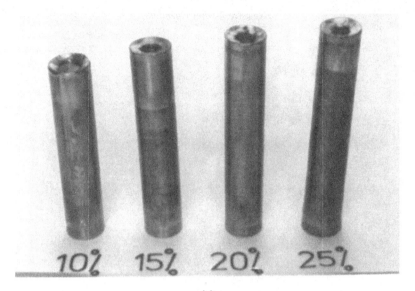

(a)

(b)

FIG. 4.59   Extruded composites of Fig. 4.58. (a) Showing percentage elongations; (b) the effect of differential elongation of bimetallic constituents of an array.

integrated with the general structure and consequently the mechanical properties of the composite are expected to improve. The numerical differences between the differential deformation of the core and that of the composite as a whole, are generally low, and in consequence the extrusion of a pre-welded array of rods can proceed under the condition of near uniform deformation.

Figure 4.60 and Table 4.12 provide information about the effects of percentage reduction, r, or extrusion ratio, R, and of the copper volume fraction on the value of the extrusion pressure for a range of employed die angles α.

The interplay between the effects of friction and inhomogeneity, that is unnecessary macroshearing in the pass, is clearly demonstrated.

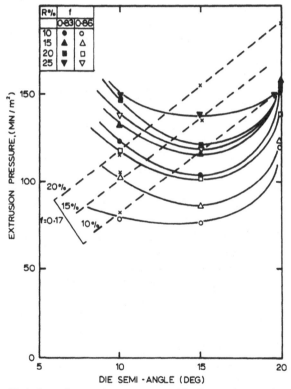

FIG. 4.60 Variation of extrusion pressure with die semi-angle for a range of extrusion ratios R, and copper volume fraction f in hydrostatic extrusion of copper/low carbon steel implosively prewelded array of rods.

TABLE 4.12
Mechanical Properties of Hydrostatically Extruded Arrays of Bi-metallic Rods

Test No.	Material		f	r (%)	Average weld strength (MPa) after		%Δ
	Flyer	Core			Welding	Extrusion	
(a) Die semi-angle $\alpha = 10°$							
1	MS	Cu	0·08	45	146	204·4	40
2	MS	Cu	0·17	40	130	158·6	22
3	Cu	MS	0·92	20	162	205·0	26·5
(b) Die semi-angle $\alpha = 15°$							
4	MS	Cu	0·08	15	170·0	214·0	25
5	MS	Cu	0·08	45	166·0	225·0	35·5
6	MS	Cu	0·17	40	123·5	154·3	25
7	Cu	MS	0·83	38	190·0	265·0	39·5
8	Cu	MS	0·92	10	124·0	145·0	17
9	Cu	MS	0·92	20	130·5	181·2	28
10	Cu	MS	0·92	27	182·0	244·0	34
(c) Die semi-angle $\alpha = 20°$							
11	MS	Cu	0·08	15	166·0	209·5	26·2
12	Cu	MS	0·92	10	165·0	190·0	15·1
13	Cu	MS	0·92	20	165·0	213·6	29·4
14	Cu	P	0·92	40	175·5	217·0	23·6

Although, for a given f, the extrusion pressure, as expected, increases with r, the increase depends on the die angle. As the effect of inhomogeneity increases with α and the influence of friction reduces at the same time, the extrusion pressure rises with the reduction of r, and with the increasing content of the steel, i.e. with the reducing f. The difference in the levels of extrusion pressure for a given reduction, die angle, and two extreme values of f, is of the order of some 65%. Figure 4.60 shows that an optimal die angle can be defined for a given value of R, and a given f. The optimal value of 2α ranges from some 8° to 30°, being lower for the higher copper content when the pressure is less than 200 MPa. The pressure rises to some 600 MPa for the optimal extruding conditions required in the case of $R = 1\cdot8$, when $f = 0\cdot08$. Again, this pattern of interdependence of process parameters reflects clearly the relationship existing between friction and inhomogeneous shearing deformations discussed earlier.

A reference to Table 4.12 gives an insight into the dependence of the quality of the weld on the basic process parameters. It follows from the table that although the shear strength of the weld does not depend either on the material order of the flyer/core configuration or on the volume fraction, the hydrostatic extrusion by consolidating the structure increases the strength of the component in each individual case. The increase in Δ ranges from 15 to 40%, and the failure usually occurs in the weaker material, i.e. copper in this case, and not at the welded interface.

The mechanical properties of the welded and then extruded specimens, are defined by the distribution of the hardness. A representative set of plots is given in Fig. 4.61. In general, welding itself produces a small degree of plastic deformation of the constituent components of an array and this is reflected in the lower levels of hardness as compared with the post-extrusion conditions. (N.B. Since hardness tests cannot be carried out before and after extrusion on the same surface, any comparison between the pre- and post-extrusion conditions is of necessity only approximate.)

The effect of plastic deformation, associated with extrusion, is considerable, particularly for low values of f. Irrespective of the metal combination, for higher reductions of, say, $r = 40\%$ and for $2\alpha = 10°$, the increases in the hardness of the flyer range from 8 to 21%, and in the case of the core from 6 to 18%. On the whole, it seems that apart from the upper value of HN, the changes in the hardness of a specimen are numerically somewhat lower in the core than in the flyer. Clearly, however, the increase in the hardness levels may indicate a slight loss of ductility on or near the welded interfaces where peaks in the distribution curves tend to occur.

4.3.4 Tubular Transition Joints

4.3.4.1 Manufacturing systems
As indicated in Section 4.2.2.3, interest in transition joints has been centred mainly on the situations in which transport of fluids, at differential temperatures, and often of corrosive properties, is involved, and in which differential properties of the conveying systems themselves have to be developed. Typical examples of these are provided by hot water/superheated steam systems in boilers, by fluidised beds used in conjunction with the combustion of coal, and, generally, by chemical apparatus and plant. On the cryogenic site, aluminium/stainless steel combinations are frequently required in, for instance, distillation columns, in shafts and in extended parts of flow valves, etc. On the other

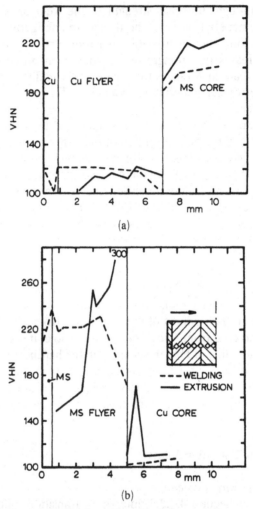

(a)

(b)

FIG. 4.61   Hardness distribution after implosive welding and hydrostatic extrusion respectively in the composites of Fig. 4.60.

hand, quite a different problem is posed by the requirements in the construction of long vacuum systems in which intersecting storage rings must be able to contain pressures sometimes as low as 10−12 torr.

The efficiency and quality of such systems depend on the degree of integrity of the transition joints and it is therefore in this connection that the implosive, metal-to-metal technique, used successfully in the manufacture of multilayered plate and tubular components, will be found particularly advantageous. Whereas, however, the 'standard' method of manufacture relies on machining the joint out of a solid bimetallic plate or block, the axisymmetrical welding techniques developed by, among others, Kameishi *et al.*,[61] Grollo[62] and Blazynski and Yacoub,[63,64] permit the use of tubing of the same diameter. The consequent saving on material and production costs is substantial.

Depending on the desired final properties of the welded assembly, a bimetallic, or exceptionally multimetallic, joint can be conveniently manufactured in one of four possible ways. Three of these are shown in Fig. 4.62, with System IV originating directly from System III by combining two units in series. In Systems I and II, welding is carried out with the assemblies arranged vertically (as in Fig. 4.62), but in Systems III and IV a horizontal arrangement is adopted.

A two-tube joint is formed by initially cutting the tube faces at an angle (θ) to allow the welding to take place on the lap interface (Fig. 4.63) (square-faced assemblies cannot be welded face-to-face), and by employing a third, sleeve tubular element. In any case, this additional element is often incorporated since the transition joint is likely to be required to withstand high pressure due to the transported fluid, and the increased welded area ensures a higher degree of tightness. The systems illustrated in Fig. 4.62 incorporate this idea. For low pressure flows, a direct two-tube composite is however quite satisfactory and the sleeve proves unnecessary.

In all of the cases, the 'to be welded' assembly is partially enclosed in a cylinder containing the explosive charge and, at the same time, the bore of the tubular joint is rigidly supported by a mandrel. In the vertical systems, the upper tube (and in the horizontal, the RH tube) is referred to as the 'tube', with the other component being known as the 'base tube'.

Although the contact between the two, in System I, is maintained on the lapped interface, welding can only be effected between the sleeve and the outer surfaces of the two components. The system is therefore suitable for short lengths of tubing and relatively low pressures only.

When, in addition to the tube/base tube/sleeve joint, welding is also required on lap interfaces, System II can be used quite effectively. Two charges are employed here, the upper one providing a sufficient axial thrust to create a welding jet on the lapped surface. Again, because of the necessity to support rigidly the base tube, and to prevent buckling, this particular method is used only for the manufacture of short

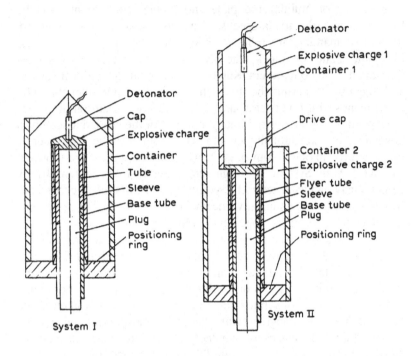

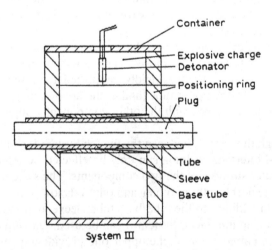

Fig. 4.62   Diagrammatic representation of three basic tubular transition joint welding systems.

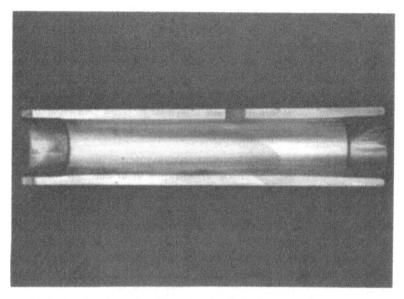

FIG. 4.63   Section through a bimetallic implosively welded (sleeved) tubular transition joint.

assemblies, which will be subjected to higher internal, working pressures.

In contrast to Systems I and II, System III gives welds on both the lapped and sleeve encased surfaces. It is suitable for long lengths of piping, although, naturally, short lengths can also be accommodated. In this latter case, the outer sleeve — which, in addition to providing an extended welded area, acts also as a stiffening and stabilising component during welding — can be easily machined off, leaving a single, non-step diameter line. In some piping configurations, this procedure is found to be advantageous.

System IV, as already pointed out, is an extension, in series, of System III, in that two joints incorporating, if required, three different materials, can be formed simultaneously. Where acoustic impedances of the constitutive materials differ greatly, two separate units, comprising System III configurations, are used since each is likely to require a different level of energy.

4.3.4.2 Metallurgical aspects

The four welding systems will produce differing interface morphologies which will be associated with both the actual geometry and the levels of energy delivered. An insight into the variations in metallurgical properties is provided by, among others, the experimentation of Blazynski and Yacoub.[63,64] These show that in all the systems employed, the wavy type of weld, and its amplitude, depends on the dynamic angle of impact between the respective elements of the assembly. In System I, typically, the amplitude is high and the frequency low (Fig. 4.64(a, b)), but the weld interface is clean (N.B. the black outline in the figure is due to etching) since no trapped jet is present and therefore no intermetallic alloy exists.

System II joints are more likely to contain intermetallic alloys at the interfaces, usually as a result of jet entrapment in between the wave crests and troughs.

Systems III and IV are generally characterised by small amplitudes and high frequency waves which make for strong joints of a relatively high integrity. A large magnification of a portion of a welded interface (Fig. 4.64(c)) shows a clean metal-to-metal interface of small wave amplitude, and, also, a change in the orientation of the grain, indicating the effect of the plastic flow imposed by the detonating charge.

The magnitudes of the respective wave parameters are particularly affected by the energy dissipated. A higher than required level of energy is likely to result in the presence of a wavy form with an intermetallic alloy or phase vortices. This is because the waves are produced at a rate of some $10^{-6}/\text{s}$ and therefore large areas will exist in which the rate of shearing is sufficiently high to produce frictional heating effects. These, in turn, cause melting of thin layers of the material. Although the melted zone is immediately cooled by the bulk of the surrounding metal, a case structure is produced, locally or continually, which may sometimes show cavities formed on cooling. System II produces some 5% of welds showing this condition, but System III appears to give fully satisfactory results. Reasonably sound Cu/Cu welds with melt layers 30 μm in thickness can be noticed; an observation in some conflict with Kreye's[65] contention that in similar metal combinations the layer will not exceed 15 μm.

In general, when molten metals at temperatures well above their solidification points come into contact with cold surfaces, showers of seed crystals are formed at the outer surfaces of the welded alloys and growth takes place in all directions until contact is made with the neighbouring crystals. Very small equi-axed, or chill, crystals are then formed. If the welding temperature is high, these crystals can re-melt and form a layer. Otherwise, the crystals can be identified as inclusions of one

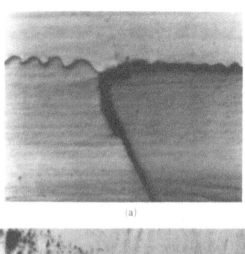

(a)

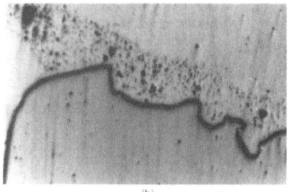

(b)

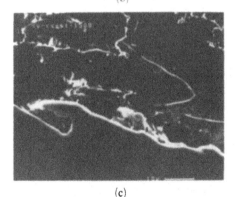

(c)

FIG. 4.64 Typical welding interfaces in implosively welded tubular transition joints. (a) Brass/brass/low carbon steel (sleeve, tube, base tube) — System I (×100); (b) detail of the weld (×150); (c) copper/copper/low carbon steel (sleeve, tube, base tube) — System III, $\theta = 40°$, base weld detail.

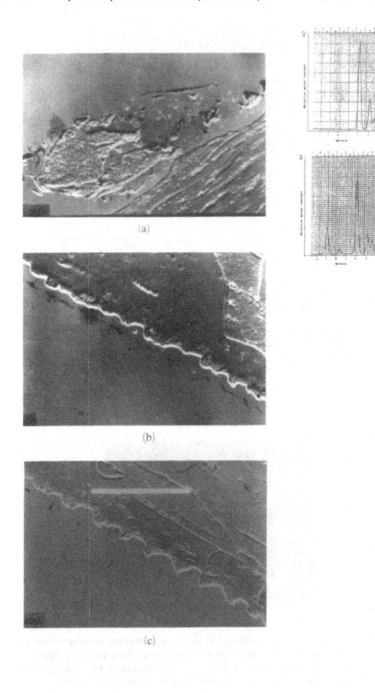

(a)

(b)

(c)

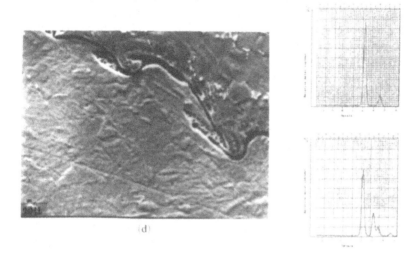

FIG. 4.65 Detail of weld interfaces in tubular transition joints. (a) System II. Steel particles in brass/low carbon steel, $\theta = 60°$, identified by the accompanying X-ray spot analyses; (b) secondary reflection waves in System II. Copper/low carbon steel, $\theta = 40°$; (c) as above, brass/low carbon steel, $\theta = 50°$; (d) System III. Steel particles in a copper/low carbon steel joint, $\theta = 60°$, identified by the accompanying X-ray spot analyses.

material in the matrix of another, or as intermetallic alloys similarly present in the basic, constituent matrices.

Both Systems II and III produce some chill crystals — as opposed to the continuous layers of metal already discussed — examples of which are provided by Fig. 4.65. Figure 4.65(a) shows fine grain areas of steel in brass in a System II Br/MS, $\theta = 60°$ joint. Steel particles vary in magnitude from 0·1 to 10 μm. Electron scan microscopic X-ray spot analysis carried out in the middle of the wave line (lower plot), shows the presence of some 30% MS, 85% Cu and 22% Zn (these percentages are relative to the total amounts of the respective metals). A change in the position of the spot by 3 μm towards brass, reveals the presence of some 92% Cu and 34% Zn, with no steel in this area (upper plot).

A similar situation is noted in System III where, for example, in a Cu/MS, $\theta = 60°$ joint, particles of steel in copper are also observed (Fig. 4.65(d)). These are of the same order of magnitude as in System II and appear as almost continuous layers. Spot analysis, made again in the middle of the wavy line, shows some 85% Cu and 45% MS (upper plot).

A shift in the spot position by 1 μm in the direction of copper, indicates the presence of Cu only (lower plot). This simply means that the melting rate of copper and brass is higher than that of steel; partly as a result of the difference in melting points and partly due to the geometries of the two systems.

An interesting phenomenon, the creation of a 'secondary wave' of metallic particles of one material in the matrix of the other, arises in System II only. The presence of a rigid anvil produces a reflection wave caused by the transformation of kinetic energy of the impacting tube into potential energy, at very high pressure, and back into kinetic energy of elastic oscillation and plastic deformation. This results, in turn, in detachment of some elements of the tube and in their transition into the matrix of the base. The actual mechanism of this process relies on the fact that the mass velocity of the plastic compression wave is higher than the collision point velocity (which remains subsonic). However, it will reduce ahead of the collision point because of the energy losses caused by plasticising of the formerly uncompressed material. At the next free surface (in most cases that of the thinner or lower strength flyer) the compression wave can be released into the plasticised region, freezing the quasi-liquid material. This phenomenon is illustrated in Fig. 4.65(b, c), where Fig. 4.65(b) refers to a Cu/MS, $\theta = 40°$ joint, and Fig. 4.65(c) to a Br/MS, $\theta = 50°$ assembly. In both cases, the secondary wave consists of particles of either copper or brass, but not of steel.

4.3.4.3 Mechanical properties

Central to the assessment of the joint quality are the problems of tightness and strength. In general, the tightness of the joint is related directly to the welding system and to the lap angle θ which is equivalent to the initial, but not dynamic, angle of impact. Since in System I the internal pressure is resisted only by the sleeve, the efficiency of the joint depends only on the quality of the weld on the sleeve bore. For 25·5 mm o.d., 3 mm wall-thickness tubes, and sleeves varying from 1·5 to 2·0 mm in thickness, the range of pressures sustained without leakage lies between 25·5 and 35·5 MPa.[63]

In System II, where the lap angle is of importance, the best results are obtained with $\theta = 40°$ when no leaks occur up to pressures of about 43 MPa, but are less satisfactory with $\theta = 50°$ (3% of joints leaking at 36 MPa), and $\theta = 60°$ (5% of joints leaking at 32 MPa). No leaking is found in Systems III and IV for pressures in excess of 45 MPa.

From the point of view of the strength of the weld, a criterion of acceptance requires that the weld be at least of equal strength to that of the weaker joint material. The bonds with wavy interfaces generally meet this requirement; the exceptions being the joints with significant volume fractions of intermetallic alloys. The mechanical properties of the welds were tested on tensile specimens (in which shear is induced) generally 2–4 mm in depth, 10 mm in width and varying in length from 57 to 120 mm. A selection of these specimens is shown in Fig. 4.66.

It is clear from Fig. 4.66(a) that no weld exists on the lap surface of a System I joint (as already pointed out earlier), and that with $\theta = 90°$, the weld between the sleeve and the tube is incomplete. In general, in System I joints, failure tends to occur within the weaker material and close to the welded surface rather than at the interface itself.

In some combinations of copper and brass, particularly in System II (Fig. 4.66(a)), failure occurs during machining thus indicating a low shear strength. The explanation lies in the formation of a narrow band of strain-hardened material near the interface which is responsible for a local increase in strength. This is sufficient to move the fracture line slightly away from the interface. The resulting brittleness causes the failure of the joint when the vibrations produced by the machining operation are imposed on the assembly.

The basic soundness of the System III joint is indicated by Fig. 4.66(b), in which the three lap angles give satisfactory welds. The numerical results of all of these tests are given in Table 4.13 and Fig. 4.67. Tensile tests were carried out using both the tube/base tube/sleeve, and sleeveless (machined off) assemblies, as indicated in the table. In sleeveless specimens, the problem of shear does not arise, but the results obtained with full, three-component test-pieces clearly depend also on the shearing strength of the sleeve/tube interfaces, $\theta = 40°$ represents the best setting.

Consistent results are observed in Systems II, III and IV where the increase in the strength of the transition joint is associated with the increase in the welded area of the lap surfaces. In all of the cases considered, the strength of the joint is, on average, less than the arithmetical mean of the strengths of the constituent materials. Figure 4.67 and Table 4.13, give the values of the maximum stresses and strains to fracture, but indicate primarily the basic soundness of the welds obtained by the smoothness of the respective stress/strain curves. The conditions of testing are shown in each case in the figure, and the selection of the curves produced is made with a view to bracketing the

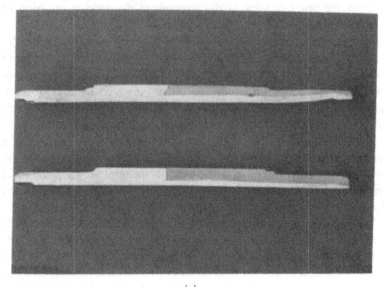

(a)

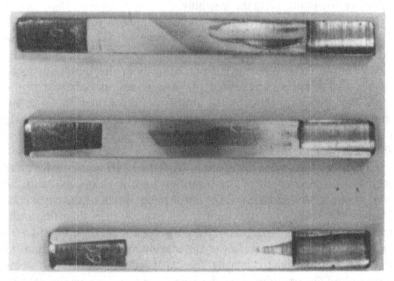

(b)

FIG. 4.66   Tensile specimens machined out of tubular transition joints. (a) Top, System II, $\theta = 60°$, bottom, System I, $\theta = 90°$; (b) System III, $\theta = 40$, 50 and 60°.

TABLE 4.13
Material Tensile Strength

Material	Strain (ε)	Maximum stress σ (MPa)	Angle (θ°)	Systems
MS	0·046	612·5	—	Before
Br	0·064	415·1	—	explosive
Cu	0·137	198·0	—	welding
MS	0·027	675·8	—	I Sleeve only
Br	0·019	537·2	—	in strip
Cu	0·042	301·6	—	form
MS–Br	0·0071	419·7	60	
	0·0042	436·3	50	II ⎰ Without
	0·0061	499·0	40	
MS–Cu	0·013	205·5	60	⎰ sleeve in
	0·021	219·4	50	II
	0·028	286·5	40	⎱ strip form
MS–Br	0·0091	470·1	60	
	0·012	481·9	50	II ⎰ With
	0·0135	506·3	40	
				⎰ sleeve in
MS–Cu	0·019	255·1	60	
	0·020	273·0	50	II ⎱ strip form
	0·025	299·6	40	
MS–Cu	0·017	192·4	60	III–IV
	0·018	215·9	50	Without sleeve
	0·023	236·4	40	in strip form
MS–Cu	0·019	275·3	60	III–IV
	0·022	287·9	50	with sleeve
	0·035	309·1	40	
MS–Cu	0·022	310·0	60	III–IV
	0·027	316·0	50	Without sleeve
	0·033	335·0	40	in tube form

extremes of the range investigated. A tubular specimen used in System IV is shown diagrammatically in Fig. 4.67(c).

4.3.5 Special Type Composites

The two main types of composite structures that do not easily fall into the already discussed groups, are the tube-to-tube plate systems, and heat exchanger tube plugging. A full description of the various approaches used will be found in Ref. 66, but the basic ideas are outlined in Figs 4.10–4.12.

The application of explosive welding to the manufacture of tube-to-tube plate systems (mainly for power stations) is developed and used on a large scale. Bimetallic joints, showing the familiar wave type interfaces, are normally obtained, and a high degree of joint integrity, combined with anticorrosive properties is maintained. The success of the operation depends, to a considerable degree, on both the ligament thickness,[67] and the geometrical pattern of the charge application.

Explosive plugging of defective circuits in heat exchangers has also been well developed and finds its main area of application in nuclear

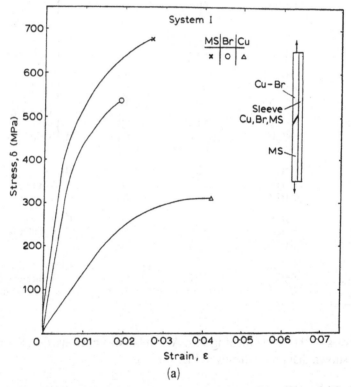

(a)

FIG. 4.67   Tensile stress/strain curves for tubular transition joints. (a) System I — low carbon steel/brass/copper joint; (b) System II — low carbon steel/copper or low carbon steel/brass joint with or without a sleeve; (c) System III — copper/low carbon steel combinations.

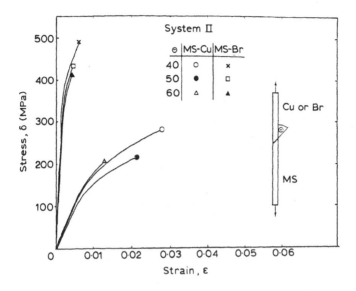

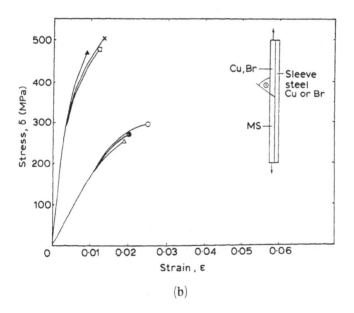

(b)

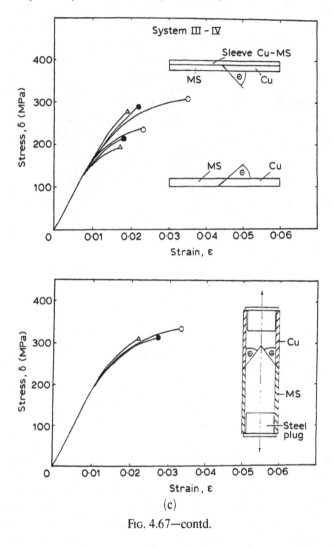

(c)

Fig. 4.67—contd.

plant. The conventional methods of fitting plugs with mechanically induced interference, or fusion welded are unacceptable in nuclear power plant because the high degree of leak tightness required cannot be guaranteed. An additional practical problem facing the maintenance engineer in these situations, is the very limited and extremely difficult access to the affected parts of an exchanger, and the consequent difficulty in either efficient insertion of a 'conventional' plug, or, later, the inspection of the area. It is for this reason, as well, that explosively welded plug — inserted in a TV-controlled remote operation — and detonated *in situ*, has a marked advantage over a conventional method. Normally, a bimetallic, wavy weld surface joint is produced that prevents leaking of the tube.

REFERENCES

1. CHADWICK, M. D. and JACKSON, P. W. In: *Explosive Welding, Forming and Compaction*, Ed. T. Z. Blazynski, Applied Science Publishers, London, New York, 1983, p. 219.
2. JACKSON, P. W. UK Patent Application No. 24299/78.
3. HARDWICK, R., BROWN, D. W. and NOWELL, D. G. *Proc. 10th Int. Conf. on High Energy Rate Fabrication*, Ljubljana, Yuguslavia, 1989.
4. CHADWICK, M. D. *Proc. Select Conf. on Explosive Welding*, The Welding Institute, London, 1968.
5. CAIRNS, J. H., HARDWICK, R. and TELFORD, D. G. *Proc. 3rd Int. HERF Conference*, University of Denver, Colorado, Paper 2.3, 1971.
6. CROSSLAND, B., BAHRANI, A. S. and TOWNSLEY, W. J. *Proc. 3rd Int. Conf. on Pressure Vessel Technology*, 1977, p. 971.
7. BAHRANI, A. S., HALLIBURTON, R. F. and CROSSLAND, B., *Int. J. Pressure Vessels and Piping*, **1** (1973), 17.
8. KOWALICK, J. F. and HAY, R. *Proc. 2nd Int. HERF Conf.*, University of Denver, Colorado, 1969.
9. CARPENTER, S., WITTMAN, R. H. and CARLSON, R. J. *Proc. 1st Int. HERF Conf.*, University of Denver, Colorado, 1967.
10. BLAZYNSKI, T. Z. and DARA, A. R. In Ref. 5, Paper 8.3.
11. CLELAND, D. B. In Ref. 1, 159.
12. SAHNOVSKAYA, E. B., SEDYKH, V. S. and TRYKOV, Yu. P. *Svar Proiz.*, **18** (1971), 34 (in Russian).
13. HAMMERSCHMIDT, M. and KREYE, H., *Proc. 7th Int. HERF Conf.*, Ed. T. Z. Blazynski, University of Leeds, 1981, p. 60.
14. BAHRANI, A. S. PhD thesis, The Queen's University, Belfast, 1967.
15. TRUEB, L. F., *Metall. Trans.*, **2** (1971), 145.
16. CZAJKOWSKI, H. In: *Proc. Int. Conf. on Use of High Energy Rate Methods for Forming, Welding and Compaction*, Ed. T. Z. Blazynski, University of Leeds, Paper 14, 1973.
17. COOK, M. A. *Science of High Explosives*, Reinhold, New York, 1958.
18. SZEKET, A., INAL, O. T. and ROCCO, J., *Proc. HERF-1984 Conf.*, Eds I. Berman and J. W. Schroeder, ASME, New York, 1984, p. 153.
19. JOHNSON, T. E. and POCALYKO, A. *High Energy Rate Fabrication*, Eds M. A. Myers and J. W. Schroeder, ASME, New York, 1982, PVP, Vol. 70, p. 63.
20. BEDROUD, J. PhD thesis, University of Leeds, 1976.
21. PATTERSON, R. A. In Ref. 18, p. 15.
22. KELOMET, ICI Nobel's Explosives Co, Ltd, Stevenston, Scotland.

23. ANDERSON, D. K. C. *Proc. 6th Int. HERF Conf.*, Haus der Technik, Essen, FRG, Paper 1.12, 1977.
24. PRUEMMER, R. *Proc. 7th Int. HERF Conf.*, Ed. T. Z. Blazynski, University of Leeds, 1981, p. 186.
25. LOYER, A., TALERMAN, M., HAY, D. R. and GAGNON, G., III *Int. Symposium on Explosive Working of Metals*, House of Technology, Pardubice, Czechoslovakia, 1976, p. 43.
26. POLHEMUS, F. C. In Ref. 9, Paper 1.3.
27. LOYER, A., HAY, D. R. and GAGNON, G. *Proc. 5th Int. HERF Conf.*, University of Denver, Colorado, Paper 4.3, 1975.
28. WILLIS, J. In: *Explosive Welding*, The Welding Institute, London, 1975, Ch. 10, p. 40.
29. SHAFFER, J. W., CRANSTON, B. H. and KRAUS, G. In Ref. 26, Paper 4.12.
30. DABROWSKI, W. In Ref. 22, p. 218.
31. REECE, O. Y. In Ref. 5, Paper 2.1.
32. WYLIE, H. K., WILLIAMS, J. D. and CROSSLAND, B. In Ref. 5, Paper 2.2.
33. SLATE, P. M. B. and JARVIS, C. V. *J. Inst. Met.*, **100** (1972), 217.
34. VELTEN, R., *Proc. 4th Int. HERF. Conf.*, University of Denver, Colorado, Paper 8.4, 1973.
35. YOBLIN, J. A. and MOTE, J. D. In Ref. 23, p. 161.
36. E. F. INDUSTRIES, Inc., Louisville, Colorado.
37. IZUMA, T. and BABA, N. In Ref. 24, Paper 2.14.
38. ANDERSON, D. K. C. In Ref. 15, Paper 18.
39. MACCLEOD, N. A., British Patent No. 765 305.
40. BLAZYNSKI, T. Z. and COLE, B. N. *Proc. High Pressure Technology Association Conf.*, The Queen's University, Belfast, Paper 11, 1968.
41. BLAZYNSKI, T. Z. In Ref. 1, p. 289.
42. BLAZYNSKI, T. Z. and DARA, A. R. In Ref. 34, Paper 2.
43. APPLEBY, E. J., *J. Appl. Mech.*, **31**, Ser. E (1964), 654.
44. WYLIE, H. K. and CROSSLAND, B. In Ref. 24, Paper 2.4.
45. BLAZYNSKI, T. Z. and MATIN, M. In Ref. 27, p. 164.
46. PRUEMMER, R. and HENNE, R., *Explosive Welding, Forming, Plugging and Compaction*, Eds I. Berman and J. W. Schroeder, ASME Century 2 — Emerging Technology Conferences — PVP, Vol. 44, 1980, p. 87.
47. BEDROUD, Y. and BLAZYNSKI, T. Z. *J. Mech. Work. Technol.*, **1** (1978), 311.
48. SANSOME, D. H. *Design of Tools for Deformation Processes*, Ed. T. Z. Blazynski, Elsevier Applied Science, London, New York, 1986, p. 73.
49. BLAZYNSKI, T. Z., *Plasticity and Modern Metal-Forming Technology*, Ed. T. Z. Blazynski, Elsevier Applied Science, London, New York, 1989, p. 355.
50. TOWNLEY, S. and BLAZYNSKI, T. Z. *Proc. 15th MTDR Conf.*, Eds S. A. Tobias and F. Koenigsberger, Macmillan Press, London, 1975, p. 407.
51. TOWNLEY, S. and BLAZYNSKI, T. Z. *Proc. 17th MTDR Conf.*, Ed. S. A. Tobias, Macmillan Press, London, 1976, p. 467.
52. ARNOLD, R. R. and WHITTON, P. W., *Proc. Inst. Mech. Engrs*, **173** (1959), 241.
53. BLAZYNSKI, T. Z. and EL-SOBKY, H. In Ref. 25, Paper 4.6.
54. BLAZYNSKI, T. Z. and EL-SOBKY, H. In Ref. 46, p. 69.
55. BLAZYNSKI, T. Z. and EL-SOBKY, H., *Metals Technol.*, **7** (1980), 107.

56. BEDROUD, Y., EL-SOBKY, H. and BLAZYNSKI, T. Z., *Metals Technol.*, **3** (1) (1976), 21.
57. BLAZYNSKI, T. Z. and BEDROUD, Y. In Ref. 24, Paper 2.12.
58. BLAZYNSKI, T. Z. and BEDROUD, Y. *Proc. 18th MTDR Conf.*, Macmillan Press, London, 1977, p. 85.
59. BLAZYNSKI, T. Z. and MATIN, M., *J. Mech. Work. Technol.*, **6** (1982), 291.
60. BLAZYNSKI, T. Z. and MATIN, M. *Proc. 4th Int. Conf. on Production Engineering*, Japan Society for Precision Engineering and Japan Society for Technology of Plasticity, Tokyo, 1980, p. 731.
61. KAMEISHI, M., BABA, N. and NIWATSUKINO, T. In Ref. 23, p. 205.
62. GROLLO, R. P. In Ref. 5, Paper 8.4.
63. BLAZYNSKI, T. Z. and YACOUB, K. A. In Ref. 18, p. 1.
64. BLAZYNSKI, T. Z. and YACOUB, K. A. In Ref. 17, p. 293.
65. KREYE, H., WITTKAMP, I. and RICHTER, V., *Z. Metallkunde*, **67** (1976), 141.
66. CROSSLAND, B., *Explosive Welding of Metals and its Application*, Clarendon Press, Oxford, 1982.
67. WYLIE, H. K. and CROSSLAND, B. In Ref. 15, Paper 15.

Chapter 5

Forming of Composites

5.1 CLASSIFICATION OF FORMING SYSTEMS

Although dynamic forming operations may not, at first, be regarded as the most obvious ways of producing composites, the high degree of diversification of the existing techniques has made them an accepted adjunct to the welding and compacting routes.

Composite prefabricates and finished components usually possess enhanced mechanical properties — resulting from the passage of shock waves — and compare well, in their response to working conditions, with the conventionally processed materials.

The 'starting' material assemblies are either of a free variety — with no mutual prior bonding or attachment — and are therefore likely to form a mechanically bonded composite only, or of a prewelded (fusion or explosion) variety, in which case a metallurgically unified bond will be retained. A third type of possible operation, although suffering from the size limitation, is that of the combined explosive bonding and forming.[1,2]

Relatively high levels of energy are required and these are supplied primarily by chemical explosives (see Chapter 3) or electrical discharge apparatus (see Chapter 1). In either case, the discharge of energy takes place in separation from the target, i.e. at a stand-off distance, and is transmitted to it through the intervening medium. In electro-hydraulic systems, the transmitting medium is invariably water, whereas in electromagnetic systems the discharge takes place in the air. Chemical explosive systems operate in both media, with water being used as a pressure and noise attenuating agent. The particular applicability of explosives to forming arises from the fact that practically any shape, in any size, can be obtained, without necessarily having recourse to any machinery. The forming tools consist of female dies which are comparatively light — depending of course on their constructional material — and which do not normally require any foundations because the inertia of the tool mass is sufficient to counteract the generated forces. The advantage of these methods over the conventional ones lies in the possibility of producing very accurately toleranced components in, if required, very complex shapes. The saving is effected by the avoidance of capital costs of presses, tooling, etc. The dies for stand-off operations are often made of cheap materials such as epoxy, concrete and Kirksite, especially if they are intended for a small number of large components, e.g. radar dishes, and can therefore be regarded as disposable.

For reasons of safety, noise containment and easy application of the transmitting medium, particularly water, forming operations are carried out in specially constructed tanks. An unconfined explosive forming facility must be so designed as to provide an environment in which an explosive charge can be effectively detonated, at a required stand-off distance from the target, and in which the effects of detonation are safely contained — even after a prolonged series of firings. Basic design principles are detailed in, among others Ref. 2.

Unlike the explosive welding contact operation, the detonation of a charge or an electrical discharge in a transmitting medium, will result in a relatively complex sequence of events. The sequence associated with an underwater explosion has a decisive influence on the mode of the dynamic response of the target. The detonation of a chemical-explosive charge is accompanied by a high rate of propagation of the detonation wave, in these cases of the order of 7000 m/s, and a high pressure, high temperature gas bubble is created in the medium. The high pressure gives rise to a shock wave which, in turn, is transmitted through the medium of water to the target. A rapid increase in pressure, reaching a peak almost instantaneously, is observed followed by an exponential type of decay. The expansion of the gas bubble continues, being accompanied by an outward accelerated flow of mass of water. The decay of the pressure pulse to ambient and below ambient levels, together with the effect of the external pressure, eventually causes the gas bubble to contract. Owing to energy losses, the contraction does not reach the original bubble size, but produces a somewhat larger diameter than that of the original bubble.

During the contraction phase, the gas pressure begins to increase again, and, eventually, the increase results in the formation and propagation of a radial outward, distinctly non-shock, pressure wave. This wave is usually referred to as the primary bubble pulse, and, together with the original shock wave effect, constitutes a major source of energy flux that is delivered to the target. Depending on the geometry and properties of the system, damped bubble pulsation may continue, but will not be of any consequence in the actual forming operation.

From the point of view of the target, the incident primary shock wave imparts outward radial velocity to it and if the target is secured to, for instance, a die, it will begin to deform. The reflection of the wave back into the medium, together with a rarefaction wave produced by the motion of the target, constitutes a secondary wave. The drop of the gas bubble pressure to a negative value causes the water to cavitate, and thus, temporarily, further loading of the target will cease. Phase one of the forming operation is thus completed.

As the contraction of the gas bubble begins, the inward flow of water associated with it, will cause the disappearance of cavitation and, in due course, will allow the primary bubble energy pulse to reach the target. Further deformation of the target then takes place.

In a more general case therefore the deformation of the target may be said to have been produced by a double energy flux. The conditions required for this to occur are a large volume of the transmitting medium, that is capable of forming the bubble, and the choice of a sufficiently large stand-off distance of the charge (H'). Considerable complications in the dynamic response of the target can be expected if the diameter of the primary bubble is less than H', and, consequently, the bubble collapses against the target. If air is used as the transmitting medium, only the primary bubble pulse will be generated.

The magnitude of the energy flux produced by the charge and available at the target, can be assessed on the basis of the Kirkwood–Bethe theory[3] which indicates that the peak pressure, impulse and energy should be a function of the quantity $(M^{1/3}H')$. Functional dependence of this type has been substantiated empirically and, for small charges, is expressed as[4]

$$E_c = \eta a_1 M^{1/3} (M^{1/3}/H')^{\alpha_2} \qquad (5.1)$$

where E_c is expressed per unit of area of incidence, η is the coefficient of the efficiency of distribution, and α_1 and α_2 are constants defining the properties of the explosive.

Since the energy flux produced by the charge is propagated radially outwards, only a small proportion of the total energy produced will be radiated in the direction of the target. The quantity actually available depends on the stand-off distance and, therefore, on the magnitude of the solid angle subtended at the charge. In the case of relatively small charges and distances, the energy produced by the primary shock wave is estimated at 23% of the total[4,5] and, where applicable, the energy associated with the primary bubble pulse at 21%. It follows therefore that η for the air-cushion system is 0·23 and that its value for the water-cushion system is 0·44.

The pulsation time t, i.e. the time that elapses from the initiation of the detonation to the first bubble minimum, and the maximum bubble radius r_M are functions of the explosive charge and ambient pressure. For a TNT charge of weight W (lb), detonated at a depth H (ft),[6]

$$t = 4\text{·}36\ W^{1/3}/(H+33)^{1/3}\ (\text{s}) \tag{5.2}$$

and

$$r_M = 12\text{·}6\ W^{1/3}/(H+33)^{1/3}\ (\text{ft}) \tag{5.3}$$

The explosive charge will also determine the intensity and configuration of the generated shock wave. For instance, for a spherical charge the peak pressure will be proportional to the cube root of the charge weight

$$p_1/p = (W_1/W_2)^{1/3} \tag{5.4}$$

where W_1 and W_2 are the respective charge weights.

In addition to the weight and shape of the charge, peak pressures are also influenced by the type of transmitting medium. Using TNT as the explosive, Cook[4] has shown that, in air, peak pressures for rectangular and cylindrical charges are given respectively by the following empirical expressions:

$$p = p_0 \left(\frac{13\text{·}5}{z} - \frac{769\text{·}9}{z^2} + \frac{36\,280}{z^3} + 1 \right) \tag{5.5}$$

and

$$p = p_0 \left(\frac{11\text{·}34}{z} - \frac{185\text{·}9}{z^2} + \frac{19\,210}{z^3} + 1 \right) \tag{5.6}$$

where $z = 3\text{·}967\ H/W^{1/3}$, H is the distance of the charge from the target in feet, W is the weight of charge in lb and p_0 is the ambient air pressure. When water is used as the transmitting medium a reasonably good approximation for pressure (p) at a point is given by the expression

$p = p_M e^{-t/\phi}$, where p_M is the peak pressure, ϕ is a constant depending on the characteristics of the charge, and t is the time. The total impulse delivered can be expressed as an integral of pressure and time $(I = \int p \, dt)$. Rinehart and Pearson[7] quote the following expressions as giving very good approximations in the case of small charges:

$$p_M = A(W^{1/3}/H)^D \qquad (5.7)$$

$$I = BW^{1/3}(W^{1/3}/H)^F \qquad (5.8)$$

$$E = CW^{1/3}(W^{1/3}/H)^G \qquad (5.9)$$

where A, B, C, D, F and G are constants depending on the explosive, and E is the energy flux passing through a unit area of fixed surface normal to the direction of propagation wave. A similar type of equation relating pressure to velocity (v) of detonation of the explosive has been proposed by Roth,[8] here

$$p_M = 155(W^{1/3}/H^{1.15})\sqrt{v} \qquad (5.10)$$

These values of pressures can be used in conjunction with eqns (2.2) and (2.3).

The effect of scale in electromagnetic forming, when using geometrically similar coils, was investigated by Al-Hassani *et al.* and analytical expressions defining the conditions were proposed.[9]

Three basic techniques characterise the stand-off forming operations. These are:

- free forming (cups, flanging and deep drawing),
- cylinder forming (sizing and extrusion),
- bulkhead forming (specified shapes in sheet and plate).

The three forming systems are shown diagrammatically in Fig. 5.1. In all of them, water is normally used as a transmitting medium. In a closed system, where low explosives (Chapter 2) are employed, to sustain the pressure for a longer period thus increasing the impulse delivered to the target, and in the open one, operated by high explosives, to moderate the very high peak pressure, and, then, to maintain an average high working pressure for periods longer than a few microseconds. Closed systems are used for the production of small components where the available stand-off distance is inadequate to permit the use of high explosives. As already indicated, forming operations can be carried out either in a tank sunk in the ground and filled with water, or in an empty tank with the water contained in, say, a polythene bag (in the case of a

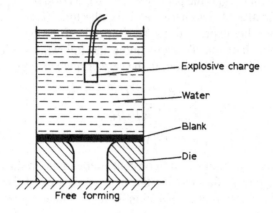

Free forming

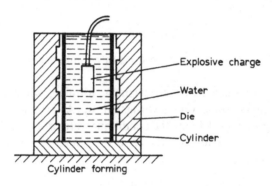

Cylinder forming

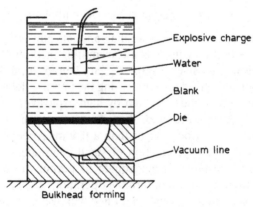

Bulkhead forming

FIG. 5.1   Basic explosive forming systems.

cylindrical component) or in any suitable disposable container. The first method is used for large components, requiring large charges, where the confinement of explosion within a large volume of water is necessary. In such a case, the space between the undeformed metal and the die has to be sealed off and evacuated to enable full deformation to take place. In the second case, it may not be necessary to evacuate the air, provided that the die is fitted with a suitable arrangement of ventilating holes.

With more complex shapes and/or large deformations, it is normally necessary to carry out a series of firings to allow the metal to fill the die without fracturing. The procedure consists simply of inserting a new charge into the die without disturbing the set-up, and in carrying on until full deformation is achieved.

A very important aspect of the dynamic forming of metal is the die design. The near absence of springback, so attractive in the high-energy rate forming, depends on a proper balance of tool profile, avoidance of sharp edges and deep narrow grooves, and on the provision of reasonably smooth transition sections.

Where a good clean mechanical bond between two or more coaxial tubes of dissimilar metals is required, for instance to combine anticorrosive properties of the assembly with good heat transfer properties, then free forming with a strip charge can be employed.

The range of engineering alloys processed dynamically includes: aluminium and its alloys, stainless steel, magnesium and some of its alloys (Mg–Al, Mg, Th), titanium and its alloys with aluminium, vanadium and manganese, refractory metals (molybdenum, tantalum, tungsten, niobium), copper and its alloys, and special alloys such as Stellite, iron–nickel, nickel–copper and chromium–nickel, and cobalt–iron type. Carbon and low alloy steels are less often used, mostly because of their low formability in these conditions.

An insight into the material response to high-strain rate processing, as opposed to either isostatic or straightforward conventional cold forming, is provided by the investigations of van Wely[10] on a range of steels, and of Orava and Kunthia[11] on TMCA 50 A, and α and β Ti–6Al–4V alloys.

In low carbon, AISI 304-L and HY 100 steels, the cold forming and explosive characteristics differ in some respects.

Cold forming	*Explosive forming*
(i) Tensile residual stresses on outer surfaces.	(i) Compressive residual stresses on outer surfaces.

(ii) Very high strain-hardening   (ii) Strain-hardening more
                                                    uniformly distributed.
(iii) Similar levels of toughness.
(iv) The same degree of increase in stress corrosion and hydrogen
       embrittlement.

                                              (v) Aging at low temperatures.

A comparison between the relevant values of the yield stress is given in
Table 5.1.

5.2 FREE FORMING

5.2.1 General Comments

Free or die-less forming is used in a variety of applications ranging
from the formation of semi-spherical, semi-elliptical and semi-
cylindrical dishes, domes and tubular components to Roots-blower type
shaped rings. The basic characteristics of these operations are simplicity
combined with reasonably, but not very high or final dimensional
accuracy.

With the exception of tubes and rings which, naturally, have to be in
their respective geometrical forms before processing, all the other
components are formed from sheet, possibly prewelded in the case of
composites, and are simply supported in open dies, as shown in Fig. 5.1.
The rims of the blanks are either firmly attached to the die surface or
left free to move and slide over the die edge when the charge is
detonated. The method of fixing, if any, will alter, on the one hand,

TABLE 5.1

Comparison between Yield Stresses (MPa) of Cold and Explosively Formed
Steel and Titanium Alloys

Material	Cold formed		Explosively formed	
	Tangential	*Radial*	*Tangential*	*Radial*
Low carbon steel	350	395	434	421
AISI 304L	409	444	497	513
HY 100	725	735	781	781
Ti-50A	346		353	
Ti-6Al-4V	918	854	901	857

frictional conditions obtaining, and, on the other, it will determine the degree of flange distortion, including the degree of wrinkling. The latter results from the lateral, plastic instability which often develops in these operations. Since relatively large deformations are obtainable in free-forming operations, these are often utilised as the first step in the sizing and final stages of component forming that have to be performed, eventually, in evacuated, shaped dies.

The amount of deformation, the level of residual stressing, and the desired shape of the component will depend on the amount of energy delivered at the target, and on the geometry of the charge. The latter is of a considerably higher importance in free — as compared with the shaped die — forming, because the resulting component profile will be governed by that of the generated and transmitted shock wave.

It is for these reasons that analytical assessments of charge size, shape and of the associated energy flux distribution have been the subject of a number of investigations which resulted in a variety of mathematical models. The energy transfer and/or dimensional analyses are the two most often applied techniques in connection with the problems attending the design of charges. The simplified theory of plasticity, involving the notion of the generalised strain and stress (flow stress) is used to predict the final geometry of the deformed component, and, in some cases, the critical conditions that may lead to plastic instability.

5.2.2 Mathematical Modelling

5.2.2.1 Domes

The mathematical modelling of the process of dome forming, usually tank heads or dishes, depends to a degree on the thickness of the original material. Both thin and thick sheet analyses rely mainly on the concept of energy transfer, but differ in their final recommendations.

(a) *Thin sheet forming.* The approximate analysis, due to Alting,[12] predicts the magnitude of the charge required to form a dome from a circular blank of diameter B, by means of an open, circular die of diameter D. The material is specified by the standard constitutive equation giving the flow stress σ as

$$\sigma = K\varepsilon^n \tag{5.11}$$

where K is a material constant, ε is the generalised strain, and n is the strain hardening exponent.

The analysis distinguishes between the plastic work done in deforming part of the blank of diameter D_D, and the work performed in deforming the flange to a diameter D_F.

The total energy of deformation of diameter D_D is then given by

$$U' = \pi D_D^2 hK[2/\sqrt{3} \ln(D_D/D]^{(n+1)}/4(n+1)(n+2) \tag{5.12}$$

The energy of flange deformation is

$$U'' = \pi(B^2 - D_D^2) hK\{1/\sqrt{3} \ln[(B^2 + D_D^2)/(D_F^2 + D^2)]^{(n+1)}\}/4(n+1) \tag{5.13}$$

The total energy U is then

$$U = U' + U'' \tag{5.14}$$

The energy required from the charge is

$$We(1 - \cos \phi) U/2\eta \tag{5.15}$$

where W is the charge weight, e is the specific energy, η is the efficiency of discharge, and ϕ is the solid angle subtended by the charge at the top die bore. For $H/D \le 0.5$

$$\eta = 4.02 - 2.83(H/D) \tag{5.16}$$

A rather complex analytical model of the deformation of a dome has been produced by Yamada *et al.*[13] This is based on multishot electrohydraulic deformation of thin lead sheet. Details of the mathematical expressions are not reproduced here, but Fig. 5.2 is included to indicate

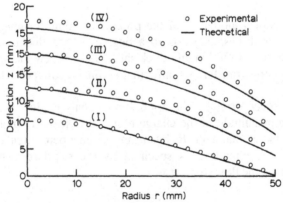

FIG. 5.2 Comparison between experimental and predicted dome profiles produced in 1 mm thick lead sheet by successive hydroelectric pulses (I, II, III, and IV) (after Ref. 13).

basic agreement with the proposed theory. The authors point out that the first pulse tends to produce a conical dome profile, but that the successive pulses smooth out the profile to semi-spherical.

(b) Thick sheet forming. The analysis of the forming of thick sheet domes from clamped circular blanks is due, among others, to Foral.[14]

To estimate the total energy delivered to the blank by the charge, use is made of the modified expression[15] for impulse I.

$$I = BW(F+1)/3H^F \qquad (5.17)$$

where B and F are explosive constants, W is the charge weight (in lb), and H is the distance from the charge to any blank element (in ft).

The total energy U delivered to the membrane is

$$U = 2B^2 W(F+1) g/h_o \rho \pi D^2 L^4 /(D/12)^{2F} D^4(F+1)$$
$$\times [1/(L/D)^{2(F+1)} - 1/(\{L/D\}^2 + 0 \cdot 25)^{(F+1)}] \qquad (5.18)$$

where D is the diameter, L is the vertical stand-off distance, h_o is the initial blank thickness, and ρ is the density.

The validity of this expression was tested on aluminium alloy membranes deformed to paraboloids. For TNT charges, for which $F = 0 \cdot 89$ and $B = 1 \cdot 46$, large stand-off distances produced a uniform initial membrane velocity.

5.2.2.2 Rectangular sheet forming

Again, analytical considerations depend very much on whether thin or thick material is used. In either case, however, two basic forming systems can be employed, because, in addition to the underwater forming (Fig. 5.1), an air/water cushion system is often introduced.

In the latter case, the initial shock wave attenuation is achieved by using, say, a polythene bag filled with water and carrying the charge, but the deformation of the target sheet proceeds, unimpeded, in the air. In this system, there is, of course, no secondary gas bubble.

The sequence of events, normally associated with a standard water system, is shown in Fig. 5.3, with Fig. 5.3(e) showing the change in the profiles resulting from the variations in the magnitudes of line charges used. An example of a circumferential strain distribution across the sheet is provided by Fig. 5.4, which refers to an aluminium alloy. For a given charge length, the strain decreases outwardly from the centre of the sheet, but an increase in charge length, made in equal increments, does not result in a uniform increase in the strain at a given radial

(a)

(b)

(c)

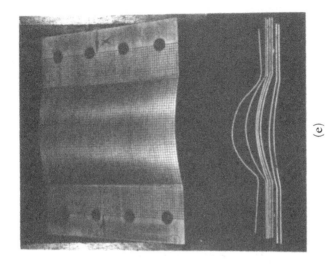

(e)

(d)

Fig. 5.3   Normal sequence of rectangular aluminium alloy sheet forming operations. (a) Forming die with sheet in position; (b) die sheet, explosive and water tank; (c) detonation of the charge; (d) sheet after forming; (e) development of sheet profiles.

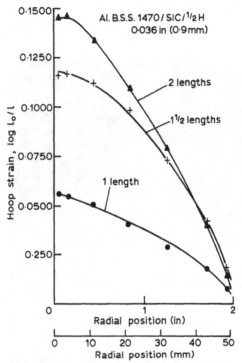

FIG. 5.4   Dependence of the distribution of circumferential strain in the sheet of Fig. 5.3 on charge length.

position. Otherwise, the pattern of distribution is symmetrical with respect to the longitudinal axis, and the magnitude of the strain near the clamped edges is not materially affected by the charge length.

(a) Thin sheet forming. The analysis of the process was developed by Blazynski,[16] and refers to the two systems described above.

In a single-pulse, air-cushion system (Fig. 5.5) the kinetic energy imparted to the sheet by the shock wave is (in the usual notation):

$$E = \rho h_o a L v^2 \tag{5.19}$$

The primary shock wave generated does not only impart an energy flux to the diaphragm, but is also responsible for the destruction of the plastic water container. In the absence therefore of a bubble pulse, the mechanism of deformation is substantially modified. Initially, the shock wave produces a high velocity of the sheet. On cessation, however, of

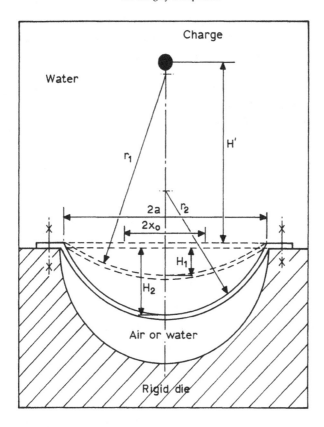

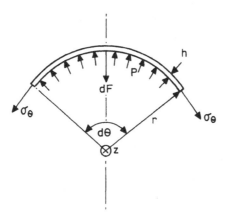

Fig. 5.5   Basic system for the forming of thin rectangular sheet.

the pressure pulse, coincident with the end of the period of deformation, the energy associated with the velocity is dissipated in overcoming the inertia effects of the moving target. The sheet is eventually brought to rest. The total polar deflection H_2 is thus obtained in two stages, i.e. during the period of delivery of the shock pulse (H_1), and a further increase during the purely inertial flow.

The magnitude of H_1 is found from the expression:

$$H_1 = v t_D \tag{5.20}$$

with the time of detonation being given by

$$t_D = L/v_D \tag{5.21}$$

where L is the charge length, and v_D is the velocity of detonation.

On analysing the situation, it is found that during the inertial movement the velocity of the diaphragm is given by

$$v^2 = 1 \cdot 73 \, K/(n+1) \{ [B + 1 \cdot 15 \ln\langle 1 + (H_2/a)^2 \rangle]^{(n+1)}$$
$$- [B + 1 \cdot 15 \ln\langle 1 + (H_1/a)^2 \rangle]^{(n+1)} \} \tag{5.22}$$

where B and K are the constants in the constitutive material equation

$$\sigma = K(B + \varepsilon)^n \tag{5.23}$$

The velocity v and deflection H_1 are obtainable from eqns (5.19) and (5.20), and so H_2 can be calculated.

The problem of transient tensile instability, leading to local yielding and possibly fracture, is of considerable importance in forming thin sheets. The strain likely to cause its onset in an air-cushion system is defined by the following equation:

$$5 \cdot 19 \, \varepsilon^2 + (5 \cdot 19 \, B - 2 - 4n) \, \varepsilon - 2B = 0 \tag{5.24}$$

In a double-pulse water-cushion system, the energy requirement is

$$E = E_W + W_P \tag{5.19a}$$

where E_W is the energy associated with the displaced mass of water, and W_P is the plastic work done in deforming the sheet.

$$W_P = 1 \cdot 73 \, a h_o L K /(n+1)(n+2) \, \varepsilon_\phi [(B + 1 \cdot 15 \, \varepsilon_\phi^{(n+2)} - B^{(n+2)}] \tag{5.25}$$

and the circumferential strain ε_ϕ is determined from

$$\varepsilon = 1 \cdot 15 [1 - (x_o/a)] \, \varepsilon_\phi \tag{5.26}$$

The total energy then becomes

$$E = W_{\mathrm{p}} + 0.5\, aHL\rho_{\mathrm{W}} v^2 \qquad (5.19\mathrm{b})$$

(b) Thick sheet. A very complex numerical solution to the determination of the profile geometry, obtained when free forming thick rectangular sheet, was proposed by Lindbergh and Boyd.[17] The derivation is general and includes cases in which the target plate, single, prestrained or a prewelded composite is either symmetrically or asymmetrically loaded. Details of the procedure adopted and methods of calculation are given in Ref. 17.

5.2.2.3 Tubular components

Free expansion of a tube presents problems similar to those encountered in sheet forming. An interesting analysis, predicting the final tube shape, including the possible tube bulging, was proposed by Yamada *et al.*[18] on the lines of Ref. 13. Good agreement between the, again rather complex, theory was obtained when hydroelectrically deforming aluminium and copper alloy tubes. Representative examples of this work are shown in Fig. 5.6.

A simpler analysis, predicting the total energy requirement as a function of pressures and velocities developed, and of the reduced mass of shell material is due to Masaki and Nakamura.[19]

A shock impulse produces a pressure *p* at the target, whose particle velocity under the impact is *u*.

$$p = 2 P_{\mathrm{M}} M\pi /[1 + (M\pi)^2]\,[M\pi \sin(\pi t/\phi) + \cos(\pi t/\phi) + \exp(-1/M\phi)]$$

$$(5.27)$$

$$u = 2u_{\mathrm{M}} M\pi /[1 + (M\pi)^2]\,[\sin(\pi t/\phi)/M\pi - \cos(\pi t/\phi) + \exp(-1/M\phi)]$$

$$(5.28)$$

$M = m/C_{\mathrm{o}}\rho_{\mathrm{o}}\phi$, *m* is the shell mass per unit area, ρ_{o} is the shell material density, *t* is the time considered, ϕ is the duration of impulse, P_{M} and u_{M} are the maxima of the incident pressure and particle velocity respectively.

Free forming of a cylinder closed with end caps and subjected to the detonation of a centrally located explosive charge was considered by Benham and Duffey.[20] Prediction of the maximum wall deformation due to both the initial impulse loading and the long-term pressure build-

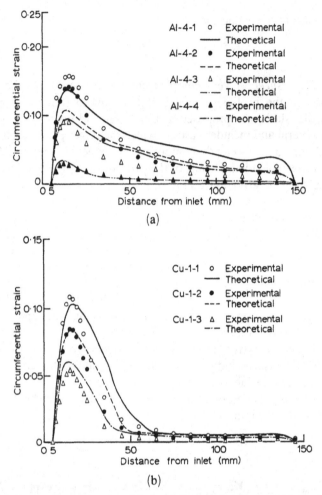

FIG. 5.6   Comparison between the theoretical and experimental profiles of freely bulged tubes. (a) Aluminium tubes deformed by condenser discharges of 4, 6, 8 and 10 kV; (b) copper tubes subjected to discharges of 10, 12 and 14 kV (after Ref. 18).

up was considered and expressions incorporating a rigid-plastic material model with strain hardening and strain rate sensitivity were proposed. The theoretical calculations based on this model compare favourably with the response of low carbon steel containment vessels, tested by means of explosive charges.

The resulting equations are complex and either an iterative technique or a computer code is needed for their solution.

5.3 AXISYMMETRICAL DIE FORMING

5.3.1 Shaping

When provision is made in a die system for evacuation of the air that otherwise will be trapped between the target and die walls, full shaping of the product becomes possible. Dished or dome-type components can be formed in bulkhead systems (Fig. 5.1), and tubular, axisymmetrical parts in solids, premachined to required profile dies. Closed or open die systems are used in the latter case (Fig. 5.7), the choice being dictated by the magnitude of the charge, and the required dimensional accuracy of the product.

The more geometrically complicated profiles can not be obtained in a single shot, unless a closed system is employed in which the shock wave is distributed more uniformly through the medium of water to all parts of the component. On expansion, the tube walls move radially outwards and, eventually, conform closely to the shape of the die cavity.

In an open system, the positions of the charges must be carefully controlled, so that the energy flux and the resulting shock wave are directed at the areas in which maximum deformations are required. The excess energy is, of course, lost to the surroundings. Procedures adopted in those cases are illustrated in Figs 5.7–5.9 on an example of a stainless steel tube formed to the shape indicated in Fig. 5.7.

In the closed system of Fig. 5.7(a), a single, centrally placed charge is sufficient to produce the required amount of deformation. This is not so in the open system (Fig. 5.7(b)) where directionality of energy flux, combined with a sequence of operations is needed. In the case considered, four firing operations are required (Fig. 5.8) and result in the desired tube shape (Fig. 5.9).

Composite, explosively prewelded tubes, of the type shown in Fig. 5.9, can be easily converted to a variety of shaped nozzles as, for instance, the components shown in Fig. 5.10. A further example of a specially designed nozzle is provided by Fig. 5.11.

The problem of dimensional accuracy arises when grooved or ribbed tubing is required, because the presence of sharp transition sections in a profile gives rise to, on the one hand, high stress concentration and possible cracking, and, on the other, incomplete filling in of the die

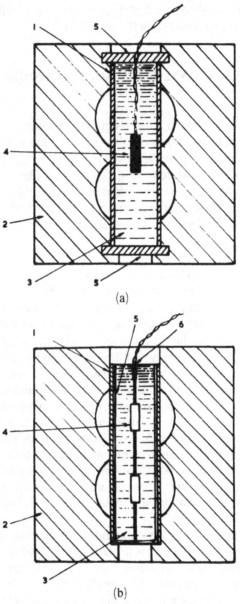

FIG. 5.7   An example of (a) closed and (b) open die systems for tube shaping. (a): 1, Tube; 2, die; 3, transmitting medium (water); 4, low explosive charge; 5, closing plate; (b): 1, tube; 2, die; 3, water; 4, high explosive charge; 5, polythene bag; 6, detonator.

Material : FDP 18 /8 TI

Initial size : 3·000 in. o.d. x 0·064 in.

Explosive content of cordite : 3·18 gm / ft

Shape of charge	Length of fuse (in.)	Total explosive content (gm)	Remarks
	10	2·65	Sizing
	10+2×6	5·83	Slow expansion of bulges
	10+2×6	5·83	Top bulge formed
	10+6	4·29	Lower bulge formed

FIG. 5.8   Sequence of operations in the open die system of Fig. 5.7.

cavity. A typical example of the difficulties that arise — in spite of the virtual absence of springback — is the forming of a ribbed tube shown inserted in one-half of an open system die (Fig. 5.12). The thinning of the tube wall in different planes of the profile, as well as the inability to obtain full conformity with the die, are indicated in Fig. 5.13. The sharper the profile of the transition section, e.g. 6–8, the greater the departure from the designed shape. On the other hand, a high degree of wall thinning is associated with the high radial deformation, as, for instance, in Sections 1–3.

The problem is, of course, magnified when deep, but small width, ribs are required. Figure 5.14 exemplifies this by showing a relatively poorly defined profile of an incompletely developed ribbing.

The bulkhead type of forming also serves as a means of introducing dimensional accuracy to axisymmetrical dome-shaped structures. By evacuating the air contained between the die and the blank, conditions for full deformation are created.

Fɪɢ. 5.9   Stainless steel tube explosively shaped in an open die system.

The usefulness of the component manufactured in this way depends to a considerable degree on the dimensions and quality of the flange. Large circumferential stresses that can set-in during the forming operation, can lead to flange buckling and therefore render the dome unsuitable for commercial applications. The tendency to buckling decreases, however, with increasing blank thickness and decreasing yield stress of its material. The effect can be further reduced by the application of lateral pressure to the flange. A theoretical analysis of this

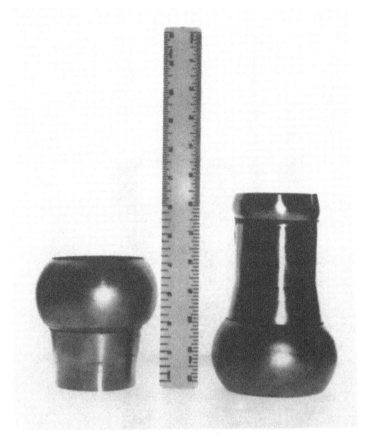

FIG. 5.10 Nozzles obtained from the tubes of the type shown in Fig. 5.9.

problem, addressed to thick-walled domes, is due to Kaplan and Boduroglu.[21]

Strain limits, determining the conditions for either tearing or flange instability arising in circular blanks formed into ellipsoidal dies, are calculated using the mathematical approach of Thurston.[22]

Although composite, prewelded or simply mechanically co-shaped components are successfully formed in the described arrangements, the ideal solution is one that combines welding and forming in a single operation. An introduction and possible solution to this problem was produced by Kiyota *et al.*,[23] when they designed a dome shaping system

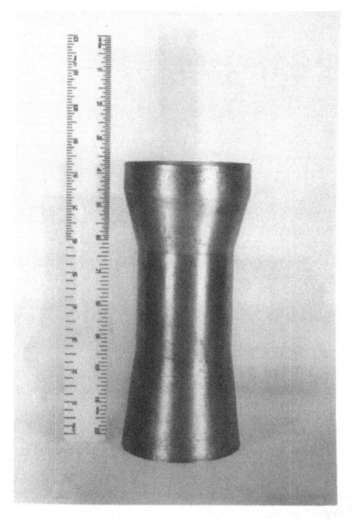

FIG. 5.11   A specially designed, explosively formed nozzle.

in which the blank to be formed was separated from the charge by an auxiliary plate and an air gap. The final design[1] incorporates a circular cutter device that shears off the auxiliary plate as it deforms through, this time, the evacuated gap that separates it from the lower blank. The shock front generated accelerates the 'flyer' until it collides with the blank and thus effects both a further deformation of the two elements of

FIG. 5.12   Forming of a sharp-transition section, ribbed tube.

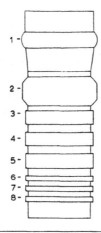

Section	Die diameter (cm)	Section diameters (cm)			Final thickness (mm)		
		Max.	*Min.*	*Mean*	*Max.*	*Min.*	*Mean*
1	9·84	9·84	9·82	9·830	1·42	1·29	1·36
2	9·85	9·85	9·84	9·847	1·29	1·19	1·24
3	8·95	8·95	8·94	8·945	1·37	1·27	1·32
4	8·95	8·95	8·94	8·945	1·37	1·32	1·34
5	8·95	8·95	8·94	8·945	1·39	1·32	1·35
6	8·95	8·70	8·66	8·682	1·47	1·39	1·43
7	8·95	8·74	8·63	8·688	1·55	1·42	1·48
8	8·95	8·61	8·49	8·549	1·52	1·47	1·49

FIG. 5.13   Dimensional accuracy obtained in the tube of Fig. 5.12.

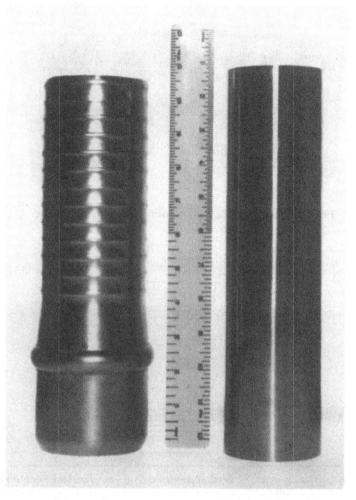

FIG. 5.14   Poorly defined tube profile consisting of small-width but deep ribbing.

the assembly and their mutual welding. This underwater explosive bonding/forming operation gives satisfactory results in a number of material combinations such as copper/low carbon steels in which the welding wave amplitudes of between 0·12 and 0·2 mm, accompanied by wave lengths of 0·3–0·4 mm, are easily produced.

In addition to profile shaping, explosive forming is also used as a substitute for tube sheet welding (Section 4.3.5). The expansion into feed water heater tube sheets of tubing up to 60 mm in diameter subjected to pressures varying in intensity between 7 and 40 MPa, is now normal practice.[24]

The Detnaform process[25,26] has been developed for the expansion of tubes into tubeplates and tubesheets, some 320 mm thick, and operates successfully on the basic engineering metal–alloy combinations of carbon/stainless steels, carbon steel/titanium, brass/carbon steel, and titanium/Inconel 625. Other combinations are naturally possible.

5.3.2 Sizing

Under the heading of explosive sizing come two main operations of manipulation of concentrically prefabricated components, and of auto-frettaging.

Prewelded half-domes or shells, intended for the manufacture of vessels for the chemical industry, are sized usually in either open or closed dies (as shown, for instance, in Fig. 5.15) to reduce the level of explosively imparted strain.

The two basic methods employed in the production of hemispherical containers are those of forming from a prewelded frustrum of a cone or from meridional strips.[27] The first technique is suitable for forming vessels from combinations of steels varying in diameters from 0·3 to 3 m, and in thickness from 2 to 6 mm. In this range of sizes and carbon or stainless steel combinations, the impact strength of the welded joints, after sizing, is, on average, some 40 J/cm².

Sizing and forming from meridional segments can be carried out using the methods of Fig. 5.16. In Method 1 (Fig. 5.16(a)), applicable to

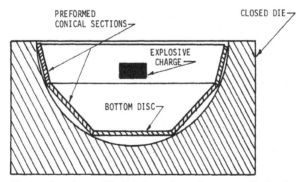

FIG. 5.15   Dome forming with fabricated preforms to limit the level of imposed strain.

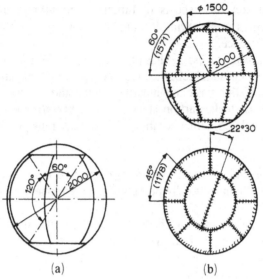

FIG. 5.16   Schematic representation of the division of a spherical vessel into segments. (a) Method 1; (b) Method 2 (after Ref. 27).

vessels of up to 2 m in diameter and around 15 mm in shell thickness, six meridional segments and two hemispherical domes are used. For vessels within the 3–10 m diametral range, Method 2 (Fig. 5.16(b)) is recommended because it offers enhanced mechanical properties of the finished shells.[27]

Explosive autofrettaging relies on the enhancement and exploitation of material properties associated with the induced stress systems. Its use is of special interest in complex shape tooling subjected to particularly severe working conditions. Although many examples can be quoted, that of forging dies has been found to be of special industrial interest.

Forging dies are subjected to large transient internal pressures that can easily give rise to eventual fracture of the material. The fracture is associated with the propagation of radial cracks that originate in the die bore. Compressive circumferential pre-stressing of such dies reduces the danger of fracture, but is difficult to generate. Although mechanical autofrettaging, by means of shrunk rings that induce compressive stresses in the bore, is possible, in many cases it cannot be used on cylindrical dies and, in any case, the level of stressing produced is low. Hydraulic pressurisation of the bore, beyond the yielding point, is economically prohibitive since it involves sophisticated equipment.

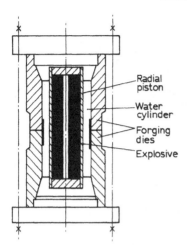

FIG. 5.17 Explosive autofrettaging of
a forging die.

However, large residual compressive circumferential stresses can be induced in cylindrical, thick-walled dies by the radial piston method based on the use of energy dissipated when an internal, explosive charge is detonated and pressure is generated. The method, suggested by Kaplan *et al.*[28] is shown diagrammatically in Fig. 5.17.

To reduce the occurrence of secondary yielding, which would lower the residual stress level, the concept of a radial piston separated by a 'cylinder' of water from the die to be stressed, has been conceived. The piston acts as an attenuator of the rate of energy dissipation and, as a result of this, the energy of the detonating charge reaches the die in the form of an uniform pressure wave, rather than in a series of shock waves. Full control of the duration and level of stressing is thus maintained.

Details of the relevant analysis are provided in Ref. 28, but it should be stressed that the analysis provides the means of examining the response of both the radial piston and the component being stressed. Since the piston has to expand some 10–20% without fracturing, the choice of its material is rather important. This is illustrated in Fig. 5.18. The correlation between this simplified theoretical model of the operation and the experimentally determined values is shown in Fig. 5.19. The discrepancy is due partly to the fact that the analysis assumes a plane strain condition that may not necessarily apply, and partly to the onset, occasionally, of secondary yielding that reduces the stress. The level of stressing is however satisfactorily high.

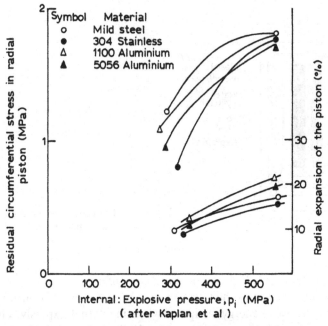

FIG. 5.18   Variation of circumferential residual stress in the piston of the system of Fig. 5.17, with material and the induced pressure.

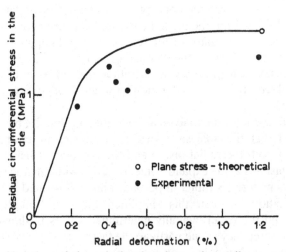

FIG. 5.19   Variation of the calculated and experimentally assessed circumferential residual stress in the die of the system of Fig. 5.17, with radial deformation.

5.3.3 Joining

The production of aluminium alloy drill pipes for the gas and oil extraction industry involves the manufacture of joints, not only tight enough to prevent leaks, but, at the same time, strong enough to produce satisfactory connections with steel couplings. Whereas conventional methods rely on screwing the two components together at elevated temperature and then producing a tight shrink-fit connection on cooling the assembly, a simpler and more cost effective method is available when explosive expansion is utilised.[29]

In this method, the aluminium tube is positioned axially within a steel coupling (Fig. 5.20) with the explosive charge inserted into the bore. The coupling is separated from the tubular die by an annular gap which is filled with water when the whole assembly is submerged in a water container. On detonation of the charge, the aluminium tube expands to the surface of the bore of the coupling, but the latter is prevented from deforming radially outwards by the practically incompressible water annulus. The single, outer steel die suffers less possible deformation and damage than a split one, e.g. Fig. 5.12, the normally labour intensive assembling operations are reduced and, from the operational point of view, the noise problem is eliminated.

Re-expansion of damaged and consequently leaking joints in boiler drums and headers is satisfactorily accomplished by techniques which

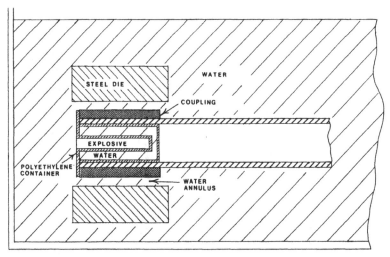

FIG. 5.20 Expansion of an aluminium drill pipe into steel coupling (courtesy ICI Explosives, Nobel's Explosives Co. Ltd, Stevenston, Ayrshire, Scotland).

resemble the approach described above. Since the amount of bulging is low, the hardness of low carbon expanded tubing is only slightly higher than the peak hardness found after conventional roller expansion. Mechanical twins present in the ferrites of such materials are due, as is usual in these cases, to the passage of shock disturbances and their effect is to increase slightly the level of hardness.[30]

5.4 LINING

Purely mechanical forming of composites with a view to providing thin anticorrosive and/or heat resistant shields, has been developed in a wide range of applications.

One of the earliest, but still well in use, examples is provided by the plating of the interior of a vessel,[31] shown in Fig. 5.21. An explosive charge, either concentrated or in the form of foil, is detonated within the 'plating' vessel (black outline) which, in turn, is positioned within, but is separated from, the vessel to be plated. On detonation of the charge, the plating component collapses onto the bore of the outer vessel and effects the sealing of its surface from the corrosive substances that may be present under normal working conditions.

This type of approach to the protection of equipment used in the chemical industry is of particular value in cases of severe service conditions such as elevated temperatures, combined with abrupt tem-

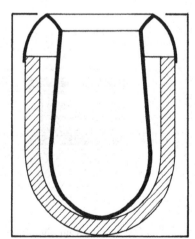

FIG. 5.21 An explosive system for the lining of a vessel with anticorrosive plating (plating vessel shown in dark outline).

perature gradients. Because it is the liner, and not the plated shell, that experiences the effects of the shock pressure and of a high degree of deformation, the residual stress field generated in the liner prevents the gap between it and the shell from forming when the temperature of the latter changes rapidly. The liner conforms closely to the shape of the shell or vessel, but no metallurgical bond (as in explosive welding) is normally present.

One of the more often used metallic combinations is that of a titanium liner in a steel shell.[32] Depending on the profile of the latter, the liner may have to be prefabricated to as near the shell profile as possible, before explosive forming is initiated. To avoid any permanent deformation of the shell, it is important that the energy balance is preserved. For cylindrical components, the total available energy of the explosive E_0 is related to the required forming energy E_F and the strain energy of the material E_S by the expression:

$$E_0 - E_F \leqslant E_S \qquad (5.29)$$

and

$$E_F = 2 R_F L h_0 [K /(n + 1)] \exp(-0.9 \varepsilon) \varepsilon^{(n+1)} \qquad (5.30)$$

where $\varepsilon = (R_F/R_0 - 1)$, R_0 and R_F are the initial and final liner diameters, h_0 is the liner thickness, L is the final length of the cylinder, and K and n are the material constants.

Explosive cladding of expended gun barrels with steel liners to provide stock for remachining, constitutes another variant of the lining technique.[33] Originally, 125 mm naval gun barrels have been processed, as shown in Fig. 5.22, by Grollo. Annular explosive charges were used as indicated in the figure and the guns were successfully relined. The induced compressive residual stresses varied from 300 to 650 MPa depending on the actual working conditions.

Equally, of course, similar techniques can be applied to any machine component subjected to rapid increases in pressure, such as, for instance, cylinders of extrusion presses and autofrettaged duplex cylinders.

A more exotic application, again in the area of chemical processing, is that of lining copper/steel components with tantalum by prewelding the triple combination and then expanding it into the desired final shape.[34]

Magnetic lining operations involving the shrinking of metallic rings onto non-metallic fixtures are carried out, on industrial scale, by means

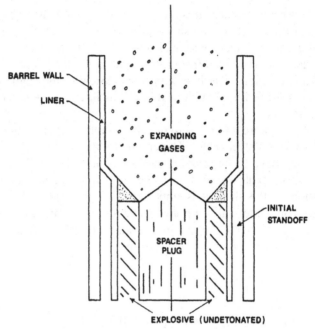

FIG. 5.22   Explosive lining of 125 mm gun barrels (after Ref. 33).

of machines such as, for example, Magneform.[35] Assemblies of electric rotors and arbors in different materials can be made easily, as well as those of parts for the car industry.

5.5 SPECIAL APPLICATIONS

The described forming techniques lend themselves easily to special types of applications requiring either an unusual material combination or uncommon geometry. The range of possibilities is very wide, but the three selected examples discussed below illustrate well the practical opportunities that may exist.

5.5.1 Multilaminate Rings

Multilaminate rings are manufactured out of explosively prewelded duplex or triplex sheet by explosively free forming (as in Fig. 5.1) the blank into hemispheres which are then sectioned to form collars. Aluminium/titanium/steel transition rings of this type form part of forged steel fittings for a helicopter tie down on aluminium flight decks of ships. Problems of corrosion are avoided.[36]

Other material combinations will furnish further sets of possibly corrosion and/or heat resistant transition joints displaying sufficiently high levels of residual stresses to ensure a good shape stability under severe working conditions.

5.5.2 Geodesic Lens

An example of a special 'one-off' component that can be produced is provided by the geodesic lens.[37] The lens consists of two skins of a complex configuration (Fig. 5.23) formed from two special aluminium

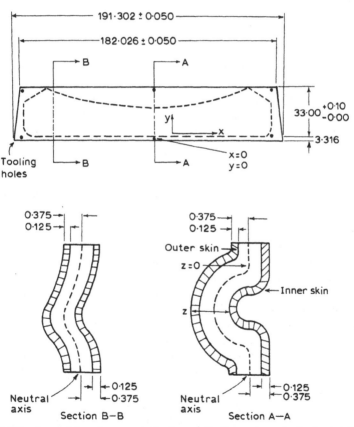

FIG. 5.23 Assembly of a geodesic lens (all dimensions in inches) (after Ref. 37).

alloys, each in excess of 16 ft in length. A closely toleranced spacing of 0·5 in had to be provided between the skins; the tolerance being 0·500 ± 0·025 in.

In view of the size of the required components, Kirksite tooling was used thus reducing manufacturing costs. Laminated 7/8 in thick blanks were formed in vacuumed (28 in gauge) die cavities. The forming operation produced a lens of a high degree of sensitivity and improved capabilities that reflected good dimensional accuracy attained.

5.5.3 Blower Impellers

Composite blower impellers form another group of special, but more frequently used components. The conventional methods of production of Roots-type blower impellers usually involve the welding together of two halves of a preformed metal sheet. There exist, of course, various techniques by means of which this is achieved, but, essentially, the method is fairly complex and expensive.

In order to both simplify the manufacture and to reduce costs involved in tooling, machining, etc., a method of free implosive forming was developed.[38] This consists of deforming seamless metal tubing to the shape of the impeller utilising the energy dissipated when suitably shaped, and positioned, high-explosive charges are detonated against the surface of the workpiece.

A high degree of dimensional accuracy can be obtained without recourse to any form of die or tool, but, when required, simple means of constraining the flow of the metal can be provided. Depending on the degree of dimensional accuracy required, two variants of the implosive technique can be employed.

The simpler method consists in providing suitably shaped explosive charges along the diametrically opposite arcs AB and A'B' (Fig. 5.24). On detonation, plastic hinges will be formed at these limiting points and the metal contained between the two arcs will travel inwards, whereas the original arcs AA' and BB' will be deformed to a shape approximating to CC'. The precise profile of ACA' and BC'B' will depend on the provision of explosive constraints along the arcs AA' and BB'. The figure shows that small explosive charges, placed diametrically opposite each other, will provide a means of imposing a constraint on the sideways movement of the metal. In fact, the manipulation of these charges, in either shape or magnitude, gives a wide range of possibilities of controlling the profile of an impeller. Figure 5.25(a) shows the difference between the profiles obtained with the same main forming

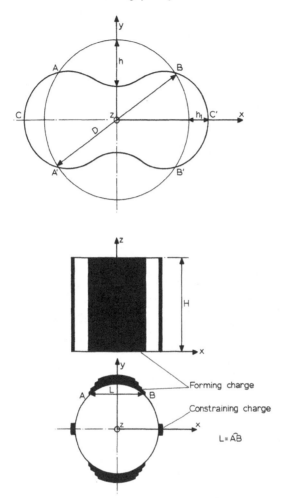

FIG. 5.24 Free explosive forming of Roots-type blower impellers (after Ref. 38).

charges, i.e. in the y–y direction, but with or without an x–x constraint. The effect of the increasing constraining charges in full-size impellers is shown in Fig. 5.25(b).

Duplex or triplex composite impellers can be formed in the same way out of either prewelded multiplex cylinders, or mechanically, coaxially preassembled tubing.

(a)

(b)

Fig. 5.25   The effect of constraining charges on the final geometry of impellers of Fig. 5.24. (a) The effect of increasing charges in forming of impeller rings; (b) increasing charges (from left to right) in full scale impellers.

The other variant of the implosive technique is based on the provision of permanent shaped metal constraints mounted on a base plate. These replace the constraining explosive charges and reduce the explosive contribution to the forming of impellers to the provision of the main charges in the y–y direction. Some technical problems are involved in the development of this type of 'tooling', but in spite of the loss of simplicity, a greater degree of sophistication than is possible with the provision of shaped tools, is advantageous in some situations.

One of the major problems arising here is that of the prediction of the magnitude and geometry of the charge needed to form an impeller of any desired size. The complexity of a rigorous theoretical approach is prohibitive and consequently other ways of establishing a relationship between the various parameters of the process have to be developed.

The successful use of dimensional analysis in the field of many engineering applications indicates the possibility of using this line of approach to the problems arising in the high-energy rate forming operations.

As with other machine components, the range of sizes of impellers can be quite extensive and therefore the provision of a scaling law provides the possibility of using the results of small scale experiments to determine the conditions required for the forming of large impellers. The successful application of this type of analysis to a more complex shape also indicates the flexibility of the technique and the obvious possibility of extending the method to components of other shapes.

A physical event, expressed as a variable y will usually depend on a number of independent variables $x_1, x_2,..., x_n$. The nature and the form of this relationship may not be known. Mathematically, however, the functional relationship between y and the independent variables can be represented by

$$y = f(x_1, x_2,..., x_n) \tag{5.31}$$

If the nature of the dependence of y is known, then eqn (5.31) will simply represent a mathematical model of a physical law. A general physical law is independent of the units employed in measurements and applies equally to the model and the prototype.

Ordinary physical phenomena can be described in terms of the four basic dimensions of mass, length, time and temperature. With reference to eqn (5.31), it is seen that the $(n+1)$ dependent and independent variables can be therefore combined into $[(n+1)-4]$ dimensionless groups. In many applications, it is found that the interpretation of the physical meaning of a non-dimensional group can be difficult, because the four basic parameters of state may not be sufficient to account for the directionality of the investigated physical effect. This difficulty can be partially obviated by the introduction of the concept of vectorial length. The directionality of an event can then be expressed in terms of L_{xx}, L_{yy} and L_{zz}, where L is the length.

With this proviso, the number of basic dimensions is increased to six, thus providing two additional equations. The number of dimensional groups is therefore given by $[(n+1)-6] = n-5$.

Equation (5.31) can be re-written in the form

$$E = f'\{(F_1), (F_2),..., (F_n)\} \tag{5.32}$$

TABLE 5.2
Averaged Specimen, Charge and Deflection Sizes

D (mm)	t (mm)	e (g)	$\dfrac{L}{s}$	h (mm)	h_l (mm)	x–x condition
100	5	5·1	3	9·0	2·0	Free
100	5	11·7	4	27·0	10·0	Constrained
155	5	14·3	8	34·0	17·5	Constrained
155	5	15·3	8	39·0	20·5	Constrained
155	10	25·0	5	22·0	17·0	Free
155	10	31·4	3	34·5	22·5	Constrained
165	5	16·1	8	65·0	17·5	Constrained
165	10	37·0	4	47·0	17·0	Constrained
275	10	63·8	9	79·0	33·0	Free

where E and F_s are the dependent and independent dimensionless groups respectively.

Based on these considerations, an equation relating the energy (e) of the forming charge (in the y–y direction), its height (s), length L, the height H of the impeller, the density ρ of its material, and the velocity of sound (v) in the material, to the original tube (ring) diameter D, and deflection h (in the y–y direction) was established. The details of the derivation and an assessment of its validity are given in Ref. 38. All the parameters appearing in eqn (5.33) are defined in Fig. 5.24.

$$e/s^{2\cdot33}H^{0\cdot33}L^{0\cdot33}\rho v^2 = 0\cdot001\,41[D/(sL)^{0\cdot5}]^{2\cdot05}](h/s)^{1\cdot41} \qquad (5.33)$$

A representative set of pre- and postforming numerical results for a ferrous metallic material combination is given in Table 5.2.

REFERENCES

1. Fujita, M. In: *High Energy Rate Fabrication PVP 70*, Eds M. A. Myers and J. W. Schroeder, ASME, New York, 1982, p. 29.
2. Thurman, A. G. and Ezra, A. A. In: *Proc. 9th Int. MTDR Conference*, Eds S. A. Tobias and F. Koenigsberger, Pergamon Press, Oxford, MS No. 19, 1968.
3. Johnson, W. *Proc. Int. Prod. Engineering Research Conference*, ASME, Pittsburgh, PA, 1963, p. 342.
4. Cook, M. A. *The Science of High Explosives*, Reinhold, New York, 1958.
5. Bebb, A. H. *Proc. Roy. Soc., Series A*, **20** (1951), 244.
6. Schauer, H. M. In: *Proc. 1st Int. HERF Conference*, University of Denver, Colorado, Paper 5.1, 1967.

7. RINEHART, J. S. and PEARSON, J. *Explosive Working of Metals*, Pergamon Press, Oxford, 1963.
8. ROTH, J. *The Explos. Engr*, **37** (1959), 52.
9. AL-HASSANI, S. T. S., DUNCAN, J. L. and JOHNSON, W. In Ref. 6, Paper 5.3.
10. VAN WELY, F. E. In: *Proc. 4th Int. HERF Conference*, University of Denver, Colorado, Paper 1.3, 1973.
11. ORAVA, R. N. and KUNTHIA, P. C. In: *Proc. 3rd Int. HERF Conference*, University of Denver, Colorado, Paper 4.1, 1971.
12. ALTING, L. In Ref. 11, Paper 6.1.
13. YAMADA, T., KANI, K., TUDA, A. and FUSE, M. Dynamic plastic deformation of a thin circular metal plate, Kyoto University Report, 1983.
14. FORAL, R. F. In Ref. 6, Paper 6.2.
15. COLE, R. H. *Underwater Explosions*, Princeton University Press, Princeton, NJ, 1948.
16. BLAZYNSKI, T. Z. *Archiwum Budowy Maszyn, Polish Academy of Sciences, Warsaw*, **20** (1973), 3.
17. LINDBERGH, C. and BOYD, D. E. In Ref. 6, Paper 6.3.
18. YAMADA, T., KANI, K., TABO, E. and INOUE, A. Dynamic plastic deformation of a cylindrical tube, Kyoto University Report, 1983.
19. MASAKI, S. and NAKAMURA, Y. *Bull. JSME*, **13** (55)(1970), 204.
20. BENHAM, R. A. and DUFFEY, T. A. In Ref. 10, SLA 73-0508.
21. KAPLAN, M. A. and BODUROGLU, H. M. In Ref. 11, Paper 5.1.
22. THURSTON, G. A. In Ref. 6, Paper 6.1.
23. KIYOTA, K., FUJITA, M. and IZUMA, T. In: *Proc. 11th MTDR Conference*, Eds S. A. Tobias and F. Koenigsberger, Pergamon Press, Oxford, 1970.
24. BERMAN, I. KUNSAGI, L. and THAKKER, B. In Ref. 6, Paper 3.1.
25. BERMAN, I. and SCHROEDER, J. W. In: *Explosive Welding, Forming, Plugging and Compaction — PVP 44*, Eds I. Berman and J. W. Schroeder, ASME, New York, 1980, p. 25.
26. SCHROEDER, J. W. In: *Explosive Welding, Forming and Compaction*, Ed. T. Z. Blazynski, Applied Science Publishers, London, New York, 1983, p. 347.
27. ANDRZEJEWSKI, H. and KRUPA, Z. In: *Proc. Int. Conference on the Use of High-Energy Rate Methods for Forming, Welding and Compaction*, Ed. T. Z. Blazynski, University of Leeds, Leeds, Paper 5, 1973.
28. KAPLAN, M., GLICK, H., HOWELL, W. and D'SOUZA, V. In: *Proc. 13th MTDR Conference*, Macmillan Press, London, 1972, p. 419.
29. HARDWICK, R., BROWN, D. W. and NOWELL, D. G. In: *Proc. 10th Int. HERF Conference*, Litostroj, Ljubljana, 1989.
30. JACKSON, P. W., LOWDON, N. and OXFORD, C. H. In: *Proc. 7th Int. HERF Conference*, Ed. T. Z. Blazynski, University of Leeds, Leeds, 1981, p. 48.
31. SIEMENS-SCHUCKERTWERKE, A. G. Berlin and Erlangen, Patent No. 1 041 999.
32. INOMATA, S., GOTO, A., YANO, K., YSUCHIMOTO, M., SHIBATA, S., FUJII, T., SAKURAI, T. and KANAMOTO, M. In: *The Science, Technology and Application of Titanium*, Eds R. Jaffee and N. Promisel, Pergamon Press, Oxford, 1970, p. 1065.
33. GROLLO, R. US Naval Ordnance Station, Louisville, Kentucky, Report No. MT-032, 1975.

34. Bouckaert, G., Hix, H. B. and Chelius, J. In: *Proc. 5th Int. HERF Conference*, Ed. R. Wittman, University of Denver, Colorado, Paper 4.4, 1975.
35. Zittel, G., Ibid., Paper 3.1.
36. Murr, L. E. and Hare, A. W. In: *Shock Waves for Industrial Applications*, Ed. L. E. Murr, Noyes Publications, Park Ridge, NJ, 1988, p. 103.
37. Lieberman, I. and Zernow, L. In Ref. 11, Paper 6.2.
38. Blazynski, T. Z. In: *Proc. 10th Int. MTDR Conference*, Eds S. A. Tobias and F. Koenigsberger, Pergamon Press, Oxford, New York, 1970, p. 511.

Chapter 6

Particulate Matter Composites

6.1 CHARACTERISTIC FEATURES

6.1.1 General Observations

The response of particulate matter to shock treatment has been under investigation for several decades, but usually in connection with metallic powder compaction. It is only in the last 20 years or so that the interest in the changes effected by the passage of a shock wave through non-metallic powder materials ceased to be purely academic. Practical, industrial considerations have given sufficient impetus to research and development into the shock treatment of powders to produce a considerable volume of literature in the area of, naturally, metallic powders, and then of ceramics, and to extend the work, more recently, to polymers and polymeric mixtures and blends.

Starting with, by now, classic papers on ceramics by Bergmann and Barrington[1] and Carlson et al.[2] among others, and developing the theme of ceramic response to shock conditions, the more recent reviews and papers of, for instance, Prümmer,[3] Gourdin et al.,[4] Prümmer and Ziegler,[5] Kondo et al.[6] and Graham et al.,[7] span the period of two decades and sum up the state of knowledge. Those publications provide further, extensive lists of references published in the intervening years, and point out three main groups of practical possibilities offered by explosive shock treatment with a special regard for the non-metallic particulate matter.

The effects of the passage of a shock wave through particulate matter can be utilised, singly or jointly, in three main areas of practical applications. These are:

- consolidation and compaction,
- structural material modification prior to further processing,
- structural and chemical synthesis.

253

The modification of structure, directly or through the medium of synthesis, together with enhanced chemical activation, consolidation of the particulate material through shock compaction, and, finally, the release of internal energy related to lattice distortion are the main attractions of the dynamic shock treatment. New materials can thus be created, composites formed and sintering, where necessary, made easier. Pre-shocked materials become more amenable to further conventional treatment by virtue of being chemically and/or physically altered or activated.

Many of these benefits, usually associated with metals and ceramics, can also be extended to polymeric powders and their mixtures with other polymeric or non-polymeric materials as shown recently by Hegazy *et al.*,[8] Blazynski *et al.*[9-14] and Blazynski.[15-17] However, whereas in the case of ceramics, thermal effects produced by shock-generated temperatures may lead to local annealing and reduction in residual microstressing, in polymers the adiabatic heat generated during the explosive compaction process is sufficient to assist the bonding operation, but its level and extremely short duration will prevent degradation of the material.

The most recent review of the effects of shock treatment of ceramics and polymers is provided by Blazynski,[15,18] with the mathematical treatment, and modelling, of compaction being supplied by, among others, Schwarz *et al.*,[19] Wilkins *et al.*,[20] Linse,[21] Lotrich *et al.*[22] and Hoenig *et al.*[23]

6.1.1.1 Compaction and consolidation

The success of the compacting process is reflected by its ability to consolidate powdered or granulated matter in a uniform manner, in the bulk of the affected material, and to bond individual grains or particles to form a coherent structure. In the conventional static or isostatic compaction, these aims are achieved, generally, through further processing of the compact, such as sintering, often accompanied by the addition of binders to create suitable chemical linkages. Such treatment is often appropriate in the field of ceramics — although even here long term thermal effects may adversely affect the properties — but is not acceptable in the case of, say, thermoplastics. The main interest in compaction of polymers lies precisely in the avoidance of heat effects associated with the conventional processing techniques. For instance, the conventional injection moulding is limited to certain materials only because the heating and cooling of the mould, the effects of high rates of shearing in small gates and runners, and the resulting molecular orientation create serious production problems. Degradation of the structure and consequently reduction in the desired mechanical and physical properties are often likely to occur. Since, however, explosive compaction is associated primarily with adiabatic heat generation, the viscoelastic nature of engineering thermoplastic materials is easily accommodated in this type of forming process. This is of particular, practical interest in the case of those plastic materials which can be produced cheaply in powder or granular form, but cannot be processed satisfactorily by conventional means. A similar argument applies to the manufacture of composite materials which consist of mixtures of different polymers, polymers and ceramics, and polymers and metal powders. Reinforcement of plastic matrices with plastic or non-plastic fibres, giving the desired directionality in properties, is, of course, another possibility offered by the dynamic compaction.

The varying interparticle collision velocities and angles produce intimate bonding that is effected through the medium of three possible shear mechanisms of friction, melting and welding.

As indicated briefly in Chapter 2, a high velocity, almost point, collision between two particles or grains produces 'hot spots' of sufficiently high temperature to give rise to friction welding (Fig. 2.20). Equally, an excess of energy, available at an interface, will produce melting and consequent adhesion through a layer of solidified melt. If the impact conditions are right — in terms of the collision point velocity, angle and stand-off distance — an explosive welding wave will be generated. Depending on the morphology or morphologies of the constituent material(s) of the compact, all three mechanisms may exist side by side, increasing the strength of the combination, but producing also non-uniformity of green density across the section. Mechanical predisposition of the powder to any particular bonding mechanism is, normally, impossible, but, naturally, a high degree of surface and grain shape regularity will improve the uniformity of the final structure.

6.1.1.2 Structural material modifications

Shock induced modification of organic and inorganic powders represents a different aspect of shock manipulation. Shock modification will affect, in the first instance, particle or grain morphology and consequently will change the specific surface area conditions and will thus favourably predispose the material to further processing. Of particular interest here are shock activated sintering, shock enhanced chemical reactivity, and enhanced catalytic activity.[24,25]

The microsecond time scale of processing has a decisive effect not only on the relatively large volume of material that can be processed, but also on its structural properties. The development of undesirable time-dependent reactions, grain growth and phase changes is suppressed or, at least, reduced. Small crystallites are likely to be formed because of the primary recrystallisation and metastability, linked to high concentrations of defects, that create conditions which differ from those obtaining in the conventional low isostatic pressure, high-temperature operations. Figure 6.1 illustrates this in the example of shock treated silicon carbide powder.

On the other hand, the availability of internal energy of lattice distortion, that is liberated in a shocked material, reduces the levels of pressure and temperature which would be normally needed if a sintering operation was required. Pre-sinter shock treatment of many materials will thus favourably predispose them for processing that might otherwise be very difficult to perform. Alternatively, the conventional

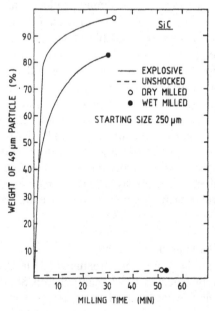

Fig. 6.1   The effect of shock processing treatment on silicon carbide (compiled from Ref. 1).

compaction of pre-shocked material may be made more efficient, in terms of a high degree of densification reached, as indicated, for instance, in Fig. 6.2. Shock processing will also, inevitably, result in thermal effects which, in the case of ceramics, are likely to produce local annealing and reduction in residual microstressing.

All these new properties, or modifications of the existing ones, have to be assessed both qualitatively and quantitatively. Since most of the changes take place on a microscale, the highly sophisticated techniques of X-ray diffraction line broadening, electron spin resonance, transmission electron microscopy, magnetisation and Mossbauer effect, and, to a limited extent, scan electron microscopy are employed.

6.1.1.3 Synthesis

A brief mention of the possibility of effecting shock syntheses was made in Chapter 2. Shock synthesis finds its practical applications in the three main areas of structural changes, general chemical compound manufacture, and, specifically, in the production of pre-shaped super-conducting materials.

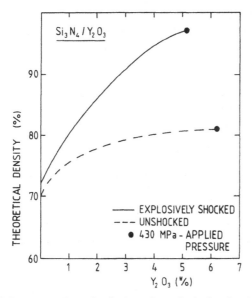

FIG. 6.2   HIP-processed unshocked and explosively shocked mixtures of silicon nitride and yttrium oxide (compiled from Ref. 26).

(a) Structural synthesis. This particular application is best illustrated on the example of industrial diamond production, although the shock synthesis of cubic boron nitride from the hexagonal phase is used commercially in the manufacture of fine, hard abrasives.[27]

The intense pressures and high temperatures generated during the passage of a shock wave through hexagonal, crystallite graphite are sufficient to alter its structure to cubic and produce diamonds (Fig. 2.6). This transformation was first observed and reported by de Carli[28] in 1961, when traces of diamond powder were detected on shocking graphite at a pressure of 30 GPa for 1 μs. Improvements in the technique used — including extension of the time of application to over 10 μs — were made, and in 1966 Allied Chemical bought de Carli's patent[29] and started the manufacture of micronsized diamond powder. Samples, some 500 g in weight, were offered on a commercial basis and used for wear resistant coatings, abrasives, fluorescent particles and catalysts.

A number of difficulties were experienced in those early experiments. One of them, the problem of the possible reversion of the formed diamond to hexagonal graphite during the long period of cooling of the bulk of the untransformed graphite material was resolved by Du Pont's improved system in which the addition of cobalt, copper, etc., to the graphite quickly lowered the temperature after the passage of the shock wave.[30] Small, rounded diamonds, free of sharp edges are thus produced and are used for polishing and lapping.

Generally, a successful structural transformation, both in terms of its stability and yield, is achieved when shock treatment is carried out at pressures above 30 and nearer 70 GPa. These must be maintained for periods in excess of 10 μs. The preferred initial density of the graphite should be 1·8 g/cm^3, and imperfections in the starting crystal form of the material should be avoided since they impede structural transformations.[3,31]

In a manufacturing variant of this technique,[32] irregularly shaped polycrystalline diamond particles are produced. These are superior to natural powder by virtue of the range of particle sizes in the mixture. This varies[32,33] from about 0·2 to 60 μm. According to Du Pont's specifications, microcrystals contained in a particle are of the order of 100 Å. An average particle thus contains a very large number of microcrystals, bonded together in a random polycrystalline structure and without any significant cleavage planes. The material obtained in this way is particularly suitable for grinding and general surface polishing of metals and ceramics, as well as for cutting purposes.

Polycrystallinity, when present, imparts particular abrasive characteristics by introducing a 'multicutting' tool of uniform properties. This ensures that flaking and, therefore, random scratching are avoided. These features are of special interest and use in the precision polishing of semiprecious gem stones, ferrites, alumina and cemented carbide parts. An illustration of this is provided by the performance of Du Pont's polycrystalline diamond material used in polishing sapphire stones for the watch making industry (Fig. 6.3).

Although synthesised diamond and c-BN powders can be used in normal applications by incorporation into suitable matrices, the densification of these materials without previous sintering is difficult in view of their strong covalent properties. However, synthesised powders when further shock treated can be successfully densified to near solid theoretical densities.

(b) Chemical synthesis. In general terms, forming of compounds and, possibly, the manufacture of entirely new material combinations are the main objectives of shock chemical synthesis, but the synthesis also provides a means of obtaining possibly unique, sometimes non-equilibrium, microstructures or otherwise difficult to produce phases. For instance, in the field of polymer production and utilisation, shock treatment may lead to, on the one hand, isomerisation, and, on the other, to

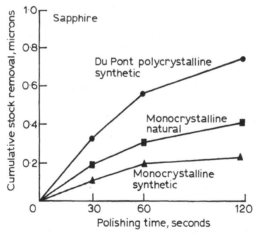

FIG. 6.3 Comparison between the characteristics of diamonds, synthesised by Du Pont methods, and conventional materials used in polishing sapphire stone gems (after Ref. 32).

polymerisation. Although these reactions proceed equally well in standard conditions, shock treatment gives a high local mobility to the bonds and affects the yield and final molecular weight of the product.[9] However, the most common application of the shock synthesis route is still in the areas of intermetallic and ceramic compounds.

A general characteristic of shock synthesised material is the lowering of transition temperature as compared, where appropriate, with the conventional treatment. For instance, copper bromide synthesised from $CuBr_2$ and Cu powders, has the transition temperature of the sphalerite–wurtzite phase of 375°C which is lower than that of conventional 396°C. Similar lowering of temperature is observed, in for example, shock synthesised niobium silicide, silicon carbide and zinc ferrite. The latter is particularly susceptible, in its structural changes, to the effects of temperature. $ZnFe_2O_4$ is explosively synthesised from a mechanically blended, in stoichiometric ratios, mixture of ZnO and haematite at pressures ranging from 7·5 to 27 GPa. The shock induced temperatures lie between 125 and 1110°C, the yield amounting to 85% at the higher temperature. The small grains present indicate primary crystallisation, but quantities of spinel ferrite are found and increase with increasing pressure. The spinel phases are either magnetic or paramagnetic and are composed of small crystallites (< 500 nm). Supermagnetic relaxation occurs often.[9,34]

Although zinc ferrite illustrates well the complexity of structures induced by shock treatment, materials like silicon nitride, titanium carbide and tungsten carbide are of much greater industrial interest. The possible development of a 'ceramic engine' depends very much on the availability of good quality Si_3N_4, free from impurities and doping agents present on the particle surfaces. An excess of these results in a lowering of the resistance to high temperature effects, as well as anticreep and anticorrosive properties. Shock synthesis provides a means of producing pure material,[9] and, further, if followed by explosive compaction will give a well consolidated and densified component. Pressures of between 20 and 27 GPa, generating temperatures of between 1500 and 1700°C, are required.

Titanium carbide is produced from a stoichiometric mixture of titanium and carbon powders at a pressure of around 27 GPa.[9] The material shocked at 22 GPa is in the form of a heavily defective powder with piled-up dislocations.

Tungsten carbide (α-W_2C or WC) results from the explosive synthesis of acetylene black and tungsten powder. Yields of up to 90%

of cermets, in pellet form, are obtained and very high hardness levels —
HV 3500 in 1 mm grain, as compared with HV 2000 in conventional
fabrication — are prevalent.[9,35]

Shock chemical synthesis combined, again, with explosive compaction begins to play a very important role in the manufacture of Type II
superconductors. These are materials in which the superconducting
state is separated from the normal by a mixed state region. The boundaries between the three states correspond respectively to the lower
(H_{C1}) and higher (H_{C2}) critical fields. The mixed state region is neither
superconducting nor fully normal since it can trap magnetic fluxes
passing through.

Practical problems arising in the production of these compounds
and, even more so, in the attempts at consolidation in the form of rods
or wires, are very severe since not only is it necessary to obtain the
material in bulk, but it is also imperative to counteract its inherent
brittleness. Niobium compounds and, more recently, ceramic oxide
materials of $T_C > 90$ K are very representative of the superconducting
materials in use. A detailed discussion of the dynamic methods of
manufacture of these materials can be found in the concluding chapter
of this book.

6.1.2 Process Parameters

Although any optimisation of process parameters depends on whether
structural, physical and/or chemical changes, or simple consolidation
and compaction are desired, the basic requirement is the generation,
propagation and dissipation of either pressure pulses or kinetic energy.
However, once the required level of either of these basic parameters is
attained, other physical and material characteristics decide the final
outcome of the operation.

In general, the response of particulate matter to shock treatment is a
function of

- The geometry of the processing system.
- The physical nature of the medium, i.e. whether gaseous, vacuum or
  solid.
- Operational temperature.
- Material characteristics:
  - particle morphology,
  - particle size and distribution in the sample,
  - tap density.

The relative importance of some of these parameters depends entirely on the nature of the material considered and, obviously, will be different in a metallic, ceramic or polymeric aggregate.

The basic structural inhomogeneity of the pre-shocked material, and therefore the inevitable variations in its response to the discrete 'jumps' of the shock wave from particle to particle, is indicative of the difficulties facing any attempts at systematisation and classification of the phenomena occurring within a sample.

The most important features of compaction — the process that, at the moment, forms the bulk of the shock-treatment operations — are a high degree of densification, and a satisfactory level of interparticle bonding. These requirements rule out, on the one hand, over or underdensification — associated with purely local conditions — and, on the other, any central core defects. Uniformity of pressure distribution across the sample to be compacted (a conical shock wave profile) is as essential as is the avoidance of hyperbolic (undercompaction), and parabolic (over compaction and central bursting and burning of the powder, Fig. 6.4) shock fronts (see Chapter 2). The near elimination of interparticle voids is always possible in a dynamic system (Fig. 6.5(a)), but even at relatively high operational pressures is unlikely to be achieved in static systems (Fig. 6.5(b)).

The degree and mode of bonding (and therefore the densification) defined by the nature of interparticle collisions, and partly by the morphology, are closely related to the velocity of impact which, in turn, depends both on the local stand-off distances and the square of the velocity of the detonation front (V_D^2). The relationship between the density of the compact and the square of the detonation front velocity — and by implication the kinetic energy — (shown in Fig. 6.6 for compacted alumina powder) is a characteristic feature of the dynamic compaction process. A set of plots of this kind establishes the optimal manufacturing conditions.

When not only the compaction or consolidation is required, but also some structure modification is needed, the relationships between particle sizes, lattice distortion, crystallite size, and the square of the detonation velocity respectively have to be established. Figure 6.7, obtained, again, in compacting alumina powders, serves as a typical example of the mutual parameter interdependence.

Of particular interest is the information about the zoning of any shock operation (Figs 6.6 and 6.7) which is a basic characteristic of any metallic,[36,37] ceramic[5] and polymeric[15] powder compaction.

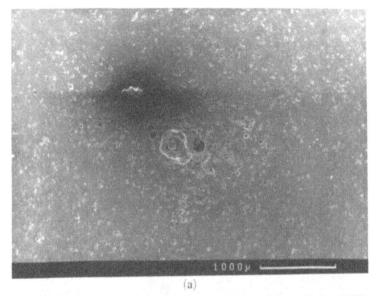

(a)

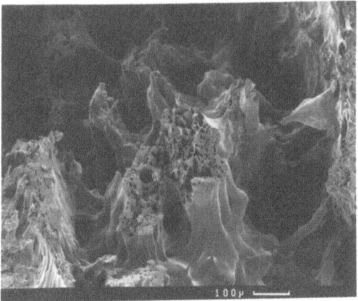

(b)

FIG. 6.4   Overintensification of the pressure pulse (parabolic shock wave front) in the compaction of a polyvinylchloride aggregate. (a) Central core burst; (b) central core melting.

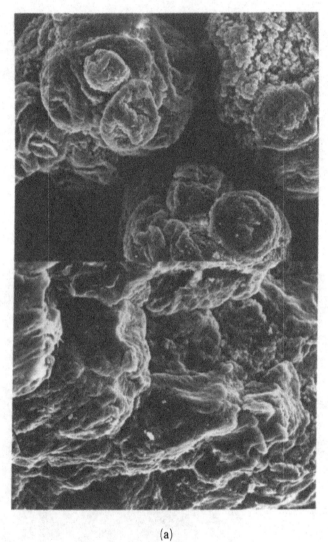

(a)

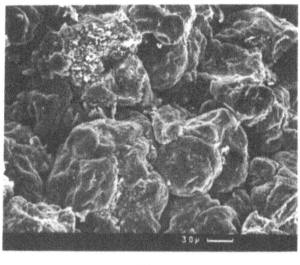

(b)

Fig. 6.5   Compaction of the PVC material of Fig. 6.4. (a) Uncompacted and explosively compacted (at about 6 GPa) powder showing a high degree of densification and void elimination; (b) statically compacted (at 350 MPa) powder with a high porosity.

At a low detonation velocity (Zone I in Fig. 6.7) for instance, the consolidation of the material is due simply to a rearrangement of particles and in this zone, lattice distortion is the same for both explosive and isostatic compactions (Fig. 6.8). In Zone II, plastic deformation and comminution are present and lead to an increase in density and, eventually, to the maximum degree of densification. Further increase in velocity and pressure (Zone III) is, however, undesirable because by producing a considerable volume of material damage it is responsible for the degradation of properties. The distortion energy generated is comparable with that found in cold-worked metals.

As already pointed out, the morphology of grain or particle surface, as well as the particle size, will play a major role in determining the operational conditions and final properties of the product. Whereas the actual morphology defines the surface particle conditions and therefore predisposes it favourably, or otherwise, to any of the mechanisms of bonding, the particle size and distribution determine the magnitudes of the interparticle stand-off distances. These parameters have therefore a

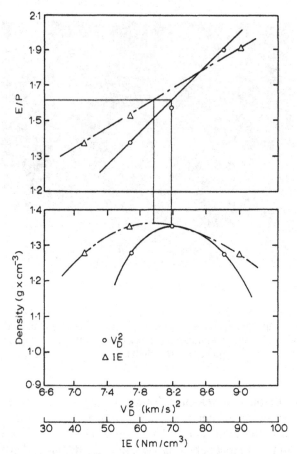

FIG. 6.6   Relationship between compact density and the square of the detonation front velocity (V_D^2) and the generated kinetic energy IE, as referred to the explosive/powder mass ratio (E/P) (data for implosive compaction of alumina powder).

direct influence on the degree of the final densification and strength of the compact. Generally, with the increasing particle or grain size, a reduction in the final density will be observed. This reflects both the influence of the morphology and particle size.

A useful illustration of the response of a particulate material is provided by, for instance, Fig. 6.9. This illustrates the effect of grain size on the densification of a PVC copolymer.[9] The degree of cohesion is seen to be larger in the narrow range of smaller particle sizes (<150

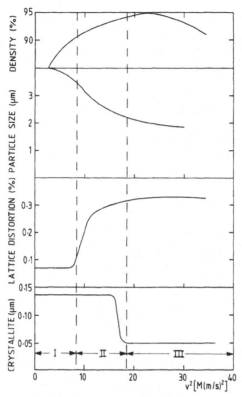

FIG. 6.7 The respective relationships between the compact density, particle size, lattice distortion, crystallite size, and the square of the detonation front velocity (V^2) (in compaction of alumina powder).

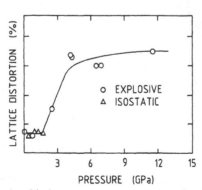

FIG. 6.8 The relationship between pressure generated and lattice distortion in explosive and isostatic compaction of alumina powder.

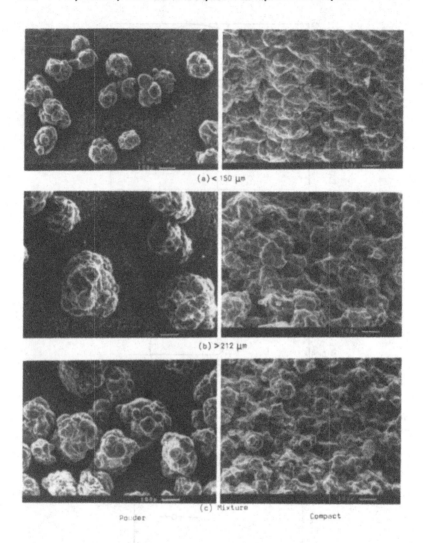

FIG. 6.9   The effect of grain size on the degree of densification in compacting ICI polyvinylchloride PC7/314 copolymer (compacting pressures between 7 and 13 GPa) (after Ref. 9).

μm) (Fig. 6.9(a)), than in the band of larger sizes (>212 μm) (Fig. 6.9(b)) which displays a tendency to slight porosity after compaction. The 'as supplied' mixture of all the sizes (Fig. 6.9(c)), shows an acceptable fracture surface with well bonded and integrated particles.

The original morphology has been altered in all cases as a result of a considerable plastic deformation of the individual grains. The closure of intergranular voids has taken place with flattening of surface nodules and regularisation of the grain shape, and the consequent high green densification. The best results are obtained with a mixture consisting of a range of particle sizes. This situation offers a great variety of impact conditions and thus provides a better means of introducing all three mechanisms of bonding. In turn, this results in higher levels of density and strength that even the lowest particle size can offer.

In structural synthesis and modification, the effect of grain morphology can be quite pronounced. Using again shock treatment of alumina powder as an example, it is seen that even at a low pressure of some 4 GPa the effect is noticeable, but increases with pressure when saturation conditions are reached at 27 GPa.[9] The effect depends not only on the square of the detonation velocity (Fig. 6.7), but also on the degree of grain coarseness. Lattice distortion is higher in fine-grained alumina (0·3–0·35%), and lower in the coarse morphology powder (0·22%) (Fig. 6.10).

The higher initial or tap powder density increases the degree of green density of the compact and therefore its ultimate strength. Further, the saving on energy required to first consolidate the material, can be utilised in actual bonding, rather than in a void closing operation. The properties of the product are thus improved as is the overall efficiency of the operation.

The effect of the geometry of the operational system is more complicated and is therefore discussed separately in Section 6.2.

6.1.3 Theoretical Background

If only mass and momentum relationships are regarded as basic characteristics of shock effects in particulate matter, then a simplified Herman's approach,[38] used by Raybould and discussed fully in Section 2.3.2, is sufficient to define the interparticle pre-melting conditions.

The question of the estimation of the porosity, e, can be approached from the static point of view and be regarded as an approximation defined fully by, for instance, Cooper and Eaton's theory.[39] In this, the relative volume compaction V_{Ce} is given by

$$V_{Ce} = (e_o - e_P)/(e_o(1 - e_P)) = (\rho_P - \rho_o)/e_o\rho_P \qquad (6.1)$$

where e_o and e_P are the loose and compacted powder porosities respectively, and ρ_o and ρ_P are the corresponding densities.

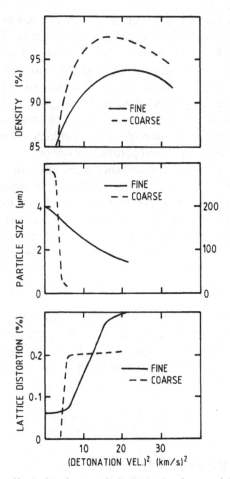

FIG. 6.10   The effect of grain morphology on density, particle size and lattice distortion of explosively compacted alumina powder (after Ref. 9).

In a static compaction process, V_{Ce} is taken to be the following function of the operational pressure P:

$$V_{Ce} = a_1 \exp(-k_1/P) + a_2 \exp(-k_2/P) \qquad (6.2)$$

where a_1 is the relative volume compaction of the first compaction stage, a_2 corresponds to the second stage, and k_1 and k_2 are constants. The first stage is defined here as the closing of interparticle voids, and the second stage, as effecting plastic grain deformation and comminution.

When considering static compaction of granulated ferrite, van der Zwan and Siskens[40] arrived at a simplified relationship between the modulus of elasticity E and the porosity of the compact in the form of

$$E = E_m[1 - (e_P/e_o)]/[1 - (D+1)(e_P/e_o)] \qquad (6.3)$$

where E_m and D are constants.

Full conditions for successful shock consolidation of particulate matter have been investigated by Schwarz *et al.*[19,41,42] who considered the effects of surface melting and resolidification, of the minimum duration of the shocked state and of the minimum amount of melted metal.

Successful consolidation is likely to materialise if the molten material solidifies and cools to below the melting point temperature T_M before the compact has recovered from the shocked condition. If this requirement is not fulfilled, the tensile rarefaction wave will fracture the compact — a situation analogous to that arising in explosive welding of multilaminates.

It follows therefore that the time of shock application t must be

$$t > t_s + t_c \qquad (6.4)$$

where t_s and t_c are the times of solidification and cooling respectively.

For the condition of no superheating of the molten material, the solidification time is given by

$$t_s = \pi D_m (LdH_m\rho)^2/16[K_m(T_m - T_1)]^2 \qquad (6.5)$$

where K_m is the thermal conductivity, $D_m = K_m/\rho c_{pm}$ is the diffusivity index, c_{pm} is the specific heat at melting temperature, H_m is the latent heat of fusion, ρ is the density, T_1 is the temperature of the particle core after the passage of the first shock wave of intensity P_1, L is the initial melted fraction assumed to correspond to spheroids of sides d.

In these considerations, normalised energy ε, and time τ were defined and used in estimating optimal compacting conditions:

$$\varepsilon = PV_o(m-1)/2c_p(T_M - T_o) \qquad (6.6)$$

and

$$\tau = 64\, Dm\, t/d^2 \qquad (6.7)$$

where V_o is the volume of bulk material, m is the powder distension or the ratio of the densities of the solid and particulate matter, T_o is the initial temperature, and c_p is the average specific heat in the T_o, T_M

range. With these stipulations, the normalised shock duration is given[42] by

$$\tau^{0.5} \geqslant 2\pi^{0.5} \varepsilon_1 (c_p/c_{pm}) [H_m/c_p(T_M - T_o) + H_M] \tag{6.8}$$

ε_1 is the normalised energy in the initial shock wave of pressure P_1.

According to these authors, the upper bound for the melt is,

$$L = P_1 V_o(m-1)/2[c_p(T_M - T_o) + H_M] \tag{6.9}$$

Consideration of temperature kinetics of metal powders has led Schwarz *et al.*[19] to conclude that:

(i) The width of the shock front is approximately equal to the particle size.

(ii) During the densification the shock energy is deposited heterogeneously near the interparticle boundaries.

(iii) The shock energy is not deposited just on the particle surface (the assumption usually made), but it affects regions of finite width.

Computer modelling of the shock consolidation phenomena is well advanced, e.g. Wilkins,[43] but for some simpler applications, these effects can be approximated by semi-empirical expressions. For instance, a consideration of the nature of the shock wave transmitted through porous materials led Roman[44] to propose a construction of suitable shock wave adiabats. The flow in the particulate matter was assumed to be hydrodynamic (i.e. no shear stresses present) giving the following relationship

$$P = \rho a^2 F/(1 - bF)^2 \tag{6.10}$$

where $F = 1 - (\rho_0/\rho)$, ρ_0 and ρ are the bulk and compact densities respectively, and a and b are the coefficients in the wave velocity U and particle velocity u equation: $U = a + bu$. The coefficient a is a function of the volume sonic velocity of the compacted material.

Raybould's investigations[45] led him to conclude that the constitutive equation for powder consolidation can be of the same type as the quasi-static pressure/density relationship

$$\rho = \rho_0\{a + (1 - a)/(1 + bP)\}^{-1} \tag{6.11}$$

where a is a constant representing the initial powder porosity, and b is a constant reflecting the stiffness.

According to Raybould, it is possible to ignore the change in the internal energy across the shock wave because, although the energy per unit area may decrease during compaction, this is probably compensated for by the stored energy of deformation. The quasistatic pressure/density relationship can be used as a reasonably good approximation to the constitutive equation, up to impact velocities of at least 500 m/s.

6.2 MANUFACTURING SYSTEMS

The generation of a shock wave of sufficient strength to produce consolidation and/or synthesis of the particulate material can be achieved by either purely mechanical means and/or by the use of high explosives as the source of energy.

In the former situation, and, up to a point, in the latter, the energy is delivered to the material indirectly, whereas with explosives, when used in implosive-type arrangements, direct transfer of energy is usual. It is mainly in this latter case that the importance of the energy/powder mass ratio, variously denoted by E/M or E/P, comes to the fore and is closely related to the geometry and mechanical properties of the powder containers which separate the to be consolidated (or simply shocked) material from the energy source.

6.2.1 Indirect Systems

The earliest indirect compression apparatus was designed and used by La Rocca and Pearson.[46] This consisted of either a single- or double-acting high-explosive actuated press (Fig. 6.11). The press was intended for the processing of relatively small quantities of material; the restriction being due partly to the limitation in the magnitude of the charge and partly to that of the clearance. The detonation of the explosive charge would result in the downwards acceleration of the upper plate

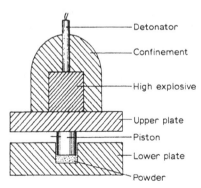

FIG. 6.11 Single acting, explosive actuated powder compression press (after Ref. 46).

and of the attached piston. Shock conditions were created when the latter impacted the particulate matter which initially would be at tap density.

Using Composition B explosive, compacts of titanium powder of green densities exceeding 95% were obtained. However, ceramic materials could not be compacted, neither could diamond synthesis be carried out.

A variant of a press in the form of a gas gun (see Fig. 1.1) was developed by Brejcha and McGee,[47] and by Hagemeyer and Regalbuto.[48] This technique consists in releasing and propelling a projectile contained in a gun-type barrel. High pressure gas, e.g. nitrogen, constitutes 'the charge' indicated in the figure. The projectile strikes a punch which, in turn, is in contact with the particulate matter. The generated accelerations and impacts (velocities of over 500 m/s) are limited and consequently the tool may not be suited for synthesis of materials requiring high levels of energy.

An improvement on the press design was provided by the uniaxial explosive compaction technique in which the only major requirement is the generation of a plane wave. The plane wave generator (see Section 4.1) is usually attached to a pad of high explosive and the pressure pulse generated accelerates a metal plate, equivalent to the punch, into a powder filled container. The method was proved to be advantageous in obtaining high density, large size semi-products (iron and SAP plates up to 520 mm²). Ceramic materials were compacted using flat plate geometry with alumina powders reaching densities of between 85 and 90% (Bogdanov *et al.*).[49,50]

The volume of post-compacting cracks, associated with the superimposition of rarefaction and reflection waves onto the primary shock pulses, was high in those systems and a remedy was sought in a more sophisticated design. In an attempt to control the wave interaction, a shock recovery system, consisting of a plane shock wave generator and a momentum trap, was devised by Sawaoka *et al.*[51,52] This technique is used for the compaction of ceramic powders in small-sample quantities. Starting powders are compacted in a stainless steel capsule, some 20 mm in diameter, the shock treatment being carried out by the impact of a flyer plate. Silicon nitride powder was thus compacted to a density of 92%, and silicon carbide to a density of 98·6%, but some cracking was still observed.

Using a similar approach, Graham[53] developed the Sandia 'Bear' and 'Bertha' recovery systems. The difference between the two lies in the sizes of the explosive plane wave generators. These are 56 and 102 mm in diameter respectively. The powder capsules are made of copper and the main casings of steel. Several ceramic powders were consolidated by Carr,[54] by first pressing the powder in the copper capsule to between 40 and 60% densities, and then compacting the samples explosively. Although a reasonable number of satisfactory compacts was obtained, some cracking continued to persist.

The idea of an indirect explosive isodynamic (hydrodynamic) compaction system was pioneered by McKenna *et al.*[55] Tungsten carbides and tungsten mixed with other carbides were compacted at pressures of between 350 and 400 MPa — to handling conditions — within 25–50 ms. In this type of arrangement, the pressure vessel is activated by a detonating charge which acts either directly or indirectly on the fluid surrounding the container.

Elaborate hydrodynamic machines were built at the Byelorussian Research Institute of Powder Metallurgy in the USSR. Titanium, stainless steel, silicon nitride, alumina/graphite composite, and molybdenum powders were compacted into semi-product shapes of crucibles, nozzles and cylindrical parts. Further sintering, where required, was possible and an indication was given that manufacturing units could be fully automated.[56-58]

A different source of energy is provided by electrical discharge. The central principle behind this technique is the sudden discharge of electrical energy from a bank of capacitors through a column of particulate matter contained in, say, a pyrex tube. As the discharge proceeds, a high transient current passes through the column and causes individual particles to heat and weld to each other. At the same time, an intense magnetic field is generated and results in the collapse inwards of the bulk of the material. It appears that compaction occurs only within a certain range of discharge energies. This depends on the dimensions of the column and on the characteristics of the processed material.[59] This technique is used in producing metal components of up to 60% relative density, but with relatively high tensile and bending strengths.

Although electrical methods have a number of advantages over the explosive techniques, for instance no ejecting pressure is required and the container may be re-used several times, they are, naturally, limited to electrical charge conducting, and magnetically responsive materials.

6.2.2 Direct Systems

Although a number of variants of the direct technique exists, the basic system is very similar to any of those discussed in Chapter 4. In its simplest form, it consists of a container filled with powder, usually at tap density, which is then surrounded (in axisymmetrical configuration) or covered with (in flat or grazing systems) explosive charge(s).

Of the two, the grazing configuration, which involves a sheet of explosive placed against a metal driving plate, is used to compact powders into flat plates. This loading method results in a detonation wave propagating in parallel to the metal surface and producing a shock pulse. The amplitude and profile of the pulse depend on the thickness of the charge and its distance from the layer of the powder. The standard procedure is to place the powder in a flat container which is located between two steel plates. The assembly is then immersed in liquid explosive. A single detonator is used in conjunction with a plane wave shaping generator.

A basic axisymmetrical system is shown in Fig. 6.12. It has been used successfully for the compaction and synthesis of metallic, ceramic and

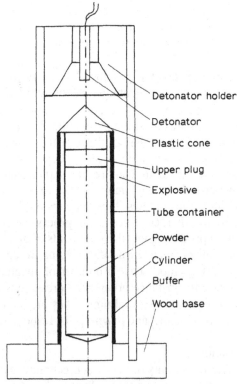

Detonator holder

Detonator

Plastic cone

Upper plug

Explosive

Tube container

Powder

Cylinder

Buffer

Wood base

FIG. 6.12   Basic axisymmetrical, direct implosive powder compacting system.

polymeric powders. A diagrammatic representation of its operation, relating to the Du Pont diamond synthesis,[32] is shown in Fig. 6.13. Charges of up to 1 m in diameter and several metres in length are used.

An alternative to the Du Pont system was developed by Kiyoto.[60] This arrangement relies on creating conditions of shock and of subsequent rapid cooling during a high velocity (appr. 6·5 km/s) collision between graphite and water. The necessary acceleration is achieved in a hollow charge system in which graphite forms the liner, and the collision pressure is of the order of 45 GPa. The yield of diamonds is about 3%.

For a given particulate matter, the two major parameters of a direct type of operation are the explosive/powder mass ratio, and the geometry and material properties of the container.

6.2.2.1 The effect of E/P ratio

The practical importance of this parameter (P/E in the early work) was recognised early on by Carlson *et al.*[2] and Leonard.[61] The investigations carried out in the 1960s led to the conclusions, summarised already in Chapter 2, that two competing mechanisms operate in dynamic powder compaction. First, there is the tendency for the pressure to increase towards the centre of the compact. Second and simultaneously, a rapid attenuation of the pressure, due to the consumption of energy caused by

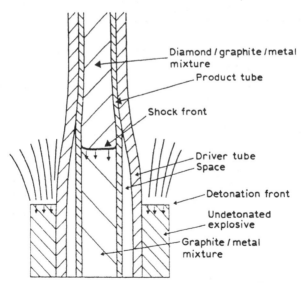

FIG. 6.13   Direct implosive diamond synthesising system (after Ref. 24).

void collapse and plastic deformation of particles, takes place. The two mechanisms have to be balanced if a constant pressure across the specimen is to be maintained. Selectively chosen and applied mass ratios are expected, according to Leonard *et al.*,[62] to ensure the existence of a balance. A linear relationship between the ratio and compressive yield strength of a number of materials compacted under identical conditions, was found to exist. Leonard's radiographic examinations indicated that failure to choose the correct value of the parameter results in either a hyperbolic shock front (Fig. 2.17(a)) and undercompaction, or a parabolic one (Fig. 2.17(c)) and overcompaction. These observations were later confirmed by Prummer[3] (Fig. 6.6) who related the E/P ratio to the square of the detonation velocity and established the detonation pressure (P_D)/velocity (V_D) empirical relationship of the form

$$P_D = 0.25 \rho_E V_D^2 \qquad (6.12)$$

where ρ_E is the density of the crystalline explosive.

A rough guide to the E/P values was also established, giving $E/P < 1$ as the sufficient level for 'soft' materials, such as aluminium, and $V_D = 2000$ m/s for full compaction, but $E/P > 2$ and $V_D > 5000$ m/s for highly alloyed steel powders (99% TD). These results reflect specific compacting geometry and range of materials.

Hoenig *et al.*[23,63] noted that ceramic compacts of densities nearing those of the solid were obtained with different E/P values, but more or less at the same detonation velocity. For instance, $E/P = 3.8$ was sufficient to compact alumina, while $E/P = 13.8$ was required to compact boron. The increase in E/P was shown to influence the quality of the compact, and so, at V_D of 4.2 km/s, and $E/P = 5.7$, alumina powder, for example, was compacted without cracking, but at a velocity of 5.5 km/s, and with $E/P = 3.8$, the central defect was present in spite of the fact that in both cases the same final density of 91% was achieved.

These results were confirmed by Peng *et al.*[64] who found that with increasing explosive mass, and at a constant V_D, the rate of pressure drop, behind the front, decreases. The decrease is believed to result in an increase in average compacting pressure and in its duration. The degree of interparticle bonding and therefore densification will thus be increased.

A further confirmation of these observations was provided, in the field of metallic compacts, by Chiba *et al.*[65] Various E/P ratios were used in compacting titanium powder and the three characteristic modes of over, under, and correct compaction were recognised as being directly associated with this parameter. Specimens compacted at ratios higher than 5·14 showed the presence of a central melt hole.

In the different material area of polymeric composites, the experiments of Muhanna[66] and Blazynski and Muhanna[67] (see Chapter 9) have led to similar conclusions. The total energy per unit volume of the container/powder system (a parameter closely associated with E/P ratio) was defined by Hegazy and Blazynski[68] as:

$$E = 2 \rho_M h d_M (V_P/d_C)^2 \qquad (6.13)$$

where ρ_M is the density of the tube container, h is the container wall thickness, d_M is the mean tube diameter, d_C is the tube bore diameter and V_P is the container radial velocity.

A more complicated expression was proposed by Wolf[69] in connection with his work on compaction of ferritic powders. This gives the required compaction energy E as:

$$E = 0·01 \, T_D \rho_P [\eta Q (M_E/M_P) - (Y_D \varepsilon M / \rho_M M_P)] \qquad (6.14)$$

where T_D is the relative green density of the powder, ρ_P and ρ_M are the theoretical powder, and tube container densities, η is the efficiency of transformation — numerically between 0·06 and 0·09 — M_E and M_P are respectively the explosive and powder masses, Y_D is the dynamic yield strength of the container material, ε is the generalised strain, M is the mass of the container, and Q is the specific energy of the explosive.

6.2.2.2 *Effect of the container*

The success of an implosive powder compacting operation, employing a direct system which incorporates an axisymmetric powder container surrounded by an explosive charge, depends on two process parameters. These are the ability to generate a conical profile shock wave, that ensures, as already noted, the uniformity of pressure distribution across the compacted powder, and the containment of the tensile release wave. Since, on the one hand, the container acts as an energy transferring medium and, on the other, as a post-shock constraint, its geometry and mechanical response determine both the nature of the pressure pulse delivered and the ability of the system to prevent fracturing of the compact.

For the given level of the available explosive charge energy, the material and wall thickness of the container are instrumental in defining the velocity of the initial impact — reflected eventually in the degree of densification of the powder — and in providing the required stiffness of the structure.

The importance of the role of the container was recognised in the late 1960s by Leonard[61] and, more recently, by Prümmer and Ziegler,[5] as well as Lennon,[70] the latter showing, for instance, that for the same load ratio, the impact velocity associated with a copper cylinder is some 12–15% higher than that for an identical steel tube. The damage suffered by the compacted specimen, as a result of the rapid release of pressure after the passage of the compressive pulse, and the inability of the structure to minimise this effect are discussed by, amongst others, Hegazy and Blazynski.[11,15,68]

Since the energy supplied by the charge is utilised in imparting a radial velocity to the container and thereby partly in deforming plastically the container and, partly, in compacting the powder, an estimate of these contributions becomes necessary. Although the development of the mathematical and computational modelling techniques is well advanced,[20,43,71] an approximate empirical approach is still needed in addition to the existing computer codes.

In either case, the response of an empty metal cylinder to the detonation of a charge provides an estimate of part of the required energy. In the course of consolidation of a cylindrical powder specimen, the total impact energy associated with the detonating charge is consumed partly by the material of the container and partly by the powder. Whereas the standard method of assessment of the available charge energy provides an approximate method of relating the quantity and characteristics of the explosive to the possible process requirements, it does not, in itself, offer a ready means of apportioning the energy between the two stages of the operation.

The amount of straining suffered by a metal cylinder subjected to implosive loading on its surface, is related directly to the level of the load. For a given magnitude of the load, the deformation of an empty cylinder is, naturally, higher than would be the case when the cylinder acts as a powder container. However, even when the latter situation arises, the actual strain imposed, although reduced, still reflects directly the load that would cause this deformation in the absence of the bore supporting powder. It is therefore legitimate to carry out experimentation on empty cylinders in order to establish empirically the relationship between strain and impact loading. When these data are obtained for different materials and geometries, the compacting component of energy can be established and further calculations concerning the condition of the powder can be made.

The basic parameters involved in the calculation of this type are the tube radial velocity V_P, the explosive detonation velocity V_D, and the $E/P = R$ ratio. When considering compaction of powder, the distribution of impact energy is defined as

$$E = E_D + E_C \tag{6.15}$$

where subscripts D and C refer to the deformation of the container and powder compaction respectively.

On detonating a charge against the outer wall of the container, the latter acquires a radial velocity V_P which is related to the impact energy E by means of the standard expression for kinetic energy:

$$E = 0.5 \, \pi(d_M h \rho L_C) \, V_P^2 \tag{6.16}$$

where d_M is the mean diameter, and L is the length of the container. The standard way of assessing the wall velocity is based on utilisation of the Gurney equation (eqn (3.21)). For ratios R less than unity, that is the situation obtaining in powder compaction, the modified[72] Gurney equation is given by

$$V_P = 0.587 \, R V_D/(2 + R) \tag{6.17}$$

In powder explosives, normally used in compaction, the thickness of the layer and the tap density of the charge will affect the velocity of detonation. Accordingly, experimental determination of the relationship between the appropriate parameters is required and can be obtained in a variety of ways, including the Dautriche method.[73]

Naturally, when an explosive charge is detonated against an empty container the impact energy E will be equal, to the first approximation, to the deformation energy E_D. The generalised strain suffered by the container is then given by the standard equation:

$$\varepsilon = 2 \times 3^{0.5} \ln(r_o/r)/3 \tag{6.18}$$

The equation reflects the changes in the geometry of the container, effected by the dissipation of energy E_D and can relate these to the mechanical properties of the container material via an appropriate constitutive equation.

The relationship between the ratio R and the reduction in the outer tube diameter has to be determined experimentally in order to avoid assumptions and approximations necessary when a purely analytical approach is used.

With the experimental and calculated data given by eqns (6.13), (6.16)–(6.18), eqn (6.15) will provide information about E_C in an actual compacting operation.

To obtain the relevant numerical data that would indicate the trends associated with both the container material and geometry, metal tubes, of wall thickness varying from 1 to 5 mm and 32 mm in outer diameter, were loaded dynamically by increasing charges of powdered ICI-Trimonite No. 3 explosive. The tube surfaces were protected by 4·5 mm thick polythene buffers. The effect of the increasing level of energy dissipated is shown, for a set of steel tubes, in Fig. 6.14. The figure shows a certain amount of surface scoring, despite the protection of the buffers, but it also indicates the uniformity of the diameter between the end plug sections, as well as the constancy of the tube length.[74]

With the explosive layer thickness ranging from 3·6 to 7·8 mm, the detonation velocity varied from 2660 to 3000 m/s. These investigations involved three basic engineering materials of EN8 steel (BS 970), commercially pure C101 copper (BS 2874), and HE30TF aluminium (BS 1474). Although 2 mm thick copper containers had been used successfully in the powder compacting experiments, the unsupported bores of these tubes could not sustain high pressures and buckled consistently (Fig. 6.15). The response of copper containers was therefore confined to 3 and 4 mm thick tubing.

Thickening of the container wall occurred with all the investigated materials, but only with aluminium containers were complete closing

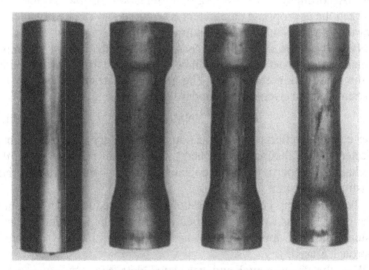

FIG. 6.14   Implosive compression of powder containers showing the effect of increasing E_D level (after Ref. 74).

down and solidification observed (Fig. 6.16). The relationships between the geometry and ratio R are given in Figs 6.17 and 6.18.

Figure 6.17 is concerned with the variation in the percentage reduction of the outer diameter — the only measurable parameter in a powder compacting system — with R. For both materials, this relationship is represented by a set of straight lines, displaced by equal distances (Fig. 6.17(a) for steel) representing equal steps in the wall thickness.

The effect of the mechanical properties of the two materials is clearly discernible from Fig. 6.17. The figure indicates a considerably higher degree of copper tube deformation for a given mass ratio R, and, also, the ability of the 1 mm steel tube to absorb more energy, and display less deformation, than even a 4 mm thick copper container. Consequently, it would be expected that the containment of the release wave by a steel-compacting system would be of a higher order of efficiency.

The relationship between the generalised strain and the impact energy per unit volume for the two materials, is given in Fig. 6.18. In both cases, the relationship is linear, for the given wall thickness, and can be used to assess the magnitude of E_D in eqn (6.15) when powder compacting. The mechanical property characteristics of the materials are again manifested in higher levels of energy required in the case of steel than copper, for the same generalised strain value. The divergence in energy consumption increases, for the given wall thickness, with the total strain imposed.

FIG. 6.15   Buckling of 2 mm thick copper tubing (after Ref. 74).

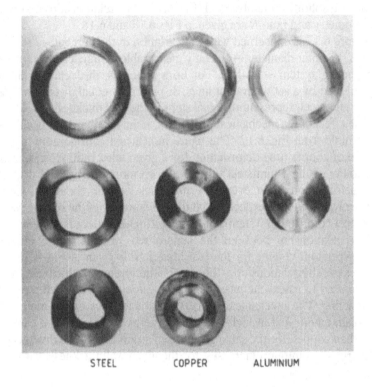

STEEL   COPPER   ALUMINIUM

Fig. 6.16   The effect of the material of the container on the degree of wall thickening for comparable values of the dissipated energy of deformation (after Ref. 74).

Within the investigated range of conditions, the correlation between strain and energy can be approximated by the following empirical equations derived from Fig. 6.18.

Steel

$$\varepsilon = (0.152 \, h^2 - 1.49 \, h + 4.02) \, E_D \qquad (6.19)$$

Copper

$$\varepsilon = (0.576 \, h^2 - 5.17 \, h + 12.5) \, E_D \qquad (6.20)$$

On substituting from eqn (6.18) for strain, the change in the actual dimensions of the tube can be assessed. If, further, the strain rate of the operation is known and the appropriate constitutive equation of the

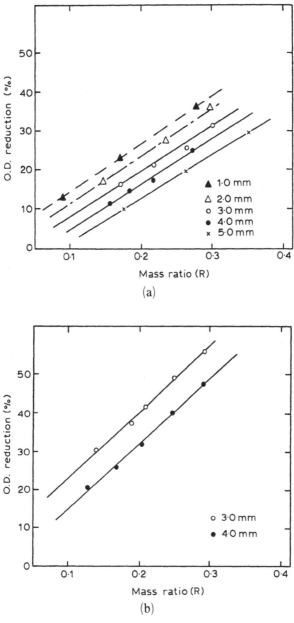

FIG. 6.17 Plastic deformation of the container in terms of its reduction in diameter and the variation in the mass ratio *R*. (a) Steel tubing; (b) copper tubing.

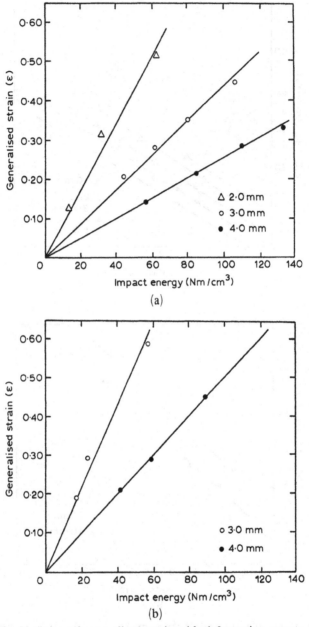

FIG. 6.18   Variation of generalised strain with deformation energy. (a) Steel tubing; (b) copper tubing.

material is available, the mechanical properties of the container can be brought into the calculations.

The actual dependence of the mechanical properties of the compact on the characteristics of the system has been demonstrated for the example of the consolidation of, among other materials, a suspension ICI homopolymer S57/116 powder.[75]

The three basic parameters characterising the response of the material to the passage of shock waves are:

- green density,
- compressive strength to fracture,
- hardness distribution.

The variation of these parameters with the container material and wall thickness is shown in Figs 6.19–6.21 for the polymeric powder compacts.

It is noted, incidentally, that in all of the cases discussed, the zoning of the density, strength and hardness curves with increasing compaction energy is present (as previously indicated) irrespective of the system used.

The rigidity of the system increases, naturally, with the wall thickness of the container tube and, consequently, the curves for the 4 mm wall show longer ranges of optimal energy levels than those for 2 and 3 mm tubing. The optimal conditions in the latter are confined to the peaks only. Unlike the aluminium containers (Fig. 6.19(a)), in which the optimal density values are reached at 30 N m/cm^3 for a 2 mm wall, 60 for 3 mm, and 90 for a 4 mm wall, copper and steel systems (Figs 6.19(b) and 6.19(c)) show a reverse tendency with the maxima reducing with the wall thickness. It is also seen that the levels of density and strength do not increase noticeably with the wall thickness in aluminium, but do so in other materials. Generally, however, the rigidity of the system is seen to increase with the strength of the container and, for the given material, with the wall thickness. Correspondingly, the density and strength of the compact will vary. Figure 6.20 shows these effects for a container 3 mm thick. In terms of green density, the maximum value of 1·35 g/cm^3 is reached at 50 N m/cm^3 in aluminium. For the same conditions, the strength of the compact increases from some 45 MPa for aluminium to 59 MPa for copper and 65 MPa for steel.

It is clear that the impact energy delivered by the system will govern the degree of intergranular bonding and the rigidity of the system will ameliorate, or otherwise, the effect of the reflected, tensile shock wave.

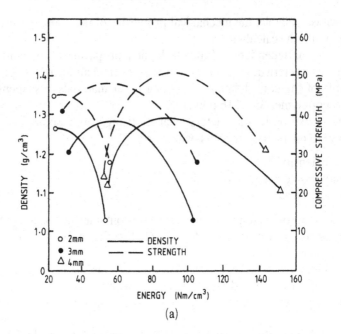

(a)

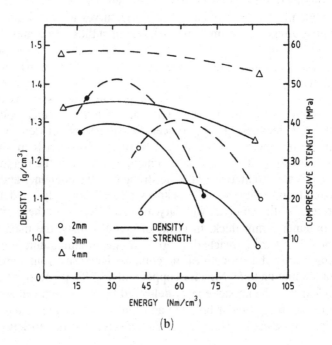

(b)

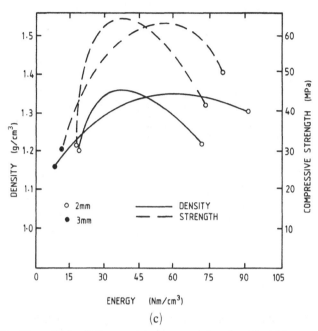

FIG. 6.19   The effect of the container material and wall thickness on the density and strength of a suspension polymer S57/116 compact. (a) Aluminium; (b) copper; (c) low carbon steel (after Ref. 75).

The increased capacity for efficient bonding is also reflected in the levels of hardness, measured across a plane section of the compact. Whereas, for a given material, hardness distribution remains practically uniform across the section, its level increases with the wall thickness of the container (Fig. 6.21). This is illustrated, as an example, in the case of copper in Fig. 6.21(a), where the difference between the 4 and 2 mm wall thicknesses amounts to some 20%. For a specific wall thickness, the level of hardness increases with the strength of the material and is the highest for steel and lowest for aluminium (Fig. 6.21(b)). The respective percentages, with respect to aluminium, are some 25 for steel and 7 for copper. Similar effects are noted in the compaction of composite materials.

6.2.3 Process Modelling

Because of the inherent complexity of the dynamic compacting process, a number of techniques has been developed in order to provide means

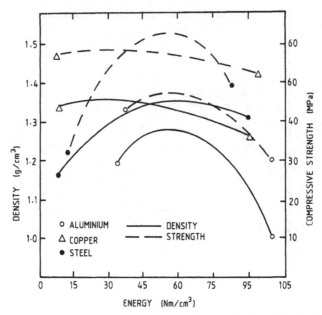

FIG. 6.20   Variation in density and strength of S57/116 (PVC) compacts consolidated in 3 mm wall-thick aluminium, copper and low carbon steel containers.

of numerical or computer simulation of the operation. Any such approach requires a solid basis of factual data and the establishment of the 'operational' boundary of its applicability.

One of the features that characterises dynamic consolidation is, as already noted, the effect of the localised deposition of energy. The prediction therefore of the possible material response to the varying energy levels, and of its distribution within the compact, is of importance in computer programming. A possible way of approaching the problem of the local microstructural changes was suggested by Gourdin[76] in his work on the dynamic compaction of Al/6%Si, steel and copper powders. A theoretical background to the assessment of some aspects of explosive compaction of porous layers, with special reference to titanium, copper and aluminium powders, was proposed by Deribas et al.[77] Methods of computing simulation of the behaviour of both the container and powder have been developed by Wilkins et al.,[20,43] while numerical modelling of the dynamic consolidation of rapidly solidified metal powders was discussed by Williamson and Berry.[71]

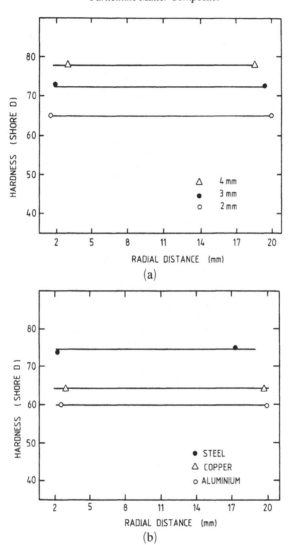

FIG. 6.21 Hardness distribution in the S57/116 compacts. (a) Varying copper tube wall-thickness; (b) 2 mm thick aluminium, copper, steel tubes (after Ref. 75).

Wilkins' simulation (HEMP-code) is used in identifying the pressure/ density/time interrelations of the materials involved. An example of this approach is provided by the simulation (Fig. 6.22) of the compaction of tungsten powder in a cylindrical configuration[3] of the basic system of

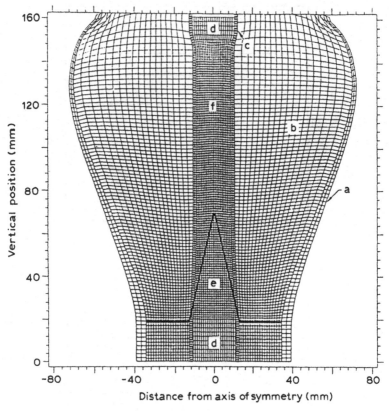

FIG. 6.22   Computer simulation of compaction of tungsten powder in a cylindrical configuration (courtesy Dr Mark L. Wilkins, Lawrence Livermore National Laboratory). a, PVC pipe; b, explosive; c, steel tube; de, steel end plugs; e, powder; f, compacted powder.

Fig. 6.12. The consolidation of aluminium nitride to 94% TD in plate, and to 97% TD in cylinder form, and of the diamond powder to solid rod were successfully accomplished using computer prognosis.[43]

6.3 MATERIAL PROPERTIES

6.3.1 Measurements

Unlike, perhaps, the case of solid composites, those originating from particulate matter are especially sensitive to the pre-processing material conditions. It is therefore of importance to consider particle size distribution and surface morphology, since these properties will eventually influence the density, strength and frictional/wear characteristics of the compound.

Measurements therefore involve assessments of both the physical and mechanical properties and in particular the density, strength, hardness levels and distribution, tribological characteristics and, where appropriate, those of creep.

6.3.2 Physical Properties

Particle size and morphology are determined prior to processing by microscopic means using both optical and, more often, electronic microscopy. The nature of post-processing (compaction) surface fractures is also examined in this way.

The apparent and tap densities of the preshocked material are determined using the standard ASTM 1895-69 test. The apparent density measurement involves the weighing of a cylindrical cup, 40 cm^3 in volume, completely filled with the tested material. The tap density is assessed by vibrating the same cup at a constant frequency while it is being slowly filled with powder. When apparently full, the cup continues to be vibrated for some 5 min and the powder is added, if necessary, until full settlement is achieved. The cup is then weighed.

The same procedure is followed when specimens for actual dynamic compaction are being prepared. All these quantities can be easily related to the theoretical density (TD) supplied normally by the powder manufacturer.

The green compact density is assessed by the method of differential weighing, either with reference to water or mercury. The standard procedure is to weigh the specimen in air (W_1), coat it with, say, vaseline to seal off the surface pores, weigh it again (W_2), and then weigh it in water (W_3). The compound density, with respect to water, is then

$$\rho = W_1/(W_2 - W_3) \qquad (6.21)$$

6.3.3 Mechanical Properties

From a practical point of view, the strength of the composite is reflected by its fracture stress in compression, and/or in bending (tension). These are determined by means of standard laboratory tests. The variation in the material properties across a specimen can be determined, with the usual reservations, by microhardness testing.

Frictional and wear-resistant properties are usefully assessed by, for instance, the pin and disc technique. In its simplest form, the method consists in rubbing some 8 mm diameter pins (made of the composite to be tested) against stainless steel discs, polished to submicron grade finish. The frictional force F is given by

$$F = T/R$$

where T is the applied torque, and R is the radius of the disc. The coefficient of friction μ is then obtained from the expression

$$\mu = F/N \tag{6.22}$$

where N is the normal load on the pin.

The wear measurement is based on the weight loss of the pin, converted to the volume loss V. The wear factor K is then defined as

$$K = V/ND \tag{6.23}$$

where D is the sliding distance.

REFERENCES

1. BERGMANN, O. R. and BARRINGTON, J. *J. Am. Ceram. Soc.*, **49** (1966), 502.
2. CARLSON, R. J., POREMBKA, S. W. and SIMONS, C. C. *Ceram. Bull.*, **34** (1965), 266.
3. PRUMMER, R. *Explosive Welding, Forming and Compaction*, Ed. T. Z. Blazynski, Applied Science Publishers, London, New York, 1983, p. 369.
4. GOURDIN, W. H., ECHER, C. J. and CLINE, C. F. In: *Proc. 7 HERF Conference*, Ed T. Z. Blazynski, University of Leeds, 1981, p. 233.
5. PRUMMER, R. and ZIEGLER, G. In: *Proc. 5 HERF Conference*, Ed. R. H. Wittman, University of Denver, Colorado, Paper 2.4, 1975.
6. KONDO, K., SOGA, S., SAWOAKA, A. and ARAKI, M. *J. Mater. Sci.*, **20** (1985), 1033.
7. GRAHAM, R. A., MOROSIN, B., VENTURINI, E. L., CARR, M. J. and BEAUCHAMP, E. K. In: *Metallurgical Applications of Shock Waves and High Strain-Rate Phenomena*, Ed. L. E. Murr, K. P. Staudhammer and M. A. Myers, Marcel Dekker, New York, 1986, p. 1005.
8. HEGAZY, A. A., BLAZYNSKI, T. Z. and EL-SOBKY, H. In: *Mechanical Properties of Materials at High Rates of Strain*, Ed. J. Harding, The Institute of Physics, Bristol, London, 1984, p. 413.
9. RAYBOULD, D. and BLAZYNSKI, T. Z. In: *Materials at High Strain Rates*, Ed. T. Z. Blazynski, Elsevier Applied Science Publishers, London, New York, 1987, p. 71.
10. BLAZYNSKI, T. Z. and HEGAZY, A. A. *Z. Werkstofftechnik*, **17** (1986), 363.
11. HEGAZY, A. A. and BLAZYNSKI, T. Z. *J. Mater. Sci.*, **21** (1986), 4262.
12. HEGAZY, A. A. and BLAZYNSKI, T. Z. *J. Mater. Sci.*, **22** (1987), 3321.

13. Muhanna, A. H. and Blazynski, T. Z. *Compos. Sci. Technol.*, **33** (1988), 121.
14. Mardani, H. A. and Blazynski, T. Z. *J. Mech. Work. Technol.*, **18** (1989), 315.
15. Blazynski, T. Z. In Ref. 7, p. 189.
16. Blazynski, T. Z. In: *Tribology in Particulate Technology*, Eds J. B. Briscoe and M. J. Adams, Adam Hilger, Bristol, 1987, p. 303.
17. Blazynski, T. Z. In: *Shock Waves for Industrial Applications*, Ed. L. E. Murr, Noyes Publishers, Park Ridge, NJ, 1988, p. 406.
18. Blazynski, T. Z. In: *Encyclopedia of Composites*, Ed. Stuart M. Lee, VCH Publ, New York, 1991.
19. Schwarz, R. B., Kasiraj, P. and Vreeland, T. Jr In Ref. 7, p. 313.
20. Wilkins, M. L., Kusubov, A. S. and Cline, C. F. In Ref. 7, p. 57.
21. Linse, Vonne D. In Ref. 7, p. 29.
22. Lotrich, V. F., Akashi, T. and Sawaoka, A. In Ref. 7, p. 277.
23. Hoenig, C., Holt, A., Finger, M. and Kuhl, W. In *Proc. 6th HERF Conf.*, Ed. R. Prummer, Haus der Technik, Essen, Paper 6.3, 1977.
24. Lee, Y. K., Williams, F. L., Graham, R. A. and Morosin, B. *J. Mater. Sci.*, **20** (1985), 2488.
25. Graham, R. A., Morosin, B., Venturini, E. L. and Carr, M. J. *Ann. Rev. Mater. Sci.*, 1986.
26. Somiya, S., Yoshimura, M., Fujiwara, S., Kondo, K. and Sawaoka, A. *Proc. Am. Ceram. Soc. 85th Annual Meeting (Basic Sci. Div. No. 135-B-83)*, 1983, pp. C51–C52.
27. de Carli, P. S., *Bull. Am. Phys. Soc.*, **12** (1967), 1127.
28. de Carli, P. S. and Jamieson, J. C. *Science*, **133** (1961), 821.
29. de Carli, P. S. US Patent No. 3 238 019.
30. Cowan, G., Bruce, N., Dunnigton, W. and Holtzman, A. US Patent No. 3 401 019.
31. Murr, L. E. and Staudhammer, K. P. In Ref. 17, p. 441.
32. Bergmann, O. R., In: *Proc. 7th HERF Conf.*, Ed. T. Z. Blazynski, University of Leeds, 1981, p. 142.
33. Balchan, A. S. and Cowan, G. R. US Patent No. 3 851 027.
34. Graham, R. A., Morosin, B., Horie, Y., Venturini, E. L., Boslaugh, M., Carr, M. J. and Williamson, D. C. In: *Shock Waves in Condensed Matter*, Ed. Y. M. Gupta, Plenum Publishing Corporation, New York, 1986.
35. Horiguchi, Y. and Nomura, Y. J. *Less-Common Metals*, **11** (1966), 378.
36. Roman, O. V. and Gorabtsov, V. G. In *Shock Waves and High Strain Rate Phenomena in Metals*, Eds M. A. Meyers and L. E. Murr, Plenum Press, New York, 1980.
37. Linse, Vonne D. In: *Proc. of the Sagamore Army Materials Research Conf. on Innovations in Materials Processing, Vol. 30*, Plenum Press, New York, London, 1985, p. 381.
38. Hermann, W., *J. Appl. Phys.*, **40** (1969), 2490.
39. Cooper, A. R. and Eaton, L. E. *J. Am. Ceram. Soc.*, **45** (1962), 97.
40. van Der Zwan, J. and Siskens, C. A. M. *Powder Technol.*, **33** (1982), 43.
41. Schwarz, R. B., Kasiraj, P., Vreeland, T. Jr and Ahrens, T. J. In: *Shock Waves in Condensed Matter — 1983*, Eds J. A. Assay, R. A. Graham and G. K. Straub, Elsevier Science Publishers B.V., Amsterdam, 1984, p. 435.

42. SCHWARZ, R. B., KASIRAJ, P., VREELAND, T. and AHRENS, T. J. *Acta Metall.*, **32** (1984), 1243.
43. WILKINS, M. L. In: *Impact Loading and Dynamic Behaviour of Materials*, Eds C. Y. Chiem, H. D. Kunze and L. W. Meyer, DGM Informationgesselschaft Verlag, Oberursel, FRG, 1988, p. 965.
44. ROMAN, O. V. In: *Modern Developments in Powder Metallurgy, Vol. 6*, Eds H. H. Hausner and W. E. Smith, American Powder Metallurgy Institute, New Jersey, 1973.
45. RAYBOULD, D. In: *Proc. 15th MTDR Conf.*, Eds S. A. Tobias and F. Koenigsberger, Macmillan Press, London, 1975.
46. La ROCCA, E. W. and PEARSON, J. *Rev. Sci. Instr.*, **29** (1958), 848.
47. BREJCHA, R. J. and McGEE, S. W. *Am. Mach.*, **106** (1962), 63.
48. HAGEMEYER, J. W. and REGALBUTO, J. A. *Int. J. Powder Metall.*, **4** (1968), 19.
49. BOGDANOV, A. P., LAZAREV, A. S., ROMAN, O. V. and FURS, V. YA. *Sov. Powder Metall. Met. Ceram.*, **7** (1973), 543.
50. BOGDANOV, A. P., LAZAREV, A. S., ROMAN, O. V. and FURS, V. YA. *Ibid.*, **11** (1973), 880.
51. AKASHI, T., LOTRICH, A., SAWAOKA, A. B. and BEAUCHAMP, E. K. *J. Am. Ceram. Soc.*, **C-322** (1985).
52. SAWAOKA, A. B. In Ref. 17, p. 380.
53. GRAHAM, R. A. In: *High Pressure Explosive Processing of Ceramics*, Eds R. A. Graham and A. B. Sawaoka, Trans. Tech. Publ., Switzerland, Germany, UK and USA, 1987, p. 27.
54. CARR, J. M. *Ibid.*, p. 341.
55. McKENNA, P. M., REDMOND, J. C. and SMITH, E. N. US Patent No. 2 648 125.
56. ROMAN, O. V., GOROBTSOV, T. V., TARASOV, G. D. and SHEKGOV, V. I. In Ref. 44, **12** (1973), 293.
57. VITYAZ, P. A. and ROMAN, O. V. In *Proc. 13th MTDR Conf.*, Eds S. A. Tobias and F. Koenigsberger, Macmillan Press, London, 1973.
58. ROMAN, O. V. and GOROBTSOV, V. G. In Ref. 17, p. 335.
59. AL-HASSANI, S. T. S. In Ref. 23, Paper 4.2.
60. KIYOTO, K. *J. Ind. Explosive Soc. Japan*, **37** (1976), 152.
61. LEONARD, R. W., *Batelle Tech. Rev.*, **179** (1968), 13.
62. LEONARD, R. W., LABER, D. and LINSE, VONNE D. *Proc. 2nd HERF Conf.*, University of Denver, Colorado, Paper 8.3, 1969.
63. HOENIG, C. L. and YUST, C. S. *Ceram. Bull.*, **60** (11) (1981).
64. PENG, T. C., SASTRY, S. M. L. and O'NEAL, J. E. In: *Powder Metallurgy — Vol. 16*, Eds E. N. Aqua and C. I. Whitman, Metal Powder Industries Federation, American Powder Metallurgy Institute, NJ, 1984, p. 523.
65. CHIBA, A., NISHIDA, M., YAMAGUCHI, T. and TOSAKA, J., *Scripta Metall.*, **22** (1988), 213.
66. MUHANNA, A. H. PhD thesis, University of Leeds, 1990.
67. BLAZYNSKI, T. Z. and MUHANNA, A. H. In: *Proc. 10th HERF Conf.*, Litostroj, Ljubljana, Yugoslavia, 1989, p. 58.
68. HEGAZY, A. A. and BLAZYNSKI, T. Z. *Int. J. Impact Engng*, **6** (1987), 63.
69. WOLF, H. In: *Proc. 9th HERF Conf.*, USSR Academy of Sciences, Siberian Division, 1987, p. 188.

70. LENNON, C. R. L. PhD thesis, The Queen's University, Belfast, 1979.
71. WILLIAMSON, R. L. and BERRY, R. A. In Ref. 7, p. 167.
72. CHADWICK, M. D. In: *Proc. Select Conf. on Explosive Welding*, The Welding Institute, London, 1968.
73. CHADWICK, M. D. and JACKSON, P. W. In Ref. 3, p. 251.
74. HEGAZY, A. A., PhD thesis, University of Leeds, 1986.
75. BLAZYNSKI, T. Z. and HEGAZY, A. A. In Ref. 43, p. 979.
76. GOURDIN, W. H. In: *High Energy Rate Fabrication — 1984*, Eds I. Berman and J. W. Schroeder, ASME, New York, 1984, p. 85.
77. DERIBAS, A. A., STAVER, A. M. and SHTERTSER, A. S. Ibid., p. 109.

Chapter 7

Shock Consolidation of Metallic Powders

7.1 INTRODUCTION

The growing demand for high-performance raw and near-net-shape formed materials has intensified in the last two decades both because of economic considerations and the need for, sometimes, unusual material properties. This is particularly obvious in the area of application of some basic engineering materials obtained initially in particulate form and represented primarily by ferrous alloys, nickel, cobalt, aluminium, copper and chromium powders.

Ultrafine copper powders, together with similar quality nickel products, are used in, for instance, the production of electronic circuits, whereas nickel and nickel oxide powders are utilised in P/M parts, welding electrodes, batteries, catalyst production, and carbide binders. Cobalt powders find their application in the production of cemented carbides, rare earth magnets and radioisotopes. Nickel and cobalt-coated composites are industrially produced, and blends of NiCrAl/ Bentonite, and Ni/Al are used respectively in aero engines and in bond coating.[1] Al/Li composites and wrought iron powders form welcome new materials in the aircraft and motor industries.[2]

Although conventional powder metallurgy techniques and methods, applied to metallic particulate matter, are very well developed and widely used, the properties of the materials produced will not be unique, but will always depend on the processing route chosen. In this respect, the dynamic processing route offers additional possibilities. The main advantages are the possible elimination of the sintering stage that can easily change and even adversely affect the properties of powders produced by rapid solidification, the possibility of processing high strength, but low ductility materials, and, finally, the availability of shock synthesis.

The interparticle bonding, often experienced in dynamic conditions, increases the degree of green density, as compared with that obtaining in static processing. Figure 7.1 gives a clear indication of this feature, in this particular case, in explosive powder compaction. The dynamic metal–powder[6] processing has developed over the last three decades or so, starting with relatively simple implosive techniques (Fig. 6.12) pioneered in the UK by the ICI,[3] and the explosively actuated presses (Fig. 6.11) by La Rocca and Pearson,[4] and by Hagemeyer and Regalbuto.[5] Aluminium, titanium, copper and tinned iron powders were compacted to high densities in the implosive systems, and, in addition, some 16 powders including chromium, cobalt, copper, molybdenum, nickel, niobium, tantalum, titanium, zirconium and stainless steels in the direct systems. Small, experimental samples are usefully produced by either light gas-guns or electromagnetic apparatus (see Section 6.2.1).

Implosive compaction can be carried out equally satisfactorily as either cold or hot. The hot processing is of particular interest when compacting high strength materials, such as alloyed steels or chromium and nickel-based superalloys and refractories. A considerable reduction in compacting pressure is usually observed and is accompanied by a reduction in the E/R ratio. For instance, the pressure required to obtain the optimal value of the density when compacting pure iron powder, reduces from 6·5 MPa in cold to 3·6 MPa in hot processing, with the corresponding drop in the E/P ratio from 1·3 to 0·8.[2]

The general compressibility of metal powders can be assessed by means of the American National Standard Test B 331-79, bearing in mind, however, that this applies strictly to uniaxial compaction.

The success of any operation depends naturally on the ability to assess, *a priori*, the required level of energy necessary, and the temperature effects in the processed powder[6] (Section 6.1.3). A simple method of estimating the required compacting pressure is that of Wolf. The result is summarised by eqn (6.14). The equation sets out the practical operational limits for specific materials. For example, iron powder compacts require an energy level of at least 5 kJ/cm³ for 95% densification, and some 2·5 kJ/cm³ for 50%. A 1% increase, within these limits, calls for an energy increase of 0·055 kJ.[7]

Although eqn (6.14) provides a ready means of an overall energy estimate, a more accurate approach is needed in those cases in which the initial particle size and prepacking material density are significant factors. This particular problem was investigated by, among others, Roman *et al.*[8] who suggested a number of empirical relationships that define compacting conditions.

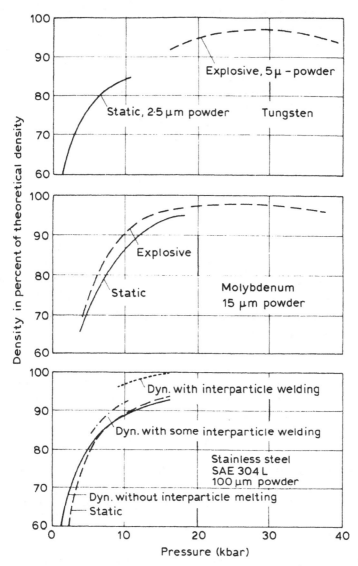

FIG. 7.1   A comparison green density and compacting pressures for statically and explosively processed tungsten, molybdenum and stainless steel powders (after Ref. 2).

Using as their basis a spherically shaped system and considering the porosity of the powder, the number of collisions, n, between the neighbouring particles is

$$n = 3 + 22\rho_1^3 \qquad (7.1)$$

where ρ_1 is the relative density.

The number of contacts per unit volume of the sample

$$N = (18 + 132\rho_1^3)\rho_1\rho_0/\pi D^3\rho \qquad (7.2)$$

where ρ_0 and ρ are respectively the solid and particle material densities, and D is the particle diameter.

The mean specific energy, E_C, available in a collision is

$$E_C = \pi D^3\rho u^2/(36 + 264\rho_1^3) \qquad (7.3)$$

where u is the mass flow behind the shock front.

It is assumed that the consolidation of interparticle voids takes place in time $t' \approx D/u$ and since this must be less than the duration of the pressure pulse, t (in most cases t varies between 1 and 10 μs) the optimum particle size is likely to lie within the 0·1–1 mm range. The total energy required to both consolidate and plastically deform 1 kg of powder is then given by

$$E_T = 0\cdot5\, u^2 \qquad (7.4)$$

A simplified form of the relationship existing between the compact density ρ_C and the kinetic energy per unit area was suggested by Lennon *et al.*[9] in the form

$$\rho_C = \rho_T - \Delta\rho \exp(-\beta E/d) \qquad (7.5)$$

where ρ_T is the theoretical density of the powder; $\Delta\rho = \rho_T - \rho_1$ with ρ_1 being the tap density, d is the container tube diameter, and β is a powder constant. A slight modification of this expression leads to

$$\rho_C = \rho_T - \Delta\rho \exp(-\beta E^\gamma) \qquad (7.5a)$$

where E is the energy per unit volume of powder, and γ is the powder constant.

To place the properties of dynamically processed composites into perspective, it is useful to consider first, in some detail, the response of the 'primary', or single materials.

7.2 PROPERTIES OF SINGLE SHOCK-TREATED MATERIALS

The basic engineering particulate materials are aluminium, copper, cobalt, iron, molybdenum, nickel, lead, titanium, tungsten and zinc. The availability of information about their response to shock processing is indicated in Table 7.1 and a set of basic Hugoniot curves for copper, nickel and stainless steel is shown in Fig. 7.2.

7.2.1 Aluminium

Because of its very high inherent ductility, the behaviour of dynamically compacted aluminium powder is similar to that of conventionally cold processed material. In their investigation of the properties of the latter, Strömgren *et al.*,[11] for instance, found that for a given particle size, the strength of the compact is proportional to the cube of the radius of the interparticle boundary. The strength depends therefore on the number of interparticle contacts (collisions) occurring during the compaction. This observation is in agreement with eqn (7.2) and with the results discussed in Ref. 8.

Computer modelling of the dynamic process has been carried out by, amongst others, Berry and Williamson,[12] using a two-spacial dimension CSQII code of Thompson and McGlaun.[13,14] Yield and fracture stresses, the elastic limit, mass density and comminuting pressure of an 6061 Al powder were assessed. Their preliminary investigation indicated that complete powder compaction is achieved in about 4 μs.

TABLE 7.1
Referenced Information on Single Metal Compact Properties

Material	Properties	
	Mechanical	*Metallurgical*
Al	2, 9, 12, 13, 14, 15, 17	8, 9, 16
Co	5	4
Cu	5, 8, 9	9, 10, 18
Fe	2, 7, 18, 20, 21	9, 19, 21
Mo	24, 25, 26, 27	10, 26, 27
Ni	9, 16	10
Pb	22	23
Ti	19, 25, 26, 29	15, 25, 28, 29
W	2, 16	2, 8

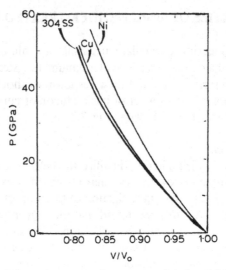

FIG. 7.2   A set of Hugoniot curves for copper, nickel and 304 stainless steel (after Ref. 10).

The ultimate compact properties vary with the compacting conditions and particle size. For instance, a fine-grained material, in the 10–160 μm range, of 2·7 g/cm^3 TD reaches only 98·7% TD after compaction from the tap density of some 44% TD.[2]

The effect of the material of the container is reflected in the nature and magnitude of the rarefaction wave which produces a 'negative' pressure.[15] A SAP powder, gun-compacted to 92–96% TD, has a UTS of between 40 and 60 MPa, which is well in excess of the 'negative' pressure of between 9·80 and 11·75 MPa produced by the reflected wave.

These quantities are influenced by the initial density and particle size distribution and therefore to avoid potential cracking of the compact, preliminary prepressing, or multistage explosive compaction, is desirable.

The influence of the container material was investigated by Bhalla *et al.*[9,16] A 60/100 aluminium powder, explosively compacted in 1–2 mm thick steel containers, failed on reaching 98·72% TD, when processed from a tap density of 49·25% TD. Starting with the same tap density, satisfactory 99·15% TD compacts were obtained in copper containers which varied in thickness from 0·4 to 0·8 mm. The effect of the grain size and compacting energy on the powder compacted in 1·5–2·0 mm thick steel containers (26 mm in diameter) was investigated on a mixture consisting of 25%–45 μm, 45%–90+45 μm, and 30%–250+90 μm. With the energy levels lying within the 34 and 182 J/cm³ range, with $\beta = 0.0023$, and $\gamma = 1.70$ (eqn (7.5a)), compacts of 76·5–99% TD were obtained.

In general, explosively compacted aluminium powders show some enhancement in their ductility (8% elongation as compared with 3% in the wrought material), an increase in hardness — 84 HV compared with 45 HV — and they display twice the fatigue strength of the wrought aluminium at 10^7 cycles.[17]

7.2.2 Cobalt

Relatively little information exists about the response of 'pure' cobalt powder, but the indication is that part of the β-phase is transformed to α or hexagonal phase. X-ray diffraction examinations confirm this finding.[4]

Compacted cobalt powder of −170, +200 particle size, shows some 93% TD (8·09 g/cm³) when processed at a relatively high energy level of 122 kJ/kg, and impact velocities of the order of 300 m/s. The corresponding average hardness amounts to 293 DPH.[5] The band of operational energies is narrow, but good quality compacts of low porosity are produced if the optimal conditions, approximating to the values given above, are created.

7.2.3 Copper

Most of the work on copper has been concerned with the metallurgical aspects. Measurements of the shock wave velocity and mass flow rate, made by means of an electromagnetic method, showed that the difference between the simplified and standard approaches amounts to 20% for porous copper processed at pressures of up to 12 GPa, but to only 10% at low pressures of up to 4 GPa.[8]

The type of material processed has a relatively low influence on the final properties. For instance, dendritic copper (−325 particle size) compacted at 270 m/s, at a 120 kJ/kg energy level, produces a compact of 99% TD (8·81 g/cm³) and of 168 DPH hardness. Irregular powder, of the same particle size, when compacted at a velocity of 245 m/s and 80 kJ/kg gives a compact of 98% TD (8·77 g/cm³) and 129 DPH hard.[5] The shock adiabat of granulated material appears to be close to that of finely dispersed powder.[8]

X-ray diffraction examination of textures of electrolytic copper powder (85% − 250 μm, 15% − 325 μm) compacted in steel containers 1 mm thick, revealed that the change from the tap density of 2·51 g/cm^3 to the final density of 8·46 g/cm^3 (94·4% TD) has a very limited effect on the texture.[18] Peripheral textures were found to be fibres (111) and (001) ⟨110⟩. Deformation of single crystals was noted, with the core of a fully compacted material having a nearly random orientation.

Larger specimens can be compacted[19] in cylindrical forms (5·5 cm diameter by 17·5 cm thick), starting with tap densities of some 5·25 g/cm^3 and reaching final densities of 95% TD with E/P ratios of the order of 6×10^{-2}.

The effect of the geometry of the container was again investigated.[9] Using electrolytic copper powder 50% − 150 + 45 μm, 50% − 40 μm, in steel containers 26 mm bore and wall thicknesses ranging between 1 and 2 mm, the varying impact velocities and energy levels produced compacts of 75% TD ($t = 1$ mm, velocity = 413 m/s, energy 77 J/cm^3), 67% TD ($t = 2$ mm, velocity = 281 m/s, energy 52·5 J/cm^3), and of 98·5% TD ($t = 2$ mm, velocity = 537 m/s, energy 285·2 J/cm^3). These experimental data were in agreement with the predictions of eqn (7.5a), taking $\beta = 0·0035$ and $\gamma = 1·32$. It is interesting to note, for comparison, that for a similar range of shock pressures (5·5–34·5 GPa), the metallurgical data for solid copper material of grain diameter of 120 μm are[10]

(i) Lower pressure — dislocation cell size 0·70 μm, difference in yield stress (before/after) 0·09 GPa.
(ii) Higher pressure — dislocation cell size 0·15 μm, difference in yield stress 0·28 GPa.

7.2.4 Iron

Shock compaction of iron powder is an early example of this type of processing. In the late 1960s an extensive examination was carried out by Butcher and Karnes[20] using a gas-gun. Specimens of porous iron 1·3, 2·2, 3·3, 4·7, 5·8 and 7 g/cm^3 in tap densities were compacted. Only partial compaction was achieved at pressures below 2·6 GPa, but complete compaction above that value would be attained in agreement with the Hugoniot theory and Gruneisen equation of state. Compact properties were in general agreement with the predictions of eqns (2.22)–(2.29).

With the dimension of the irregular void shape structure of sintered iron being of the order of 0·02 mm, the yield stresses of compacts obtained at low impact pressures were 0·12 and 0·56 GPa for the 4·57 and 7 g/cm³ density specimens.

Larger specimens (container bore 5·5 and length 175 mm) were also compacted at that time,[19] starting with 2·5 g/cm³ tap density and using $E/P = 0·12$. Final compacts, 32 mm in diameter, would attain some 97% TD.

With a view to providing compacted material in the shape of preforms suitable for cold rolling and forming, Witkowsky and Otto[21] compacted an atomised iron powder Ancorsteel 1000 (+60–325 μm) and provided also a comparison with the conventionally pressed material.

Flat compacts, 40 × 40 mm, and 60 × 60 mm, 6–25 mm thick were explosively manufactured giving green densities of between 92·8 and 99·2% TD. For E/P ratios of 0·6 and 1·1, the yield stresses of the compacts amounted to 24 and 32 MPa respectively with the corresponding Brinell hardnesses of 111 and 160.

The green densities were much higher than those obtained in conventional processing where at a pressure of some 475 MPa, the maximum attainable density was only 84%. It was noted that sintering of the compacted material within a 425–1000°C range, lowered the density to about 1·5%.

Again, the effect of the container was studied, in this case by Lennon *et al.*,[9] who examined the properties of reduced iron oxide sponge powder of the following particle size distribution: 20% − 125 + 90 μm, 30% − 45 μm, 50% − 90 + 45 μm. Starting with a tap density of 3·75 g/cm³ (47·5% TD) and using $\beta = 0·0040$, $\gamma = 1·30$ in eqn (7.5a), the following properties were established.

For a 1 mm thick container wall, the impact velocities of 300 and 400 m/s produced 66·5 and 81·0% TD compacts respectively. For thicker 3 mm wall containers, velocities of 150 and 230 m/s gave 57 and 69% TD, but the maximum final density of 98·1% TD was obtained with a container 1·5 mm thick, and at an impact velocity of 580 m/s. The effect of the container rigidity, associated with the varying shock conditions, was therefore clearly demonstrated in this case.

A compressed air-driven machine, used by Raybould,[18] demonstrated the possibility of producing preshaped compacts. Using maraging or EN24 steel dies and zinc stearate or PTFE spray as lubricants, atomised iron powder was successfully compacted. On average, densities in the 97–99% TD range were reached, with the corresponding UTS values of 1·5–2·0 GPa, and HRc hardnesses of 35–50.

More recent work[2] shows that explosive compaction of 1 μm iron powder, shocked from a tap density of 3·45 g/cm^3 (44% TD), can easily produce a 98·7% TD specimen.

Aluminium containers, 15–70 mm diameter, and 1–5 mm thick, used by Wolf[7] in explosive compaction of Fe powder, produced an interesting energy/density relationship that, geometrically, resembles a stress–strain curve. Up to 90% TD, a linear relationship exists, with the energy increasing from some 20 to 40 kJ/cm^3. However, the density of the compact remains constant, at that value, irrespective of any further increase in the energy level. Wolf concludes that 40–50 kJ/cm^3 are necessary for the optimal condition to be established.

7.2.5 Lead

Although of lesser practical importance in P/M operations, the consolidation of lead powder offers a further insight into the behaviour of shocked materials.

As far back as 1959, Cross[22] examined the possibility of hot explosive compaction at pressures of up to 350 MPa. The arrangement used was comparatively simple, consisting of a 200 mm outer diameter, 6 mm I.D. die. A low explosive was used to compact a range of powders. The object of this investigation was an assessment of the effects of particle size and oxide film thickness on the compact properties.

The presence of the oxide would affect the achievable degree of interparticle bonding and, therefore, the strength of the compact. This observation was confirmed experimentally on the basis of particle diameter/oxide thickness ratio (D/t). For $D/t = 16\cdot6$, the compressive compact strength was found to be about 13 MPa, increasing to 45 MPa for $D/t = 72$ and at a pressure of 350 MPa.

The particle size effect, associated with the influence of the intergranular stand-off distances, was very pronounced at 350 MPa, when for a 5 μm agglomerate the compressive strength was 12 MPa, rising to a maximum of 45 MPa for the 41 μm sample, and reducing to 27 MPa for the 131 μm powder.

A more recent work[23] employing a gas-gun, was concerned with the compaction of spherically shaped, 1·5 mm diameter, unsintered lead powder of an initial porosity of 30–35%. In a range of velocities between 195 and 700 m/s, a heavy deformation of the grain, with double concave and convex boundaries, was produced at higher velocities. However, irregularly shaped particles were formed in the lower velocity range and were accompanied by a large volume of voids.

A high degree of melting and its penetration into the interparticle spaces were regarded as necessary for high strength compacts.

7.2.6 Molybdenum

The basic data, relating to shock conditions of molybdenum powder and including P_H adiabat, Gruneisen coefficient, and the relative ratio of thermal P_T pressure to hydrostatic, conventional, components, were provided recently by Roman and Gorobtsov.[24] Information about the properties of shocked solid molybdenum is available in Ref. 10.

A detailed discussion of the mostly mechanical properties of molybdenum powder compacts can be found in the work of Prümmer and Ziegler.[25,26] Their investigations were concerned with the relationships between the impact pressure and green density, on the one hand, and lattice distortion on the other.

In the earlier work, a satisfactory compact density of 97·5% TD was obtained at an impact velocity of 3000 m/s and with $E/P = 0.95$. This was accompanied by some 0·14% lattice distortion. Changes in structural properties taking place in the isostatic compression were compared at a later date with those occurring under the shock conditions. Using a powder of an average particle size of 15 μm and isostatic pressures of up to 17·5 kbar, compacts were manufactured in sizes of up to 5 mm in diameter. The results of the combined investigations are summarised in Figs 7.3 and 7.4.

Figure 7.3 indicates that in terms of the green density of the respective compacts, the differences between the two conditions are relatively small. In the explosive compaction the density increases to a pressure of some 25 kbar, but a slight reduction occurs at higher pressure values. The maximum density reached is 98% TD. At an isostatic pressure of 17·5 kbar, the final compact density is 96% TD, and the density/pressure relationship is similar in character to the dynamic one.

The lattice distortion, associated with the explosive processing (Fig. 7.4) increases up to a pressure of 15 kbar, but no further increase is noted at higher pressures. The trends in the isostatic condition are similar, but in both cases the increase in density is thought to originate as a result of lattice distortion.

The metallurgical aspects of the shock compaction of molybdenum were investigated by Murr,[27] who carried out an extensive range of experiments compacting fine-grained (1–25 μm) material in 76·2 mm diameter low carbon steel cylinders. Initially, the container would be at 19°C and 1·2 Pa pressure. 98·3% TD green density compacts, showing

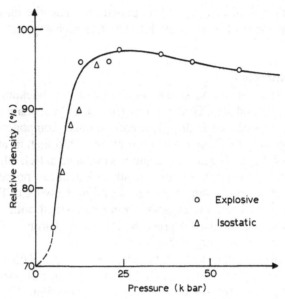

Fig. 7.3   A comparison between the green densities in the isostatic and explosive pressure compactions of molybdenum powder (after Ref. 25).

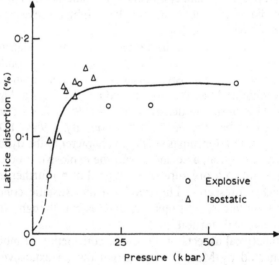

Fig. 7.4   A comparison between the degree of lattice distortion in the isostatic and explosive compactions of molybdenum powder (after Ref. 25).

signs of microcracks were obtained. The lack of full densification was attributed to the presence of these particular faults.

The compacts ranged in hardness from 300 to 400 HV10, with an average axial plane hardness of 381 HV10, and a radial plane hardness of 360 HV10. These values are shown to be in excess of sintered molybdenum extrudates (270 HV10), conventionally P/M processed bars (250 HV10), and explosively shocked and rolled 1–67 μm sheet (390–188 HV50). The only unalloyed material approximating to these levels of hardness is the shock loaded 1 μm molybdenum foil which reaches a hardness range of between 340 and 390 HV50.

The investigation shows that even after 1 h anneal at between 980 and 1200°C, the hardness is still higher than in the conventionally processed material. The explanation is sought in the development of small grain microstructure within the fine-grained consolidated material which is absent in, for instance, explosively loaded molybdenum sheet. This material displays pronounced dislocation tangles and loops.

7.2.7 Nickel

Since nickel forms the basis for a number of important metallic blends, its response to shock compaction is of considerable importance. A fair volume of detailed literature is available, with the Williams *et al.*[9,16] work being representative of the published results.

A comparison between a spherical, carbonyl, type 435 nickel powder, and a hydromechanically prepared irregular Grade 'F' material is given in Figs 7.5 and 7.6.

The particle distribution of the spherical material was 1·4% + 12, 8·8% − 120 + 240, 4·5% − 240 + 350, and 85·3% − 350, average particle sizes varied from 7 to 10 μm, and the tap density was 3·8 g/cm³ (42·7% TD). The irregularly shaped powder was characterised by 10% − 120 + 240, 28% − 240 + 350, and 62% − 350 particle size distribution. In this case, the tap density employed was 4·56 g/cm³ or 51·23% TD.

Some of the spherical material was initially prepressed to 67% TD, but this did not affect noticeably the green compact density (Fig. 7.5). Figure 7.6 shows that for a given level of impact energy delivered, a 50% increase in the container diameter is accompanied by about 4% reduction in compact density, whilst some 103% increase produces a 11% reduction.

A further investigation[9] of the properties of the spherical material — but of a different particle size distribution (85% − 45, 15% − 125 + 45)

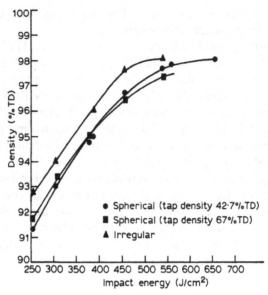

FIG. 7.5   The correlation between green compact density and container impact energy for spherical and irregular nickel powder (after Ref. 16).

and using a 26 mm diameter container of 1–2 mm wall thickness — produced a range of densities. Starting with the tap density of 4·29 g/cm³ (47·5% TD), a density of 67·8% TD was obtained at 337 m/s impact velocity and 50·5 J/cm³ energy value. Compact properties were again predicted with the help of eqn (7.5a), using $\beta = 0.0069$, and $\gamma = 1.11$.

For comparison, the properties of the solid nickel material[10] shocked over a period of 2 μs, were UTS of 0·33 and 0·56 GPa for the corresponding impact pressures of 15 and 45 GPa.

7.2.8 Titanium

The availability of spherical titanium powder, made by the rotating-electrode technique, provided Linse[28] with an opportunity to study the effect of shape, in addition to that of size distribution. Samples 38 mm in diameter of +30–200 mesh powder were explosively consolidated, but full densification of over 99% TD was obtained only with the preheated material. Full particle deformation was observed, with ductile fracture on grain boundaries. In all of the cases investigated only limiting bonding was found to exist.

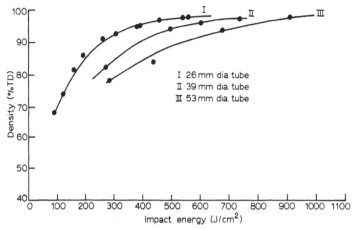

FIG. 7.6   The effect of specimen size on the green density of spherical nickel powder compact (after Ref. 16).

The effect of a spherical configuration is to reduce the amount of deformation and energy absorption, when compared with a similar cylindrical arrangement, to the extent of possibly providing insufficient balance with respect to the magnitude of the impact pressure. On the other hand, an attempt at improving the bonding properties of the agglomerate, by increasing the duration of pressure application, is likely to lead to a build-up of excessive heat in the interior of a particle and to the consequent melting and shock damage to its structure.

A comparison between the shock and isostatic compactions is given, again, by Prümmer and Ziegler.[25] At pressures of the order of 1·5 GPa, 5 μm titanium powder was successfully compacted to 99% TD in processing routes. The results are shown in Fig. 7.7, the trend being similar to that discussed with reference to molybdenum powders. A lattice distortion of 0·12% was found to result from the shocking of titanium powder to 99% TD at 2650 m/s and with $E/P = 1$, in 25 mm diameter, 1·5 mm thick, and 250 mm long steel containers.[26] However, at a low E/P ratio of 0·09, W–Ti powder, 1·53 g/cm^3 TD, consolidated in a 35 mm diameter, 175 mm long tube, would reach only 89% TD.[19]

The problem of micro and/or macro post-compacting crack development was investigated by Roman and Bogdanov,[15] who established the fact that the 'negative' or reflected pressure wave is responsible for these faults if it exceeds an average 6·9–8·8 MPa green strength of titanium compacts.

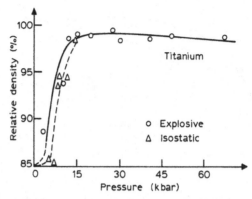

FIG. 7.7   The relationship between impact pressure and green density in isostatic and explosive compactions of titanium powder (after Ref. 25).

To rectify the problems arising out of the cracking, Chiba *et al.*[29] have designed a modified implosive pressure system. The powder to be compacted is at a tap density in the inner tubular container. The middle, larger diameter and coaxially arranged, tube forms an annular cylinder that is filled with a liquid. The third, outermost tube serves as the container for the explosive charge. Using an aluminium powder container, a copper liquid tube, and a steel charge container, titanium powder less than 44 μm (-325 mesh) was compacted with water, liquid paraffin, or an aqueous solution of gum arabic providing, in turn, the pressurising medium. No faults were detected in the $1.8-5.14$ E/P range, but overcompaction was observed for $E/P > 5.6$. The optimum results of over 95% TD were obtained with $E/P > 1.8$, and preferably at a value of approximately 3.5. This compares favourably with the unpressurised systems in which $E/P = 3.4$ reduces microcracking.

Although interparticle bonding was produced, the green compact strength was only 280 MPa which represents one half of that of the as-cast material. Further sintering is therefore recommended. It would seem from the values quoted by the authors that the nature of the pressurising medium has very little influence on the compact properties.

7.2.9 Tungsten

A comparison between the response to shock treatment of the granular and dispersed powder materials seems to indicate[8] that a higher compact strength is attained with the granular powder. A series of experiments were carried out with 1–2 mm granules and 5–10 μm dispersed powders. The granules, which are porous spheres exhibiting brittle fracture at static loads of the order of 100 N, would undergo full plastic deformation in dynamic processing at 10 GPa, giving compacts of green density of about 85% TD. At the same time, finely grained powder would show very little plastic flow and would reach only 65% TD. To improve these properties, sintering at 1250°C had to be introduced.

Tungsten powder of $4\cdot5\% - 120 + 240$, $37\cdot5\% - 240 + 350$, and $58\% - 350$ size distribution, giving $3\cdot3$ g/cm^3 TD was compacted implosively[16] in 21 mm diameter, 1–2 mm thick steel containers. With a tap density of 66% TD, final 89% and 95·2% TD densities were reached respectively in 2 mm wall-thickness containers. The lower density corresponded to $E/P = 0\cdot5$, giving 618 J/cm^2 at 890 m/s, and the higher was produced at $E/P = 1\cdot2$, which gave an impact energy of 1935 J/cm^2 at 1575 m/s. The constant β (eqn (7.5a)) was estimated at $0\cdot0015$, and for an energy of 1300 J/cm^2, the compact microhardness was about 475 Hm.

The metallurgical problems arising out of evaporation were examined by Prümmer.[2] Whereas 2–10 μm tungsten powder (19·3 g/cm^3 TD) would be compacted satisfactorily (from a tap density of 45·6% TD) to 97·4% TD, 5 μm powder in a 17 mm diameter container would be affected at a pressure of 5 GPa and an impact velocity of 3500 m/s. In this case, the converging shock pulse would raise the core temperature to above 3000°C, producing melting and subsequent recrystallisation of an area of 0·12 mm in diameter.

Again, a comparison between the isostatic and dynamic conditions has been made[25] and the results for 2·5 μm (explosive processing) and 5 μm (isostatic) particles are given in Fig. 7.8.

The data for isostatically compacted material merge with the explosively obtained values. At the highest 1·8 GPa isostatic pressure attainable, the green compact density amounted to 85% TD, but at an explosive pressure of 3 GPa, a density of 96% TD was reached. Above that pressure a deterioration in the properties was observed, with the density falling to 90% at 4 GPa. The pattern of the pressure/density relationship is tri-zonal.

7.3 PROPERTIES OF METALLIC COMPOSITES

In the broadest sense of the word, a particulate metallic composite can be produced either by shock synthesis, or shock consolidation. In some

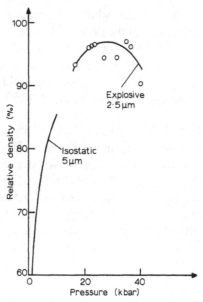

FIG. 7.8   Relationship between pressure and green density of explosively and
            isostatically compacted tungsten powder (after Ref. 25).

cases the simultaneous combination of the two processes is possible.
From the purely compacting point of view, the basic matrix materials
are aluminium, copper, iron, nickel, tin, titanium and tungsten. The
blending and/or alloying of these with other metallic powders is discussed in the following sections, and the availability of information is
indicated in Table 7.2.

7.3.1 Aluminium-based Compounds

A considerable volume of information on shock consolidation of
aluminium-matrix composites is available, but relatively little has been
published on the subject of shock synthesis. An early example of the
latter treatment is provided by the work of Otto *et al.*,[30] who in the early
1970s investigated the synthesis of aluminium niobimide (Al_3Nb).

The following review is concerned entirely with the compaction of
composite powders obtained originally in conventional manufacturing
operations.

7.3.1.1 The 2000 series

The two, industrially more interesting, representatives of the aluminium–copper series are the 2014 and 2024 alloys. Both are dynamically

TABLE 7.2
Referenced Information on Metal
Composites

Material	References
Al$_3$Nb	30
2014 Al	31
2024 Al	32, 33
Al-1·4Co	34
Al-Fe	35, 36
Al-Li	33, 37, 39
Cu-Ni	6, 16, 40
304 steels	23, 36, 43, 44
Fe-Cr-Al	28
Fe-40Ni	37
Fe-Cu	46
Ancorsteel 1000	47
Nickel aluminides	48, 49, 50
Inconel 718	37, 51
Nickel titanide	52
Niobium stannide	53
Titanium aluminides	26, 37, 54
Titanium erbides	37
Tungsten aluminides	55
W-Cu	16
Tungsten titanide	24
W-Ni-Fe	56

compactible and, in this way, make a useful addition to the existing range of materials.

(a) 2014 Al (4·49 Cu, 0·50 Fe, 0·88 Si, 0·80 Mn, 0·03 Ti). This non-rapidly solidified powder of average size fraction of about 60 μm, was successfully compacted using an electric gun incorporating an electro-magnetic accelerator actuated by a 36 kJ capacitor bank.[31] Compaction was carried out in a 19·1 mm diameter die, with zinc stearate as lubricant.

The particles of the starting material were morphologically irregular; the particle distribution being 8·5% − 170 + 200, 31·5% − 200 + 325, and 60% − 325.

Starting with a tap density of 62% TD, and depending on the energy levels and the associated impact velocities, increasing compact densities of 70, 89 and 99·9% TD were obtained for the corresponding sets of the velocity and energy values of: 250 m/s, 65 J/g, 305 m/s, 106 J/g, and 450 m/s, 231 J/g.

The original powder microhardness of 100 DPH would reach a saturation level of 160 DPH at about 125 J/g, and the UTS (determined in bend tests) would approach that of wrought 2014 Al alloy at about 170 MPa. The compacts show relatively low ductility, but this increases with the energy input.

Metallurgically, the compact structures indicate that oxides are well broken and dispersed, and that the particle deformation is concentrated in the areas of surface irregularities, especially on the particle boundaries. Cores of individual particles are not affected by plastic flow.

(b) 2024 Al ($Al_{0.934}$, $Cu_{0.045}Mg_{0.025}Mn_{0.006}$). The usual material used in the dynamic compaction[32,33] is either some 50 μm rapidly solidified process powder, or 2·16 g/cm^3 sintered product. Apart from its industrial applications, the 2024 Al alloy is of particular interest in the shock treatment field since it is being used by the Los Alamos Scientific Laboratory as a shock-wave standard material.

Gas-gun compaction of 60% crystal density powder, 85% 325 mesh, of 16·1 μm average particle size can be hot compacted in a graphite die at 600°C.

Even in the lower range of impact velocities, compact densities of 85% TD (108 m/s), 96% TD (304 m/s), and 99·9% TD (420 m/s) were easily obtained, producing an average elastic limit value of about 80 MPa.

Irregular shaped voids of mean pore diameter in the 2·2–2·7 μm range were observed in the compacts.

7.3.1.2 Aluminium/cobalt

The effect of possible thermal modification of the compact structure was studied, on the example of Al–1·4Co (by weight) alloy, by Wright *et al.*[34] Explosive compaction of commercially produced (by air atomisation) material was carried out using powder screened to − 325 mesh at both ambient (28°C), and − 50°C temperatures. Pressures of between 7 and 8 GPa were estimated to be operating in the cores of the irregularly shaped particles for periods of 5 μs. For comparison, commercially pure aluminium powder, of the same characteristics, was also consolidated. Both materials when shocked from 55% TD tap density produced compacts of 98·7% TD.

Breaking up of oxides took place and resulted in interparticle bonding by melting. Extensive plastic flow and particle elongation were observed, but no change in fine precipitate microstructure was noted. Both low and ambient temperature samples showed very similar microstructures.

The 'ambient temperature' Al–1·4Co compact had a cellular structure, with arrays of precipitates aligned on the cell boundaries. The Al_9CO_2 precipitates were not regarded as resulting from the dynamic treatment, but rather as being products of the original powder-manufacturing process. At the low temperature of $-50°C$, Al–1·4Co subgrains were still within the pre-existing cells, with the elongated cells contained, for both temperature values, within the $1·5 \times 2·5$ μm size.

The pure aluminium powder showed extensive recovery, when consolidated at 28°C, displaying a large number of equiaxial grains separated by well-defined dislocation boundaries. At the low temperature, the 2 μm subgrain might be reduced and the mean subgrain would be found to be within the 1–2 μm range.

The authors found that the hardness of the compacts was comparable with that of the cold-worked 1100 Al alloy, although the structure was not that of a cold-worked material. The fracture strength of the compacts was about 140 MPa, but the green specimens would fail at a low elongation of less than 1%. No macroductility was generated on annealing the material for 1 h in the 200–500°C temperature range — with a view to produce sintering — but fracture stress was reduced as the temperature increased.

7.3.1.3 Aluminium/iron
Compressed air-driven compacting machines have proved adequate in consolidating aluminium/ferrous powder mixtures into, sometimes, finished parts.[35,36] For instance, an atomised aluminium/steel admixed alloy powder (Al–25Fe) was compressed[35] in dies lubricated with either zinc stearate or PTFE spray, to produce 25 mm bevel gears.

Some 94–98% TD compacts displayed high tensile strength varying from 1·5 to 2·0 GPa, a high degree of toughness, and hardness levels within the 35–50 HRc range. Strain-hardened material was found in the gear-teeth, with the compact showing interparticle welding.

Again, 20% M2 steel, dispersed in an aluminium matrix powder was consolidated[36] at an impact velocity of 1000 m/s, reaching a density of 99% TD, Hv = 110, and a bend strength of 300 MPa. The material showed a good wear resistance against steel and iron, and in abrasive testing against silica particles it produced a wear rate similar to that of a medium alloy steel, or twice as good as aluminium silicon casting alloy.[36]

A rapidly solidified particulate of Al–8Fe–2Mo, made by atomisation and melt-spinning, and having a very well defined cellular structure, was consolidated dynamically, including a HIP operation, by Miller *et al.*, (of the University of Illinois, Urbana).

A single-stage gas gun, producing impact velocities of between 1200 and 1500 m/s was used. Dynamically compacted specimens were HIP'ed at 350°C for 2 h. High densities of the order of 99% TD were normally obtained, the compacts having a laminated structure with clearly defined interparticle boundaries. No evidence of any boundary melting was detected and bonding was assumed to have taken place by mechanical interlocking of flakes. Small regions of recrystallised material of very fine equiaxed aluminium grains were present, and in some areas intercellular phases would decompose into coarse Al_6Fe. However, with no Al_3Fe present, negligible thermal decomposition could be assumed. It seemed that a slight growth in Al_6Fe was the only effect of HIP.

Both the dynamically compacted and the additionally HIP treated samples displayed a Rockwell B hardness of about 50 on the transverse sections. Hardness measurements on the longitudinal sections would result in the generation of cracks caused by delamination.

According to the somewhat limited data about the explosive compaction of Al–8·4Fe–7Ce powder.[37] The UTS of these compacts reached the value of 256 MPa.

7.3.1.4 Aluminium/lithium

The saving on weight accomplished when aluminium/lithium alloys are incorporated in aircraft structures, together with the strength and stiffness characteristics, are the most attractive features of these composites.[38] The reduction in weight can be quite considerable with the aluminium matrix being capable of dissolving up to 20% of lithium by weight. Two equilibrium phases of saturated solid solution and Li_3Al are produced in conventional processing at 602°C. At room temperatures, Al–5Li alloys of a nominal density of 2·53 g/cm^3, and an elastic modulus E of 78 GPa, give an E/ρ ratio of 1·52 and a 0·1% proof stress of 400 MPa. This is due to the fact that each weight of lithium dissolved in aluminium reduces the density of the compact by about 3% while raising by 6% the value of the modulus.[39]

Ahrens *et al.*[33] and Myers *et al.*,[37] recognising the practical importance of these alloys, carried out extensive investigations into the properties of dynamically manufactured blends.

Al–3Li and Al–1·1Li rapidly solidified, single phase powders, 50 μm mean particle size were compacted in gas-guns of 1·6–4·8 GPa pressure range. Impact velocities of between 800 and 1300 m/s were generated in connection with a tap density of 60% TD. Under these conditions, a pressure of 3 GPa was found to be optimal giving 97·2% TD Al–3Li, and 98% TD Al–1·1Li compacts. Pressures above 3·7 GPa produced melting and vaporisation of lithium.

Tensile tests done on the higher lithium content compound indicated 0·2% proof stress of between 110 and 131 MPa, and a ductile failure after elongation of 0·7%.

Initially, both materials were polycrystalline, fcc, α-phase compounds of the following particle distribution (Al–3Li values are given in brackets): 0·63% (1·18%) 180–125, 1·58% (6·12%) 125–90, 8·01% (17·09%) 90–64, 21·25% (19·17%) 64–45, 31·58% (25·25%) 45–32, 28·45% (22·65%) 32–20, and 8·7% (8·51%) 20–5.

The Hugoniot data for 1·6 g/cm³ bulk density were found to be consistent with a linear shock velocity (km/s) V_s given by

$$V_s = 5·22 + 1·31\, u \tag{7.6}$$

where u is the particle velocity and the Gruneisen constant is taken as 2·12.

A little cracking is occasionally present, its existence being linked with the failure of an interparticle phase rich in magnesium. The surface concentration of an Mg/Al compound was twice that of the sample as a whole, with, at the same time, a surface lithium deposit in an Al–3Li compact being three times that in the interior of the compact.

Dehydroxylation at 227°C of LiOH and Al(OH)₃ present on the surface of the original powder is required. This treatment, followed by pre-annealing at between 400 and 500°C reduces the density of macroscopic cracks and is found to increase the efficiency of interparticle bonding in the compact. Some lowering of the weight of the structure can also be achieved by employing Al–Li–Cu powders.[37]

Gas-gun compaction of 75–80 μm, practically spherical particles with small surface nodules, produces satisfactorily strong compacts in spite of an occasional presence of molten and resolidified pockets. For example, Al–3Li–1Cu–1Mg–0·2Zr alloy produces a UTS of 282 MPa, while the Al–3Li–1Cu–0·2Zr compound gives a UTS of 265 MPa.

7.3.2 Copper-based Compounds

The available information points to the almost exclusive use of copper/ nickel alloys in the field of dynamic compaction. Equal fractions (by weight) of the spherical Type 435 nickel powder, obtained by the Carbonyl Process, and the electrolytically processed copper were compacted explosively in steel containers 26 mm in diameter, 2 mm wall thickness, and 65 mm long.[16] The particle size distributions of the two powders were:

Copper: $0.2\% + 100$, $30.5\% - 100 + 200$, $29.1\% - 200 + 350$,   and $40.2\% - 350$.

Nickel: $1.4\% + 120$, $8.8\% - 120 + 240$, $4.5\% - 240 + 350$,   and $40.2\% - 350$.

Since the investigators considered the area of the container surface in preference to the mass of the powder, the impact energies supplied are given per unit area. For the high degree of densification of 97.2% TD, an energy input of 382 J/cm² was recommended. An estimate of the required energy E_C, was obtained from the consideration of the law of mixtures (similar to eqn (4.25))

$$E_C = E_A f_A + E_B f_B \qquad (7.7)$$

where A and B refer to the two components, and f represents the volume fraction. A value of some 350 J/cm² was indicated, which compares quite well with the experimental data.

The remaining items of information are concerned with the shock treatment of copper-based alloys in the manufacture of thermocouples and the subsequent measurement of temperature rise during dynamic compactions of powders.

Schwarz *et al.*[6] produced a copper/constantan interface, by first mechanically alloying a constantan powder (55% Cu and 45% Ni, by weight). The mixture was obtained by vibratory methods of some 5 h duration, followed by annealing in purified hydrogen at 750°C for 16 h. The powder was then ground to − 200 mesh and dried in a nitrogen atmosphere.

Layers of copper and constantan powders (about 60 μm average particle size) were statically compressed at 140 kPa to a green density of 5·27 g/cm³. The green compacts were then used in a gas-gun to measure voltages associated with the passage of shock waves which were generated at pressures ranging from 1·3 to 9·4 GPa. Rapid increases in compact temperatures, of 45–81 ns duration, were observed resulting at first in peak temperatures, and then in equilibrium conditions throughout the compact in question. With the increasing shock pulses, the equilibrium temperatures rose from 425 to 1215 K.

A similar application of shock treatment was reported by Raybould.[40] Using a compressed-air-gun, capable of producing impact velocities in the 100–2000 m/s range, copper/nickel powder junctions were produced. With an average particle size varying from 0·1 to 0·5 mm, and tap densities of 40 and 45% TD respectively, shock pulses of 3 and 3·5 GPa were generated. The compacted copper/nickel junctions would register a temperature of about 750°C.

Although no information has been provided about the mechanical strength of these various compacts, some insight into the possible values can be obtained from the data on the explosively shocked solid, and admittedly different, copper alloy. A $Cu_{70}Zn_{30}$ alloy, compacted at pressures ranging from 5 to 40 GPa, produced 0·2% proof stress varying from 160 to 420 MPa, and UTS between 210 and 430 GPa. Shock pulses of 2 μs duration were used on 120 μm size particles.[10]

7.3.3 Iron-based Compounds
With very minor exceptions, interest is centred here on a wide range of high quality steel powders, but some other ferrous alloys and wire-reinforced composites have also been manufactured.

7.3.3.1 Steel compacts
304 series. Typical examples of the work carried out in this range are the investigations of the properties of shocked Type 304SS, and 304L powders. An in-depth investigation of the explosive compaction of 304SS powders was made by Korth et al.[41] Two rapidly solidified powders, one produced by centrifugal atomisation (RSR) (cooling rate 10^5 K/s), and the other by dissolved gas atomisation (DGA) (cooling rate 10^2 K/s) were compacted. In the 'as-received' state, the RSR-type powder was dendritic in structure and displayed a microhardness of 245 DPH when of the following composition: 70·5 Fe, 9·1 Ni, 18·4 Cr, 0·8 Mn, 0·7 Si, 0·6 Mo, 0·5 Cu, 0·002 S, 0·020 P and 0·05 C. The DGA-type material had a cellular morphology with a microhardness of 190 DPH and the composition of: 71 Fe, 9·9 Ni, 19 Cr, 0·03 Mn, 0·04 Si, 0·02 Mo, 0·005 Cu, 0·005 S, 0·005 P and 0·02 C. The theoretical density of both materials was estimated at 7·9 g/cm³.

The compaction (from the tap density of 58·2% TD) of 80 μm average size RSR powder produced 66·6% TD final compact, whose 0·2% proof stress amounted to 740 MPa and UTS to more than 1·05 GPA. No elongation or reduction in cross-sectional area was evident. The hardness of 350 DPH is equivalent to 50% cold-worked steel of the same category. The compaction of the DGA-type powder (40 μm average particle size) from a tap density of 54·9% TD to a final of 65·4% TD, resulted in a considerable extrusion of particles (shared with the RSR powder), melting and fuse-bonding of individual particles.

In comparison with the same powder material, hot-extruded at 900°C, the green dynamic compacts show very much higher stress values, but non-existent ductility. For instance, the conventional material has a proof stress of only 340 MPa, and UTS of 743 MPa, but its elongation and reduction of area amount to 62 and 77% respectively.

The relatively low final densities are associated with the physical difficulty of close-packing larger sized particles. To provide some insight into this problem, Staudhammer and Murr carried out an interesting investigation on the shock-consolidation of solid rods and tubes of the 304SS material. The results are discussed in Ref. 42.

The main investigations of the properties of the 304L steels focused, on the one hand, on the surface morphology/impact pressure dependence, and, on the other, on the mechanical and metallurgical features of the compacts.

For example[43] spherical morphology powders, produced by spin atomisation, of the following composition 70·82 Fe, 10·5 Ni, 18·5 Cr, 0·15 Co and 0·028 C, were explosively compacted to a density of some 74% TD. Separate consolidations of 100 mesh (150 μm), and of binary 12% − 170, 78% − 325 samples were conducted at very high impact pressures.

In the 100 mesh aggregate and at a tap density of 47% TD, extremely fine dendritic morphology was observed, with some interparticle melting pesent. The particle core pressures were estimated at 66 GPa.

The binary blend, compacted from the tap density of 57%, showed some small cracking, attributed to the solidification shrinkage of the molten regions, at the estimated surface impact pressures of 40 GPa, and core pressures as high as 100 GPa. It is, however, suggested by the authors that the lower pressure range should be sufficient to fully densify the material, if it is applied to the bulk of the specimen.

The work of Raybould[36,44] provides more detailed numerical data about the properties of 304L steel composites. Atomised 100 μm particles, when gas-gun compacted in 50 mm diameter containers, showed an increase in the final density from 60% TD at 180 MPa to 97·5% TD at a pressure of 1·2 GPa. Substantial interparticle melting (a calculated average temperature rise of 250°C) was observed. The lower tensile strength compacts of 320 MPa displayed a fracture toughness of

22 MN/m$^{1.5}$ which corresponds to that of a sintered powder material of 95% TD final density. The higher tensile strength range of 500 MPa, had the corresponding fracture toughness value of 31 MN/m$^{1.5}$, comparable with 32 MN/m$^{1.5}$ of the same, but wrought material.

The average microhardness of 420 Hv compared favourably with 345 Hv of a 50% cold-worked steel, or 350 Hv of explosively shocked solid material.

The presence of ductile fracture regions suggested good interparticle bonding. An extensive surface shearing and incidence of 'explosive welding' waves were also noted.

Since it is not only the pressure, but also the impact velocity that dictate final characteristics of the product, the effect of compacting at 400 and 1100 m/s was examined. In the first instance, specimens annealed at 1100°C, produced final densities of about 95% TD, with 160–240 Hv, and a bend strength of 1 GPa, whereas at the higher impact velocity final densities of 99% TD were attained. In the latter case, a hardness of 400 Hv and a bend strength of between 400 and 700 MPa were attained. The variation of the compact tensile strength with the dimensionless rise time, for a range of densities, is shown in Fig. 7.9.

An unsintered, high carbon chromium bearing steel powder, of 1·5 mm diameter particles, originally annealed at 1260 K and cooled down at the rate of 5 K/min, can be successfully shock treated.[23] An average chemical composition of this particular material is 9·95–1·1 C, 0·15–0·35 Si, <0·5 Mn, <0·0025 P, <0·0025 S, 0·90–1·20 Cr. The compaction is carried out within a 770–1530 m/s velocity range, at impact pressures (estimated from the Hugoniot compression curve for porous iron) ranging from 8 to 19·5 GPa.

Green compacts obtained at 8 GPa give a final density of 70% TD, whereas the annealed powder material produces compacts of 65% TD at 11 GPa, and 70% TD at 19·5 GPa. Plastic flow takes place in the pre-annealed material, but without the anneal, the particles are fragmented and fractured. Because of the insufficient levels of energy dissipated, only partial interparticle bonding occurred. Concave and/or flat shaped particle boundaries were observed, their final shape depending on the magnitude of the local deformation and the shock-wave amplitude.

The consolidation of 44–74 μm spherical particle AISI 9310 steel powder (3·2 Ni, 1·39 Cr, 0·65 Mn, 0·24 Mo, 0·1 C) of a bulk density of 7·90 g/cm^3, by energy dissipation in the range of between 240 and 770 kJ/g, in 2–3 μs shock pulses, gives a set of very useful mechanical properties.[45] Starting with a tap density of 60·6% TD, and using

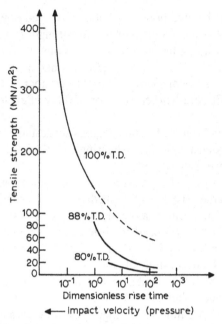

Fig. 7.9   Variation of tensile strength of 304L steel powder compacts with dimensionless rise time (after Ref. 40).

impactor plates at velocities ranging from 1500 to 3000 m/s, compact hardnesses of between 390 and 510 DPH are easily obtained. The tensile strength of these specimens can be as high as 1·4 GPa, with Young's modulus lying between 66 and 181 GPa. A practically complete interparticle bonding is attainable at 800 kJ/g and at impact pressures of between 3·6 and 19 GPa.

7.3.3.2 Special alloys

The relationship between the shock wave (V_s) velocity and compaction pressure (P) was studied using a Fe–Cr–Al 44–100 μm particle size metallic powder.[28] An increase in the shock wave velocity up to a pressure of 7·5 GPa was observed. At this pressure, the mechanism of energy absorption and dissipation appears to change as the free interparticle surface is reduced. Intensive particle bonding occurs, partly as a result of the breaking up of the free surfaces and partly because of the local surface melting, with consequently less energy being available for the acceleration of the back-surface particles.[28] The compaction of this material from a tap density of 62% TD provides a means of establishing an empirical expression linking the shock wave velocity and pressure in the form of

$$V_s = 897 + 2·30\ P^3 \quad (m/s) \qquad (7.8)$$

An interesting comparison between the properties of a Fe–40Ni explosively compacted and a sample of the same material HIP-processed and, subsequently, cold-worked, was made by Flinn *et al.*[37]

The explosive compaction results in a material having a 0·2% proof stress of between 513 and 536 MPa, UTS of 633 MPa and percentage elongation and reduction of area of 4·6 and 1·2–2·8 respectively. The hardness range lies between 209 and 214 DPH. The HIP and cold-worked compact shows a 0·2% proof stress of between 621 and 678 MPa. UTS ranging from 655 to 690 MPa, elongation of 18% and reduction of area of 54%.

Although the microhardness is comparable in both cases, the very localised yielding and plastic deformation of the former compound are responsible for the very limited melting and resolidification, and for a substantial presence of poor quality mechanical bonds. In consequence, the mechanical properties are lower than in the conventionally processed material.

Of particular practical interest is the range of Fe–Cu alloys. These possess higher strength than the pure Fe after annealing of statically consolidated mixtures, which are normally sintered afterwards at a temperature above the melting point of copper.[46] It has been shown, however, that dynamic compaction will produce a material of yet better quality that displays hardness values of between 100 and 200 Hv.

7.3.3.3 Reinforced composites

One of the very early, in the history of dynamic powder compaction, investigations into the manufacture of a wire reinforced powder matrix was carried out by McClelland and Otto[47] using an Ancorsteel 1000 iron powder and tungsten wires. The composition of the powder was 0·02 C, 0·20 Mg, 0·015 P, 0·015 S, 0·10 Si, residue Fe, and its particle size distribution was 2% − 80 + 100, 17% − 100 + 150, 28% − 150 + 200, 31% − 200 + 325, 22% − 325. The tungsten filaments were 0·127 mm in diameter and their UTS amounted to 2·48 GPa.

A two-piston, explosively actuated press was used for the compaction of matrix filament layers. On average, the volume fraction of the filament material varied, reaching some 10% for the compact, and the wire spacing was 0·40 mm. The law of mixtures was employed to assess the relevant parameters. The final green compact properties were 93% TD for the volume fraction V_f of between 1·6 and 5·1, elongation 10–26%, 0·02% proof stress 85–98 MPa, and UTS 145–269 MPa.

Some specimens were subsequently sintered in a hydrogen atmosphere at 1120°C for 1 h, and would then show a final density in the 90–95% TD range, and an average hardness of about 62 BHN. The stress–strain curves for a variety of green compacts are shown in Fig. 7.10.

7.3.4 Nickel-based Compounds

Unlike the ferrous composites, the shock treatment of nickel-based compounds has embraced both chemical synthesis (nickel aluminides) and/or compaction and consolidation; the latter mainly in the area of Inconel-type and nickel titanide alloys. Nickel aluminides are gaining rapidly in industrial importance because of their high temperature strength and resistance to corrosion and oxidation. In conventional processing boron is often added to improve ductility.

7.3.4.1 Nickel aluminides

The first shock chemical synthesis was carried out by Horie *et al.*[48,49] by explosively shocking a blended mixture of elemental nickel and aluminium powders. A nominal composition Ni–30Al (by weight)

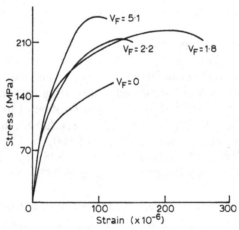

Fig. 7.10   The stress/strain curves for a range of volume fractions of tungsten-filament reinforced Ancorsteel 1000 powder matrix (after Ref. 47).

mixture of -325 mesh aluminium, and $-200+325$ mesh nickel powders, 5–15 μm and 44–74 μm average size respectively, was prepared.

The explosive treatment produced ordered Ni_3Al with $NiAl$, Ni_2Al_3 and $NiAl_3$ in the heat affected zones. Consolidation to 4·25 g/cm^3 or 60% TD in two pressure ranges of 14–16 GPa and 19–22 GPa for 4·6 μs duration, resulted in estimated temperatures of 450°C and 590°C, a microhardness of 438 DPH in the Ni_3Al areas, and a dislocation-free structure. The hardness is comparable to that of cold-rolled, or rapidly solidified Ni_3Al with boron as an additive. The hardness of the $NiAl$ component was 621 DPH as compared with 179 DPH of pure nickel powder.

The high hardness of Ni_3Al was attributed to a very small crystallite size, usually of about 5 nm. No evidence was found of any amorphous material in the mixture, and it was noted that Ni_2Al_3 and $NiAl_3$ compounds were formed, but were not obtained in fast quenching ion-implanted layers.

A further examination of the possibility of synthesising these compounds was made using a composite in which each spherical particle of aluminium powder was coated, on the surface, by 80% nickel (by weight). The deposition was made by hydrogen reduction in aqueous metal salt solution.

With both materials being 99·9% pure, and particle sizes ranging from 53 to 88 μm, satisfactory synthesis was achieved, accompanied by consolidation which led to a regular stacking of the particles with only a small deformation of the aluminium cores.

Although $NiAl_3$, Ni_2Al and Ni_3Al were present, the bulk of the synthesised material consisted of $NiAl$. Much higher impact pressures were required than in the previously discussed case.

Employing an explosively activated flyer plate for compaction, Song et al.[50] synthesised two mixtures. One, containing flaky nickel (-325 mesh) and irregular aluminium (-325 mesh) powders, and another consisting of spherical nickel and irregular aluminium (both -325 mesh) particles. Stoichiometric blending was used to give 87Ni–13Al. In all of the cases, a tap density of 65% TD was employed. Three impact pressures of 6·5, 12·5 and 16·8 GPa were applied for periods varying from 1 to 1·8 μs and produced impact velocities of 900, 1370 and 1600 m/s respectively.

Equiatomic $NiAl$ was obtained at 12·5 and 16·8 GPa, but Ni_3Al resulted at the lowest pressure. Generally dendritic structures were observed, but the lowest and highest pressures produced regular, spherical voids.

A purely compacting (explosive) operation,[16] based on a Ni–30Al mixture (by weight) prepared in a ball mill and carried out in steel containers 26 mm in diameter, 1·5 mm thick, produced 99·13% TD compacts. The explosive to container area mass ratio was 1·4, and an average energy level was 307·5 J/cm².

7.3.4.2 Inconel 718

This nickel-based, heat resistant alloy obtained by rapid solidification, was explosively consolidated in a pre-heated (525°C) state to soften the powder and reduce the incidence of cracking.[37,51]

Consolidation at the basic pressures of 3, 8 and 18 GPa generated the final density in excess of 98% TD. The 40 μm powder was packed in steel tube containers and sealed at 10^{-5} torr vacuum to minimise oxidation during heating. Some 20% interparticle melting was present after the compaction, but its fraction did not vary in the 2·5–9·8 μs range of pressure application. The quality of the compact was found to be related to the level of interparticle melting which, in this series of experiments, varied between 9·5 and 31·8%.

The 9·5% melt was associated with 3 GPa pressure and 525°C pre-heat. The UTS of the compact was 423 MPa, and the hardness reached a value of 440 Hv100. At 31·8% melt and 18 GPa (740°C pre-heat), 0·2% proof stress was 683 MPa, UTS was 855 MPa, and the hardness amounted to 320 Hv100. The highest values were obtained at 8 GPa and 525°C pre-heat. Here, the 0·2% proof stress was 877 MPa, UTS was 1·16 GPa, hardness 328 Hv100, and the interparticle melting reached 19·1%.

Extremely small grains were detected in the melt regions. The microcrystalline grains were about 0·02 μm in diameter, suggesting cooling rates of between 10^9 and 10^{10} K/s. A dislocation-free micro-dendritic structure was obtained, with some highly deformed dendrites being present.

7.3.4.3 Nickel titanide

The compaction of two-phase NiTi powder was carried out in a gas-gun by Thadhanni *et al.*[52]

A bcc Ni–45Ti, and fcc Ni–65Ti (by weight) mixture was compacted using 2–60 μm, spherically shaped, rapidly solidified material. The impact velocity was 950 m/s, and the impact energy reached the value of 316 J/g.

A unique structure modification was observed with bonding being effected by interparticle melting (estimated at 10%) and the molten material rapidly solidifying to largely amorphous and/or microcrystalline phases. Fine microcrystalline grains 0·04 μm in diameter were found in the interparticle regions. An extensive plastic deformation of the individual particle core, resulting in twinning, as well as grain elongation and a degree of recrystallisation were observed.

7.3.5 Tin Alloy

An example of a synthesised metallic tin compound is afforded by the explosive treatment of a tin and niobium powder mixture[53] to produce Nb_3Sn.

7.3.6 Titanium-based Alloys

Chemical synthesis and/or powder consolidation have been successfully achieved by shock treatment of titanium aluminides and erbides, as well as of some more complex titanium-based alloys.

7.3.6.1 Titanium aluminides

Synthesis of titanium and aluminium from stoichiometric, mechanically mixed powder ratios was carried out[54] and produced limited yields of $TiAl_3$. In addition, well ordered superstructural Ti_9Al_{23} and Ti_8Al_{24} compounds were formed.

Dynamic consolidation of TiAl34V1·3 compound (obtained by a rotating electrode process) of 50 to -200 mesh sized material by explosive tube and plate impact at both the room and elevated temperatures, led to the presence of large quantities of TiAl. The compact was found to be very hard and brittle, with no plastic deformation or interparticle melting present, but having strong interparticle bonding.[37]

Similar treatment of a TiAl14Cb21, -80 mesh, mixture resulted in the presence of Ti_3Al particles. These showed a high degree of interparticle deformation, but in spite of this, they retain cellular structure in the core. Localised melting is observed in the interparticle collision areas. Very fine grain is produced in this metastable material.

An earlier example of the explosive consolidation of a material belonging to this group, is that of the compaction of TiAl6V4 powder by Prümmer and Ziegler.[26] A metal container, 25 mm in diameter and 1·5 mm wall thickness was used and a pressure of about 2·25 GPa was generated. At an impact velocity of 2350 m/s, and $E/P = 2·1$, 99% TD compact density was obtained, with the lattice distortion of the order of 0·32%.

The variation of the compact density and of E/P ratio with V_D^2 is shown in Fig. 7.11.

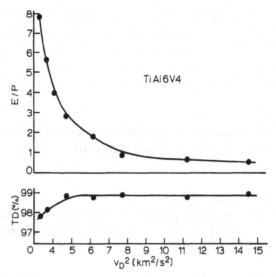

FIG. 7.11   Variation of the compact density and E/P ratio with the square of the velocity of detonation in explosive consolidation of TiAl6V4 powder (after Ref. 26).

7.3.6.2 *Titanium erbides*

The interest in titanium alloys containing fine erbia dispersions has grown in view of their effective high temperature strengthening and grain-growth limiting properties. Three such alloys have been shock-consolidated, two of these by Miller, Sears and Fraser of the University of Illinois, Urbana, and one by Meyers *et al.*[37]

The dynamic compaction of Ti–0·7Er produced extensive grain deformation, but no interparticle melting. The α'–Ti microstructure was retained during the compaction, and subsequent heat-treatment for 10 h at 600°C produced sintering in some of the interparticle bonds, as well as fine erbia dispersions. Some 99% TD compact densities were obtained with an average erbium particle size of about 17 nm. The immediate post-compaction mechanical properties were yield stress of 845 MPa, and UTS of 1·25 GPa, but after heat treatment, the yield stress was reduced to 257 MPa and UTS to 408 MPa. The same material HIP-treated gave a larger grain size of 38 nm.

The TiAl8Er0·7 alloy produced more porous compacts of between 96 and 97% TD and it retained its martensitic structure of the carbon-free powder. Some sintering of the interparticle bonds was again present after heat treatment at 600°C. Pure compaction resulted in a yield stress of 1·06 GPa and UTS of 1·075 GPa, but these values were reduced to 572 MPa and 714 MPa respectively after the treatment.

A Ti 6245 alloy (TiAl6Sn2Zr4Mo2Er1) of particle size range varying from 50 to 250 μm (obtained by a rapid solidification process) was explosively consolidated.[37] Very heavy grain deformation was observed, accompanied by a uniform dispersion of < 50 nm erbia precipitates.

7.3.6.3 Complex titanium alloys

The more complex titanium alloys are represented by Ti 17 and Ti 662 compounds. Both were implosively compacted in 50 mm diameter cylinders, 500 mm long. Compacts of up to 10 kg mass were obtained.

Both materials were produced in powder form by a plasma rotating electrode process and had an original particle size within the 50–250 μm range. Ti 17 (TiAl5Sn2Zr2Mo4Cr4) and Ti 662 (TiAl6Sn2V6Tl1Cu1) display a microcrystalline, dislocation-free grain structure after shock processing. Melted and resolidified interparticle regions are present and bonding is observed. Some spiral cracking can occur in the brittle TiAl alloy, and, occasionally, transverse cracking associated with longitudinal tensile stresses may be present.

7.3.7 Tungsten-based Alloys

Tungsten-based composites are used primarily as structural materials in working environments characterised by high temperatures and an oxidising and/or corrosive atmosphere. Aluminides and titanides are the main groups, but other material combinations are also sometimes encountered.

7.3.7.1 Tungsten aluminides

A successful synthesis of this material was achieved by Stavrer *et al.*[55] who explosively shock-treated a blended mixture of elemental aluminium and tungsten powders. The tungsten material was obtained by thermal decomposition of an acid, followed by reduction of the oxide in a hydrogen atmosphere, whereas the commercially pure aluminium powder was produced by air atomisation. Aluminium particles, 73 μm, were highly oxidised, but small 8 μm tungsten particles were oxygen-free.

The original blend had a density of 1·18 g/cm³ (25% TD) and was shocked by the impact of a flyer plate. The resulting material was mainly WAl_{12} with an admixture of unreacted aluminium and tungsten elements. The yield of the compound was less than that from a standard liquid-phase reaction and sintering at 685°C, but no difference in the chemical composition was detected.

7.3.7.2 Tungsten–copper alloy

This was obtained[16] by explosive compaction of a blend of tungsten powder, manufactured by a reducing method and electrolytic particulate copper mixed in a ball mill. Steel containers, 26 mm in diameter and 1 mm thick, were employed. The explosive mass to container tube area ratio of 3·48, and the optimal energy input of 711 J/cm^2 produced a W–30Cu compact of 94% TD. The relevant experimental numerical data were assessed by means of eqn (7.5a).

7.3.7.3 Tungsten titanide

W–15Ti mixtures were compacted[24] explosively in a gas-gun to 70–90% TD final densities. The corresponding impact pressures employed varied from 371 MPa to 1·53 GPa, and the appropriate Gruneisen coefficients from 0·020 to 0·077. Large specimens, 210 mm in diameter and 7–10 mm thick, were manufactured and, after sintering, used for magnetron sputtering.

7.3.7.4 Special application

The explosive welding of solid steel to solid W–Ni–Fe plates is considerably facilitated if an intermediate layer of a powdered W–Ni–Fe mixture is used. Peikrishvili *et al.*[56] bonded such a metallic combination at an elevated temperature of 300°C by producing a transient zone of solid solution on compaction. Depending on the circumstances, the zone can extend to between 0·10 and 0·15 mm and the solid solution associated with it can reach a hardness of about 580 HV5.

REFERENCES

1. Capus, J. M., Lund, J. A. and Clegg, M. A. *Int. J. Powder Met.*, **25** (1989), 141.
2. Prummer, R. In: *Explosive Welding, Forming and Compaction*, Ed. T. Z. Blazynski, Applied Science Publishers, London, New York, 1983, p.369.
3. British Patent No. 833 673, 1960.
4. La Rocca, E. W. and Pearson, J. US Patent No. 2 948 923.

5. HAGEMEYER, J. W. and REGALBUTO, J. A. *Int. J. Powder Met.*, **4** (1968), 19.
6. SCHWARZ, R. B., KASIRAJ, P. and VREELAND, J. R. In: *Metallurgical Applications of Shock-Wave and High-Strain-Rate Phenomena*, Eds L. E. Murr, K. P. Staudhammer, M. E. Meyers and M. Dekker, New York, Basle, 1986, p. 313.
7. WOLF, H. In: *Proc. 9th HERF Conference*, Academy of Sciences USSR, Siberian Division, Novosibirsk, 1986, p. 188.
8. ROMAN, O. V., NESTERENKO, V. F. and PIKUS, I. M. *Fizika Goreniya i Vzryva*, **15** (1979), 102 (in Russian).
9. LENNON, C. R. A., BHALLA, A. K. and WILLIAMS, J. D. *Proc. 6th HERF Conference*, Haus der Technik, Essen, Paper 6.1, 1977.
10. MURR, L. E. In: *Shock Waves and High Strain Rate Phenomena in Metals*, Eds M. M. Myers and L. E. Murr, Plenum Publishing Corporation, New York, 1981, p. 607.
11. STROMGREN, M., ÅSTROM, H. and EASTERLING, K. E., *Powder Met.*, **16** (1973), 155.
12. BERRY, R. A. and WILLIAMSON, R. L. In Ref. 6, p. 167.
13. THOMPSON, S. L., Sandia National Laboratory Report, SAND77-1339.
14. THOMPSON, S. L. and MCGLAUN, J. M. Sandia National Laboratory Report, SAND81-0987.
15. ROMAN, O. V. and BOGDANOV, A. P. In Ref. 9, Paper 6.5.
16. BHALLA, A. K. and WILLIAMS, J. D. In: *Proc. 5th HERF Conference*, University of Denver, Colorado, Paper 2.2, 1975.
17. ZUBAR, V. V., KORSHENEVSKY, A. P. and SHUGANOV, A. D. *High Energy Rate Fabrication — 1984*, Eds I. Berman and J. W. Schroeder, ASME, New York, 1984, p. 143.
18. CABANILLAS, E. and CUSMINSKY, G. In: *Proc. 7th HERF Conference*, Ed. T. Z. Blazynski, University of Leeds, Leeds, 1981, p. 242.
19. GELTMAN, G. In: *New Methods for the Consolidation of Metal Powders*, Eds H. H. Hausner, K. H. Roll and P. K. Johnson, Plenum Press, New York, 1967, p. 65.
20. BUTCHER, B. M. and KARNES, C. H. *J. Appl. Phys.*, **40** (1969), 2967.
21. WITKOWSKY, D. S. and OTTO, H. E. In: *Proc. 4th HERF Conference*, University of Denver, Colorado, Paper 9.3, 1973.
22. CROSS, A. *Iron Age*, (12) (1959).
23. TANIGUCHI, T., KONDO, K. and SAWAOKA, A. In Ref. 6, p. 293.
24. ROMAN, O. V. and GOROBTSOV, G. In: *Shock Waves for Industrial Applications*, Ed. L. E. Murr, Noyes Publishers, Park Ridge, NJ, 1988, p. 335.
25. PRUMMER, R. and ZIEGLER, G. In Ref. 18, p. 274.
26. PRUMMER, R. and ZIEGLER, G. In Ref. 9, Paper 6.2.
27. MURR, L. E. *Mater. Sci. Engng*, **57** (1983), 107.
28. LINSE, VONNE D., In Ref. 6, p. 29.
29. CHIBA, A., NISHIDA, M., YAMAGUCHI, T. and TOSAKA, J. *Scripta Metall.*, **22** (1988), 213.
30. OTTO, G., ROY, U. and REECE, O. *J. Less-Common Met.*, **32** (1973), 355.
31. IYER, N. C., FIKSE, D. A. and MALE, A. T. In Ref. 17, p. 137.
32. BUTCHER, B. M., CARROLL, M. M. and HOLT, A. C. *J. Appl. Phys.*, **45** (1974), 3864.

33. AHRENS, T. J., THADHANI, N., MUTZ, A. H., VREELAND, T., SCHWARZ, R. B., TYBURCZY, J. A., SHASTRI, S. L. M. and PENG, T. C. In Ref. 6, p. 83.
34. WRIGHT, R. N., DOYLE, T. E., FLINN, J. E. and KORTH, G. E. *Mater. Sci. Engng*, **94** (1987), 225.
35. RAYBOULD, D. In Ref. 18, p. 249.
36. RAYBOULD, D., MORRIS, D. G. and COOPER, G. A. *J. Mater. Sci. Letters*, **14** (1979), 2523.
37. MEYERS, M. M., THADHANI, N. N. and LI-HSING, YU In Ref. 24, p. 265.
38. WEBSTER, D. *Metal Prog.*, **53** (7) (1984).
39. KELLY, A. *Compos. Sci. Technol.*, **30** (1985), 171.
40. RAYBOULD, D. In Ref. 18, p. 261.
41. KORTH, G. E., FLINN, J. E. and GREEN, R. C. In Ref. 6, p. 129.
42. STAUDHAMMER, K. P. and MURR, L. E. In Ref. 6, p. 237.
43. STAUDHAMMER, K. P. and JOHNSON, K. A. In Ref. 6, p. 149.
44. RAYBOULD, D., *J. Mater. Sci.*, **16** (1981), 589.
45. KASIRAJ, P., VREELAND, T., SCHWARZ, R. B. and AHRENS, T. J. In: *Shock Waves in Condensed Matter — 1983*, Eds J. R. Asay, R. A. Graham and G. K. Straub, Elsevier Science Publishers, BV, Amsterdam, 1984, p. 439.
46. RAZAVIZADEH, K. and DAVIES, B. L. *Powder Met.*, **25** (1982), 11.
47. McCLELLAND, H. T. and OTTO, H. E. In Ref. 21, Paper 9.1.
48. HORIE, Y., GRAHAM, R. A. and SIMONSEN, I. K. *Mater. Letters*, **3** (1985), 354.
49. HORIE, Y., GRAHAM, R. A. and SIMONSEN, I. K. In Ref. 6, p. 1023.
50. SONG, I., THADHANI, N. N. and JING DING In: *Proc. 10th HERF Conference*, Litostroj, Ljubljana, Yugoslavia, 1989, p. 76.
51. WANG, S. L., MEYERS, M. A. and SZEKET, A. *J. Mater. Sci.*, **23** (1988), 1786.
52. THADHANI, N. N., VREELAND, T. and AHRENS, T. J. *J. Mater. Sci.*, **22** (1987), 4446.
53. OTTO, G., REECE, O. and ROY, U. *Appl. Phys. Letters*, **18** (1971), 418.
54. HORIE, Y., HOY, D. E. P., SIMONSEN, I., GRAHAM, R. A. and MOROSIN, B. In: *Shock Waves in Condensed Matter*, Ed. Y. M. Gupta, Plenum Publishing Corporation, New York, 1986.
55. STAVRER, ST, GOSPODINOV, V., DOICHEV, Al. and RADEV, R. In Ref. 50, p. 123.
56. PEIKRISHVILI, A., JAPARIDZE, L., GOTSIRIDZE, G. and CHIKHRADZE, N. In Ref. 50, p. 333.

Chapter 8

Shock Treatment of Ceramic Powders

8.1 GENERAL OBSERVATIONS

To a much higher degree than the metals, the response of the ceramic particulate matter to shock treatment reflects its sensitivity to the varied effects of the passage of a shock wave through an agglomerate. With the focus of attention directed, these days, more and more towards the many industrial applications of ceramics, the ability to respond easily to a relatively simple and cheap manufacturing technique is of considerable practical importance.

Chemical, structural and combustion (involving two or more complex powders) syntheses produce new ranges of materials, whereas, consolidation combined, where necessary, with further sintering, leads to the manufacture of near-shape, or basic shape pre-fabricates.

Oxides and non-oxides of basic ceramic materials, but particularly those of the inorganic refractory type, are of specific interest because they form the bases for a variety of parts in the manufacture of integrated electronic circuits, ceramic capacitors, mechanical seals, and, as cermets, in high-duty cutting tools.

Syntheses, whether chemical or structural, give rise to a variety of materials ranging from the 'exotic' high-pressure NH_4Br–$CsBr$ phase, through the reactions between tin and chalcogens, europium and chlorine ($EuCl_3$), neodymium oxide and neodymium, to, for instance, zinc ferrite, aluminium and silicon nitrides, titanium oxides, borides and carbides, and a range of ferrites. Mixtures of the oxides of lead and zirconium, zirconium and iron, and titanium and samarium serve as further examples of combustion synthesis. Structural synthesis manifests itself primarily in lattice distortion in composites like alumina, beryllia, zirconia, magnesia, silicon and boron carbides, and silicon and titanium nitrides.

In general, the combination of elemental powders into new substances $(Ti + C \rightarrow TiC)$, oxides into more complex composites $(Al_2O_3 + SiO_2 \rightarrow Al_2SiO_5$, or $Re_2O_3 + ZrO_2)$, the forming of solid solutions $(Al_2O_3 + SiC \rightarrow Al_2O_3$ structure), oxidation $(Cr + Cr_2O_3 \rightarrow Cr_3O_4$ and $Nd_2O_3 + Nd \rightarrow NdO)$, decomposition $(NiO \rightarrow Ni + O)$, or shock activation are the general characteristics of shock syntheses.[1,2]

Of particular practical importance is the manufacture of composite cermets of TiC–TiN, Al_2O_3–TiC, and Si_3N_4 bases, as is that of metallic glasses and superconductors. Although these materials come under the general description of ceramics, the complexity of their respective structures and methods of manufacture entitle them to a separate, and more detailed, discussion. This will be provided in Chapter 10.

The consolidation of powders and their mixtures represents the second, but a very important, facet of shock treatment. Although, as was indicated in Chapter 7, final compact densities of near 100% are common in metallic composites, an average theoretical density of 95% is more common in ceramic agglomerates. This is explainable in terms of the reduced plastic deformability of ceramic materials and the consequent localised resistance to the transmission of the shock wave. A degree of comminution is thus required and this is achieved through the medium of brittle fracture. In some cases, notably with alumina and boron carbide, the mechanism of adiabatic shear plays a major role in stimulating particle comminution (Fig. 8.1).

Whereas the silicon compounds, SiC and Si_3N_4 in particular, are acquiring an ever growing importance in the design of ceramic engines, turbines and mechanical seals, alumina and beryllia have made inroads into the area of electronics where they are used as substrates for film circuits. On the other hand, large plates, discs and similarly shaped pre-fabricates can be easily manufactured out of vacuum refined titanium, molybdenum, niobium and tantalum silicides whose consolidation is followed by sintering, albeit at a reduced temperature and pressure.

Although, admittedly, some of the dynamically produced composites are, as yet, of somewhat limited interest, the fact that they can be so obtained and, more importantly, that the shock treatment modifies their structure and properties is potentially significant. Table 8.1 gives an extensive list of the compounds produced or used in shock treatment, but only the well-documented and industrially important materials will be discussed in detail.

In general terms, however, the success of any of these operations depends both on the knowledge of the likely properties of the processed

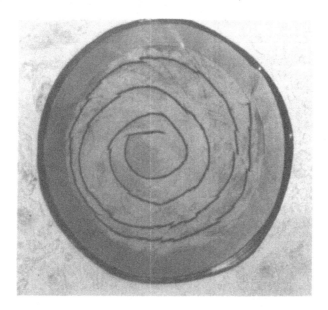

FIG. 8.1   The incidence of adiabatic shear bands in explosively compacted alumina powder. 0·1 mm thick niobium foil serves as an indicator of the degree of shearing (courtesy Dr R. Prümmer).

materials, as these develop during the operation, and on the ability to model the course of the operation. The first problem is solved when the Hugoniot curve, appropriate to the material and operational conditions, is available, such as for instance Fig. 8.2. The modelling of the operation can be carried out either by computer simulation of the type developed by Wilkins,[3] or by the application of the CSQ plane-wave code[4] which, if used in the axisymmetric mode, leads to the three-dimensional HULL code. The latter provides a closer insight into the pattern of behaviour of the powder during dynamic compaction.

Since a marked difference exists between the response to shock treatment of the oxide and non-oxide based composites, the following discussion is sub-divided into these two distinct groups.

8.2 OXIDE-BASED COMPOSITES

In view of their respective practical importance, the oxides of refractory materials, such as alumina, ferrites and rutiles, form the basis of a

TABLE 8.1
Chemical and Structural Shock Effects in Ceramics

Composite	References
(a) Oxide syntheses	
Al_2O_3	8, 9, 11, 12, 13, 14, 15, 16, 17
Al_2O_3–SiO_2	1
Al_2O_3–ZrO_2	20
Al_2O_3–Ti	18
$BaTiO_3$	1
$BaFe_{12}O_{19}$	7
BeO	22, 23
$CaCO_3$	23
CuMgO	19
$CuFe_2O_4$	24
$CoFe_2O_4$	24
$NiFe_2O_4$	24
Fe_2O_3	15, 24
Fe_3O_4	15
$PbZrO_3$	24, 25, 26
$PbTiO_3$	1
$Pb(Ti, Zr)O_3$	1
MgO	7
$MgSiO_3$	1
MnO_2	15
$MnSiO_3$	27
SiO_2	29
TiO_2	2, 31
ZnO	32
$ZnFe_2O_4$	32, 33
ZrO_2	20
(b) Oxide compaction	
Al_2O_3	8, 9, 10, 11, 12, 17
Al_2O_3–SiO_2	1
Al_2O_3–ZrO_2	20
Al_2O_3–Ti	18
Al_2O_3–AlN	19
$BaFe_{12}O_{19}$	7
BeO	21
Fe_2O_3	15, 24
Fe_3O_4	15
MgO	7, 21
MnO_2	15
SiO_2	5
SnO_2 + Al, C, Si	30
TiO_2	31
UO_2	7, 21
$ZnFe_2O_4$	32, 33
ZrO_2 + rare earth	34

Composite	Synthesis	Compaction
(c) Non-oxides		
AlN	12, 14, 31, 35, 36, 37	12, 35, 38
B_4C		7
BN	18, 39, 40	39
$MoSi_2$		18, 21
$MoSi_2$-BN		21
ZrB_2-$MoSi_2$-BN		21
Nichrome-BN		18
Nb_3Sn	41	
Nb_3Si	42	
SiC		43–46
Si_3N_4		12, 47, 48
TiB_2		14, 37, 49
TiC	14, 31, 37, 50	19, 31
WC, WC_2	30	
W-3OUO_2		21

number of investigations. This does not mean, however, that some of the less known oxide based materials, e.g. beryllia, magnesia and urania, which are slowly coming into their own in a range of applications, should be excluded from this review.

Apart from the obvious attractions of chemical synthesis and compaction, it is structural modifications that so often make shock treatment preferable to conventional manufacturing techniques. In this context, the residual straining of the lattice, and the changes in the specific surface area acquire great significance. These effects are normally related to the operational pressure and it is for this reason that their variation with this particular parameter, in the case of a few selected refractory materials, is indicated in Figs 8.3 and 8.4.

8.2.1 Aluminium Oxide – Al_2O_3

With diverse applications, ranging from electronic systems substrata through tribologically abrasive substances[6] and admixtures with polymers, to brass drawing dies that give tools lives almost equal to that of a WC:Co and diamond inserts, alumina has been a widely studied material. In spite of being difficult to consolidate, even under dynamic conditions, structural modifications resulting from shock treatment are of considerable practical interest. A large volume of literature is

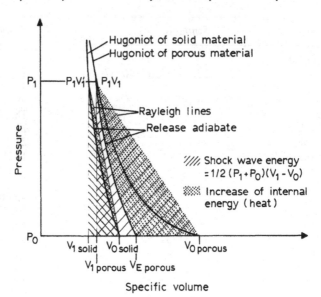

Fig. 8.2   Basic Hugoniot curves for the explosive compaction of particulate ceramic materials.

available and the following comments are meant as a resumé of this information.

The main thrust of the investigations has been the modification of structure, followed by the mechanical properties of the compact. Most of this work was carried out in direct, usually tubular, explosive systems, but smaller samples, intended for more sophisticated physical tests, were occasionally obtained in flyer-plate, momentum trap recovery assemblies.

The earliest experimentally obtained data are due to Bergmann and Barrington[7] who in 1966 consolidated α'-Al$_2$O$_3$ powder. An X-ray line broadening suggested 560 Å as crystallite size and the calculated, from the line broadening, increase in surface area was estimated at 2·7 m^2/g; the actually measured increase amounted to 3·1 m^2/g, or some 29%. The compact density was low at 63·3% TD, but it was an improvement on 56·8% TD of the unshocked material.

The bulk of the investigative work in the 1970s was done by Prümmer and Ziegler,[8,9] and was supplemented by Hoenig *et al.*[10] The former authors were mostly concerned with the relative response to shock of fine and coarse-grained alumina and, for comparison, of fine and coarse materials, but milled for 157 h.

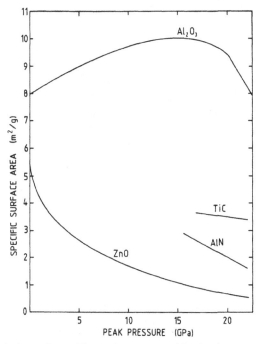

FIG. 8.3   Variation of specific surface area with shock pressure for some
refractory materials (after Ref. 5).

The original fine-grained powder had an average particle size of 3·5
μm and a crystallite size of 0·14 μm, with a lattice distortion of
0·075%, dislocation density $1·5 \times 10^{10}/cm^2$, a distortion energy of 0·146
J/g, and a BET surface of 1·2 m^2/g. The coarse-grained alumina had
particles of 3 μm, consisting of single crystals.

On explosive compaction at 2300 m/s in SAE 1010, 25 mm diameter, 1·5 mm thick containers, a high degree of activation, increasing
with pressure, was recorded. The activation of the fine powder was
higher than that of the coarse material. The crystallite sizes were
changed to 0·05 and 0·15–0·30 μm for the two powders, with the lattice
distortion values being 0·35 and 0·22% respectively. The corresponding
dislocation densities were $1·2 \times 10^{11}$ and $3·3 \times 10^9/cm^2$, and distortion
energies increased to 3·22 and 1·25 J/g.

The fine and coarse milled compacts showed the same crystallite size
of 0·07, the same dislocation density of $6·1 \times 10^{10}/cm^2$, but had
different lattice distortions of 0·17 and 0·065%, and distortion energies

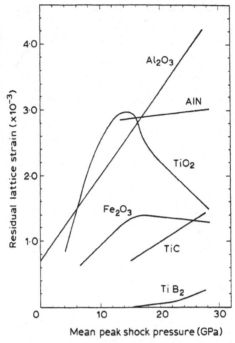

FIG. 8.4   Variation of residual lattice strain with shock pressure for some refractory materials (courtesy Dr R. A. Graham).

of 0·75 and 0·109 J/g respectively. This critical increase in lattice distortion is associated with the detonation velocity of 3690 m/s for fine alumina, 2240 m/s for the coarse powder, and 4690 m/s for the material that was sintered after shock treatment (Fig. 8.5).

A summary of these investigations is provided by Fig. 8.6 which also draws attention to the tri-zonal nature of powder response. In Range I, the density of the compact increases, but is not accompanied by any noticeable structural changes. In Range II, a large increase in both the density and lattice distortion takes place, but the crystallite size is reduced, whereas Range III is characterised by a decrease in density and lack of structural changes.

The highest compact densities of 99% TD were obtained for the coarse powder at a pressure of 4·1 GPa. The required E/P ratios depended on the type of powder and the detonation velocity (Fig. 8.7).

A continuation of this investigation by Hoenig et al.[10] added some numerical data and also produced a hydrodynamic model of the

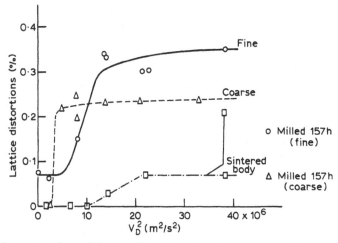

FIG. 8.5 Variation of lattice distortion with the square of the detonation velocity for fine, coarse and milled for 157 h, explosively compacted alumina powder (courtesy Dr R. A. Prümmer).

operation. This was based on the constitutive equations for alumina, a Hugoniot curve (Fig. 8.8), and a plane-wave compaction computer code.

Significant interparticle melting and consequent bonding in α-Al_2O_3, tabular, Grade T-61 powder were present at pressures in excess of 8 GPa. The powder (10–177 μm particle size) was explosively flat compacted in 50·8 mm diameter, 9·5 mm thick discs from 62% TD tap density at 4900 m/s, and at an average pressure of 6 GPa. The final densities lay in the 94–98% TD region, but 90% TD compacts were the only ones free from microcracking. The microhardness of the shock treated material was about 2000 kg/mm^2 which compares favourably with 1580 kg/mm^2 of conventionally hot-pressed (at 1810°C), 95% TD material.

The information was further supplemented by Hoenig and Yust[12] who shock treated and consolidated the same material at detonation velocities of 4200 and 5500 m/s, and corresponding pressures of 3·6 and 7·7 GPa, generated by E/P ratios of 5·7 and 3·8.

Although the final density of the compact (starting from 55 to 65% tap density, and compacting in 25·4 mm diameter, 1·65 thick and 180 mm long steel containers) was 91% TD in both cases, overcompaction took place at 7·7 GPa pressure. Here, all particles were reduced in size with their central core regions being below 10 μm. A very fine-grained,

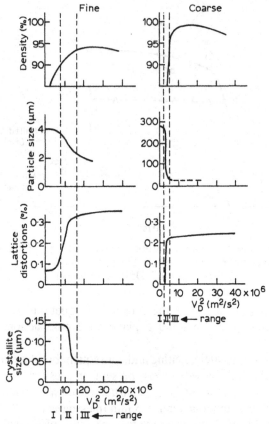

FIG. 8.6   The relationship between the square of the detonation velocity and physical properties of fine and coarse grained, explosively shocked alumina powders (courtesy Dr R. A. Prümmer).

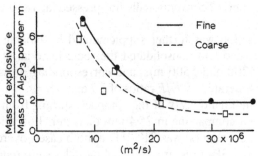

FIG. 8.7   *E/P* ratios for the compaction of fine and coarse alumina powders (courtesy Dr R. A. Prümmer).

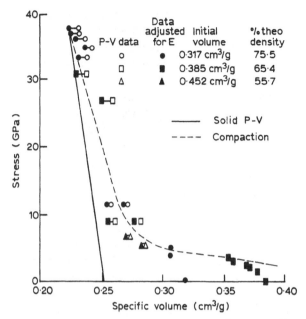

FIG. 8.8   Hugoniot data for porous α-Al$_2$O$_3$ material (3·97 g/cm^3), plotted on the zero-reference plane (after Ref. 11).

but densely faulted and strained polycrystalline matrix was generated, with some recrystallised, strain-free grains and columnar growth present at crack boundaries. The green compacts showed a micro-hardness of some 2070 KHN 100, and after heat treatment for 36 h at 1375 K in air, a hardness of 2280 KHN.

The mechanism of densification and the associated metallurgical phenomena were examined by Beauchamp *et al.*[13] The investigated particulate material, consisting of α-Al$_2$O$_3$ and metastable phases, was explosively compacted from a tap density of 56% TD at pressures ranging from 13 to 26 GPa to 90% TD. The accompanying maximum adiabatic temperature rise was 400°C.

The original material had been plasma spheroidised and contained some 17% of particles with internal voids. The parent phase (10–20 μm) was spinel thinly distributed between the monoclinic (0·3–0·5 μm thick, and 2–10 μm long) phase. The latter had some 5×10^7/cm^2 uniformly distributed dislocations, and contained also a few twins. The α-phase was present in the form of 1–10 μm grains, with low 10^6/cm^2 density dislocations.

The shock treatment did not produce any phase transformations, but resulted in severe deformation of the monoclinic phase, a high dislocation density of $10^{11}/cm^2$, and the appearance of many fine twins. There was no twinning or cleavage present in the α-phase, and the dislocation density matched that of the monoclinic. This range is comparable with the effect produced by severe cold-working of metals.

It was found that the densification was affected by plastic deformation involving dislocation slip in the α-alumina, and by slip and twinning in the metastable phases. Unlike the results of Hoenig *et al.* no fracture or interparticle bonding were observed.

The variation of the specific surface area, first noted by Bergmann and Barrington, was confirmed by the investigations of Lee *et al.*[14] On shock treating Grade RC-HP-DBM, 3·14 μm particle alumina at pressures between 4 and 22 GPa that produced peak temperatures of 100 and 350°C respectively, it was found that the specific surface area responds in a well-defined manner. Initially, this parameter increases from the value appropriate to the unshocked powder to a maximum in the region of 10 GPa pressure. The peak change amounts to about 20% of the initial 8 m^2/g value. The pattern is shown in Fig. 8.3.

On the purely compacting side of the problem, the investigation confirmed that at low pressures only comminution of particles obtains, but interparticle bonding takes place at higher shock pressures.

The effect of shock on the residual lattice strain has been investigated by Graham and his co-workers[15,16] on samples of Linde C, and Reynolds alumina powders. Starting with tap densities of between 40 and 60% TD, the lower 4–8 GPa pressure range produced residual strains of the order of $2·5 \times 10^{-3}$, whereas the 20–27 GPa pressures gave a strain of $4·5 \times 10^{-3}$. The crystallite size was reduced linearly as the pressure increased (Fig. 8.4). Again, the bulk temperature of the compact varied from 100 to 450°C, with the particle edge temperature being estimated at around 800°C.

A question that often arises in shock compact manufacture is that of the likely effect of repeated shock wave passages. The effect of this type of treatment on the microstructure was studied by Akashi and Sawaoka[17] on 99·98% pure alumina 5–10 μm platelets prepacked at 440 MPa into 12 mm diameter, 5 mm thick stainless steel capsules. A flyer-plate system generated shock pressures of between 21 and 29 GPa at 2500 m/s impact velocities.

After a single compaction, the powder, remaining white, suffered very little particle reduction, but cracks and subgrain microstructures, consisting of particles 0·1 μm in diameter, were present. The second densification changed the colour to grey and considerable reduction in particle size occurred. The agglomeration of particles produced a 1 μm polygon microstructure consisting of fine 50 nm crystallites. The change in colour appears to be caused by non-stoichiometric alumina resulting from the presence of oxygen vacancies that are produced by the passage of the shock wave.

It was noted that in this first stage extremely high dislocation densities were generated in crystalline alumina above the Hugoniot Elastic Limit. Crystal particles were transformed into amorphous-like structures. With the shock and residual temperatures being 1300 and 1000°C respectively (at 21 GPa), the polycrystalline structure of the compact is the result of recrystallisation of the amorphous-like phase. The two temperatures rise to 1750 and 1500°C after the second compression and thus create conditions for bonding of the fine 50 m crystallites. The authors point out that the final compact structure is bound to be affected by the rate of cooling and shrinking of the capsule.

Blends of alumina with other metallic or non-metallic oxides are also produced and, although as yet of limited interest, they are indicative of the possible future developments.

An interesting example of combustion synthesis is the aluminium silicate obtained by shock treating a mixture of alumina and silica[1]

$$Al_2O_3 + SiO_2 = Al_2(SiO_3)_2 \qquad (8.1)$$

This type of induced reaction could be utilised eventually in, say, the manufacture of Sialon. On the other hand, the compaction of sintered, porous granules of α-phase alumina coated with titanium powder (2–5 μm particle size) shows promising results with its 85% TD density (quite high for granulated material) and considerable, fracture-free, plastic deformation.[18]

A situation that leads to a uniform distribution in a binary shock-compacted blend arises when the dimensions of the velocity non-equilibrium regions, behind the incident and reflected shock waves, do not exceed the sizes of the particles involved. This, in turn, is likely to happen when either the respective mass difference is small, or when the interparticle pressures are high. A 50/50 mixture of alumina and AlN powders, explosively compacted to 54% TD, illustrates this effect. It was found[19] that under those conditions no separation of the two substances was present.

The fabrication of a zirconia-toughened alumina composite was recently investigated by Murat *et al.*[20] High purity Reynolds-HP-DBM α-alumina and MEL.SC15 zirconia powders were blended in an attritor mill in distilled water prior to compaction. Mixtures containing 10 or 20% (volume) of zirconia were compacted at pressures ranging from 5 to 20 GPa. Hot explosive compaction at 500 and 1000°C was also employed.

The retention of the tetragonal zirconia phase was considerably increased in the preheated (20°C) compacts. Only low densities of between 60 and 70% TD (50–55% TD tap density) were obtained in the lower pressure range, but no microcracking ensued. Pressures of the order of 20 GPa produced higher compact densities, but were also responsible for the onset of cracking.

Shock consolidation followed by sintering at low, in comparison with conventional, temperatures of between 1200 and 1475°C, increased the final densities to above 90% TD, irrespective of the operational shock pressure.

At 5·7 GPa, a 10% ZrO_2 mixture reached a value of 300 GPa for the Young's modulus, after sintering at 1300°C for 8 h, and had a 31% retention rate. At the same pressure, a 20% mixture showed only a 13% retention and a low value of $E = 197$ GPa.

The variation of hardness with the pre-shock temperature and heat treatment is shown in Fig. 8.9(a). Figure 8.9(b) shows that at pressures above 20 GPa, the alumina particles remain equiaxed, but the zirconia elements deform heavily and form a film around the alumina grains. The authors attribute the high degree of retention of zirconia to enhanced recrystallisation associated with shock treatment.

8.2.2 Barium Composites

The two barium composites of considerable industrial interest are barium titanate ($BaTiO_3$) and barium ferrite ($BaFe_{12}O_{19}$). Barium titanate is conventionally obtained by pressing and sintering and has a melting point of 1680°C, and a density of 6·03 g/cm^3. Because of the high values of the dielectric constants, it is used in powder capacitor materials, in ultrasonic generators, and in shock-wave parameter measuring transducers. For practical applications, it is much more easily obtained by explosive synthesis,[1] of the type

$$TiO_2 + BaCO_3 = BaTiO_3 \qquad (8.2)$$

that involves the mixing of the two constituent powders.

The effectiveness of post-shock sintering of $BaFe_{12}O_{19}$ powder depends on the activation of the material. This, in turn, is related to the operational temperature. For a sintering time of 3 h, the density of the compact increases by about 8·7% at 1250°C to 5 g/cm^3, and by 9·1% at

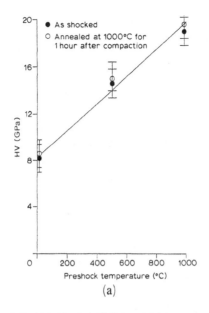

(a)

(b)

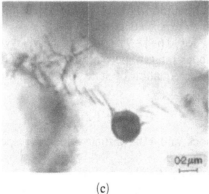

(c)

FIG. 8.9 Characteristic features of zirconia toughened alumina powder. (a) Variations of hardness with pre-shock temperature in 10% (volume) ZrO_2 mixture compacted at 21·4 GPa; (b) SEM micrograph of the above compact after hot explosive compaction at 500°C; (c) TEM micrograph showing dislocation loops punched out of an intergranular zirconia particle in a sample compacted at 5·7 GPa and sintered at 1475°C for 75 min (courtesy of Dr O. T. Inal).

1100°C to 4·2 g/cm² in comparison with the unshocked, but sintered, material.[7] The increased reactivity is linked to the available lattice distortion energy.

8.2.3 Beryllium Oxide — BeO

As an electronic substrate with a high melting point of 2550°C, solid density of 3 g/cm³, and Knoop hardness of 1250, beryllia is basically preferable to alumina. Its thermal conductivity of 2·18 W/cm°C exceeds that of aluminium oxide (0·17 W/cm°C), but it suffers from the disadvantage of being toxic. The conventional manufacturing technique is by hot pressing, but the dynamic approach offers an alternative.

In this respect, an early interest in the material was shown by Carlson *et al.*[21] who in 1964 compacted explosively pure beryllium and, then, beryllium oxide powders. Beryllium powder, 75 μm average particle size, was densified from a tap density of 31·7% TD to 96% TD and formed into rods and tubes. Beryllia, 49 μm particle size, was similarly compacted into fluted tubes of 96% TD, from a tap density of 68·3% TD.

A more detailed examination of the dynamically consolidated 0·6 porous beryllia was carried out by Yaziv *et al.*[22] Shock loading of plate impacts provided numerical data for the determination of the response of this polycrystalline material. A zero porosity condition was reached at a pressure level of 9·6 GPa, accompanied by a spall strength of 0·2 GPa for the elastically stressed material. The spall strength was found to be decreasing above HEL. The elastic region of experimentation was confined to pressures between 3·3 and 8·5 GPa, with HEL being reached at 8·65 and 9·4 GPa, and the plastic zone at 8·95 GPa. The corresponding impact velocities ranged from 200 to 477 m/s, and from 512 to 575 m/s. The authors have produced a spall model for high HEL solids, based on their experimental data.

Hugoniot data for 0·6–18·9% porosity beryllia were obtained by Marsh[23] for the shock range of between 10 and 100 GPa, and led to the establishment of the equation of state for fully dense material in the form

$$U = 8·6 + 1·2u \tag{8.3}$$

where U is the shock, and u the particle velocities respectively.

8.2.4 Iron Compounds

Prime examples of shock-induced structural changes are the two oxides of iron, magnetite (Fe_3O_4), and haematite (Fe_2O_3). Haematite, used among other applications in the dynamic synthesis of zinc ferrite, is of particular interest because in its normal condition it is an antiferro-magnet with zero-net magnetisation in the perfect lattice. The investiga-tions of Graham *et al.*[24] show that shock-induced deformation produces defects in α-haematite which, in turn, are responsible for the onset of disorder in the atomic substrata and the consequent changes in their parallelism or antiparallelism. Conversion to magnetite of some 1–3% of the material, at temperatures ranging from 600 to 800°C, changes its physical properties, inducing a degree of magnetisation. The main factor influencing its onset appears to be the residual lattice strain, which is detectable at even very low shock pressures of 4 GPa (strain of 6×10^{-4}). As Fig. 8.4 indicates, this lattice defect reaches a peak at about 17 GPa ($1\cdot4 \times 10^{-3}$), and then declines slowly, at 30 GPa, to a value of $1\cdot2 \times 10^{-3}$. The reduction is associated with the annealing effect produced by a relatively high shock temperature (bulk tempera-tures of up to 400°C, with edge temperatures reaching 650°C) attending a high pressure treatment.[15]

The BET specific surface area change is generally low, not exceeding a factor of two. For instance, a 99·5% pure, 200 nm average particle size α-haematite powder, dynamically compacted in the 7·5–22 GPa range to a density of 2·88 g/cm³, changes its specific surface area from the original value of 5·37 to 6·65 at 8 GPa, 6·93 at 17 GPa and 2·69 m²/g at 22 GPa.

Magnetite, used in the syntheses of cobalt, copper and nickel ferrites also shows significant structural changes. A commercially pure powder, used by Williams *et al.*[15] exhibited a large residual strain in its original, unshocked condition. Shock densification to 2·20 g/cm³ produced further increases in the broadened diffraction profile and in the con-sequent residual lattice strain.

The original specific surface area of 10·92 m²/g was reduced to 6·64 m²/g at a pressure of 7·5 GPa (bulk temperature 450°C, edge tempera-ture 525°C) and, further, to 1·31 m²/g at 27 GPa (bulk temperature 650°C, edge temperature 1150°C). The reduction is, again, explainable in terms of the relative ease of annealing and of interparticle bonding.

8.2.5 Lead-based Compounds

The shock-induced enhancement of chemical reactivity is well demon-strated by the synthesis of lead zirconate.[24–26] In its original condition, zirconia powder has a low specific surface area of 8·5 m²/g which does not change when the material is shocked in a 20–27 GPa pressure range. However, the behaviour of the shocked zirconia is that of the unshocked material of twice the specific surface area. The usual procedure is therefore to shock the zirconia first, and then to react it with the lead oxide

$$PbO + Zr_2O \rightarrow PbZrO_3 \qquad (8.4)$$

The reason for the enhanced reactivity of shocked zirconia with lead oxide lies with the considerable reduction in the crystallite size, combined with the formation of the tetragonal phase particles. This, in turn, is influenced by the reduction in the reaction temperature of zirconia linked to the rise in shock pressure. The temperature falls from some 830°C at 10 GPa to 790°C at 27 GPa. At the lower pressures, localised particle size reduction is observed, but at the higher pressures a strain-free tetragonal phase, recrystallised to 30 nm, is present, and lattice strains of 3×10^{-3} set in. The conversion of the original monoclinic to tetragonal phase occurs at a rate of 10% at 20 GPa, and 20% at 27 GPa.

Other mixtures, displaying similar characteristics, are also synthesised by shock treatment.[1] Of these, lead titanate, and a complex ferroelectric ceramic lead zirconate, titanate composite are of interest.

$$PbO + TiO_2 \rightarrow PbTiO_3$$

$$PbO + TiO_2 + ZrO_2 \rightarrow Pb(Ti, Zr)O_3 \qquad (8.5)$$

8.2.6 Magnesium Oxide

In conventional applications, magnesium oxide is characterised by a high melting point at 2850°C, a relatively low density of 3·6 g/cm³, a Knoop hardness of 370, Young's modulus of 280 GPa, and, at a temperature of 600°C, by a thermal conductivity of 59 W/m K. These properties make it an ideal material for work at elevated temperatures, as an additive to alumina and zirconia, and a binder in the compaction and sintering of difficult to sinter amorphous nitride-type ceramics.

Shock consolidation of the powder produces both a satisfactory degree of densification (97·3% TD in the case of 49 μm particle size material),[21] and structural modifications that enhance its properties, with the size of the coherently diffracting domain estimated[1] at 160 Å.

A comparison between the increase in the specific surface area calculated from the line broadening was made by Bergmann and Barrington on the assumption that this was produced only by small crystallite size.[7]

With X-ray line broadening, as a crystallite, the unshocked material showed values in excess of 2500 Å, and the shocked powder some 350 Å. The calculated broadening would suggest less than 6 m^2/g for the unshocked sample, and 48 m^2/g for the shocked. In reality, the two values were 3·5 and 4·2 m^2/g (or a 20% increase). It appears therefore that a large proportion of the line broadening is due to lattice strain and not predominantly to the reduction in crystallite size.

Shock treated material exhibits a rapid response to sintering at much lower than conventional temperatures, and reaches a density of 2·75 g/cm^3 (76·4% TD) after 4 h at 1210°C, and a density of 80·5% TD at 1390°C. The corresponding unshocked compact data are 72·2% TD at 1210°C, and 75·5% TD at 1390°C.

The oxide can be easily shock-synthesised with silica, as follows

$$MgO + SiO_2 \rightarrow MgSiO_3 \tag{8.6}$$

8.2.7 Manganese Composites

Manganese dioxide (MnO_2) undergoes changes in its specific surface area with as little as 10% of the material being affected by shock treatment.[15] In its original condition, a 99·9% pure powder shows a sharp diffraction pattern. Its specific surface area is assessed as 0·94 m^2/g, and at pressures of 20 and 27 GPa it densifies to 2·09 and 3·14 g/cm^3 from 40 and 60% TD tap densities respectively.

At the lower pressure, the bulk temperatures are 500 and 125°C (for the 40 and 60% tap densities respectively), and the specific surface areas change correspondingly to 3·09 and 3·96 m^2/g. At 27 GPa, the two temperatures are 650 and 450°C, and the specific surface areas become 2·14 and 2·39 m^2/g.

Those very substantial increases in specific surface areas are clearly functions of pressure and tap density, and they appear to follow a pattern similar to that of alumina (Fig. 8.3). However, this material is characterised by numerically smaller values than those of alumina.

The response of manganese dioxide to shock is also unusual because of the emergence of metal atoms with reduced valency, as, for instance, in the synthesis of $MnSiO_3$, where free electrons are supplied by the dielectric–metal type conversion.[27]

8.2.8 Neodymium Oxide

The NdO material is used in the solid state for the colouring of glass lasers and consequently its properties in shocked and unshocked conditions arouse some interest. The effect of shock treatment on the activation rate and particularly on the rate of hydration was studied by Adadurov *et al.*,[28] and the synthesis by Batsanov.[29] In shock compression Nd_2O_3 was converted to the corresponding compound of bivalent neodymium

$$Nd_2O_3 + Nd \rightarrow NdO \qquad (8.7)$$

Following various different treatments, the hydration rate of the pure (99·97%) material was measured in a temperature range of 22–23·5°C. Three conventional techniques were used and their effects were compared with those produced by shock loading.

With the material calcined at 1200°C and cooled, the incubation period was observed to have been between 150 and 350 h. When the oxide was fragmented at 1200°C for 2 h, and then held in moist air for 260 h, the incubation period was reduced to 60 h. Finally, the material treated as in the latter case, but cooled in a desiccator, showed an incubation period of 20 h. By comparison, the shock treated NdO powder was typified by a rapid increase in the water content and the lack of any incubation period. Generally, a marked increase in the effectiveness was noted in the shocked powder and was reflected in the lattice properties.

8.2.9 Silica — SiO_2

In addition to being an additive, as already indicated, to alumina, magnesia, zirconia, silicon carbides and nitrides, and, more recently, polymers (Chapter 9), silica has properties similar in some sense to those of pure silicon ($E = 1·30$ GPa, thermal conductivity 1·47 W/m°C) which is used in electronic circuits, and in the manufacture of lenses.

Shock-consolidated silica agglomerate undergoes a change in lattice dimensions of between 0·013 and 0·067 Å, but its isomorphous transformations yield unstable modifications, as evidenced by subsequent heating.[29]

The investigations by Blazynski and Hegazy[5] in 1985, carried out at pressures reaching 24 GPa, showed that the irregular, average size 15 μm particles (Fig. 8.10(a)) deform plastically with relatively little comminution at pressures of up to about 17 GPa (Fig. 8.10(b)).

The material was found to be very sensitive to the effects of the release wave — associated also with the container properties — and at pressures exceeding the optimal value of 18 GPa the onset of microcracking was observed.

(a)   (b)

FIG. 8.10   Explosive compaction of silica: (a) uncompacted particulate
material; (b) after compaction at 18 GPa.

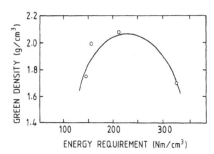

FIG. 8.11   Variation of green compact density with shock energy in an
explosively   consolidated   silica
sample.

The dependence of densification on the peak pressure (Fig. 8.11) shows a marked similarity to the behaviour of alumina (Fig. 8.6) in that three distinct zones are discernible. Low densification due mainly to particle re-arrangement, at pressures lower than 17 GPa, the optimal conditions of comminution and interparticle bonding in the range of 17–20 GPa, and the degradation of properties beyond that pressure can be identified.

The peak green density attained was $2·08$ g/cm^3, or approximately 89% TD. With interparticle bonding present in the compact, subsequent sintering could be avoided.

8.2.10 Tin Compounds

The shock compression of stannic oxide (SnO_2), either on its own, or as a mixture with aluminium, carbon and silicon powders provides an interesting insight into the effect of initial porosity and temperature of the agglomerate on the final volume fraction.

The experiments of Staver[30] involved the use of 5 μm SnO_2, 5 μm aluminium, 30 μm carbon and 30 μm silicon powders. The compactions were carried out at pressures reaching 15 GPa and were accompanied by a considerable degree of comminution.

The blend of the stannic oxide and aluminium powders was characterised by the increase in the tin content as the isobar–isostatic potential, and particle size, were reduced. The same effect was observed with an increase in the shock pressure.

The ratio of the volume of the oxide and of tin in the container, before and after the compaction, expressed as a percentage, was found to be a function of both the pressure and initial porosity of the sample. For instance, within the 10–15 GPa pressure range, the SnO_2 ratio has a value of $1·67$, but for its mixture with carbon, this value is raised to $4·16$, whereas for a blend with aluminium powder the ratio increases to 24. The compact of stannic oxide and silicon displays a different pattern of behaviour. Up to a pressure of $7·5$ GPa, the ratio is maintained at about $5·8$, but it then rises rapidly and reaches a value of about 60 at 15 GPa.

The variation of the ratio is also influenced by the initial temperature in that in the range of between 77 and 300 K the oxide ratio increases only slightly from $2·8$ to 3, but the corresponding (SnO_2 + Si) mixture ratio changes from 5 to $5·8$, and at a higher temperature of 477 K it reaches a value of $14·8$.

Repeated compressions of the sample of a SnO_2–Si blend tend to reduce the rate of decrease of the ratio when no fracturing is present, but they increase it slightly when fractures occur. The authors suggests that this may be the result of the existence of microareas of concentrated thermal energy.

8.2.11 Titanium Oxides

The high melting point ($1840°C$) of titanium oxide, combined with its high density of $4·26$ g/cm^2 and Knoop hardness of about 1000, make it particularly suitable for high temperature applications. TiO_2, in both its rutile and anatase forms, responds strongly to shock treatment, including densification, showing increased chemical activity and structure modifications.

The extensive investigations of Graham *et al.*[2,31] provide a large volume of information about this material in its powder form.

Shock-modified rutile has enhanced catalytic properties which are not linked to point defects. The catalytic activity at 450°C, studied in connection with the oxidation of CO in a flow reactor, increases by up to five orders of magnitude, depending on the pressure, and remains unchanged for a matter of hours. Shock pressures of up to 27 GPa reduce the particle size through the process of deformation, but no significant change is observed. The crystallite size varies with the crystallographic direction and the presence of small grains in the shocked material appears to be mainly due to primary recrystallisation.

Shock-induced changes in the specific surface area are relatively small, but the lattice distortion is considerable. In both the rutile and anatase forms, the distortion is detectable at pressures as low as 4 GPa, but the saturation level reached depends on the material. The residual lattice strain (Fig. 8.4) shows an increase from 0.8×10^{-3} at 4 GPa to a peak of 3×10^{-3} at 14 GPa. The strain then reduces rapidly to about 1.5×10^{-3} at a high pressure of 28 GPa. The reduction is attributed to the fact that the shock-induced temperature affects the microstrain through the process of annealing. The strain is reduced at high pressures and the corresponding high temperatures. Thus, powders of the lower tap density of, say, 43% TD show less strain (1.4×10^{-3} at 29 GPa) at higher than at lower pressures in comparison with, for instance, a 55% TD sample (1.7×10^{-3} at 29 GPa). Because of the tetragonal symmetry of the structure, the strain is anisotropic with the directional differences as high as 50%. Annealing effects are felt at temperatures as low as 320°C, but the full benefit is not likely below 1100°C. Defect configurations, causing the strain, are present at the catalytic activity temperature.

A high concentration of dislocations, deformation twins in 2 μm particles, deformations and annealing twins in 5 μm particles, and cleavages are present. The concentrations are so high that only in the low pressure range of 5–7 GPa is observation possible. However, no crystallographic shear defects exist.

Paramagnetic defects also occur in large concentrations, but are reduced at the catalytic temperature and are not therefore associated with that activity. The paramagnetic resonance, absent in the unshocked white powders, is present as either an isotropic or anisotropic defect that changes the colour to grey and alters the *g*-factor. At 20 GPa, the isotropic *g*-factor is 2·0029 and the anisotropic one is either 1·965 or 1·937. The corresponding concentrations are 3×10^{16} and $3 \times 10^{19}/$ cm^3. The resistivity of the shocked material reaches some 10^{-4} Ω/cm.

Chemical or elevated temperature treatments of non-shocked material are incapable of producing the high deformation — uniformly distributed within the grains — and the degree of grain refinement that are observed in the shock-consolidated material. Conventional processing results in low deformation and causes crystallographic shear.

8.2.12 Uranium Dioxide

Regarded usually as a bivalent uranyl radical in uranium salts, UO_2 ($\rho = 10·88$ g/cm^3) had been used widely in photography and for making fluorescent glass. It was only in the late 1960s that its use in vibrational compaction and swaging in fuel element fabrication became of interest, and it was in this context that shock treatment of the powder was introduced.[7,21]

The basic idea was to agglomerate micronised UO_2 powder into a high-density coarse-particle product that, after suitable low-temperature hydrogen treatment, could be used to produce a stoichiometric oxide of large particle size. The explosive compaction of the powder resulted in a slight increase in the crystallite size from the original less than 1 μm to about 2 μm, at a pressure of 55 GPa. Localised heating was regarded as the likely explanation for the change. The ratio of oxygen to uranium of 2·08 in the original material, remained practically unchanged after the compaction, showing just a slight reduction to between 2·052 and 2·075.

A comparison between the green densities of the shocked and 'as received' material indicated considerable differences. A tap density of 29% TD produced an agglomerate density of 57·9% TD after shock densification, whereas a 20% TD tap density gave a 44·9% TD in the conventionally compacted material.

In sintering, the HIP-ed material, granulated to 5600 μm and initially at a tap density of 58·8% TD, produced a 95% TD compact of bonded structure after 3 h at 1150°C and at a low pressure of 70 MPa. The extremely large particles of the pre-shocked material gave the same HIP results as the very finely-grained powder processed conventionally under the same conditions.

8.2.13 Zinc Compounds

Explosive densification[14] of zinc oxide (ZnO) powder of 3·14 μm mean particle size, provides a means of investigating the change in its specific surface area. Starting with a tap density of 3·01 g/cm^3 (55% TD), the compaction of the powder produces peak temperatures of 225 and 400°C for the 7·5–22 GPa shock-pressure range. As a result of inter-particle bonding, the specific surface area is seen to decrease with increasing pressure (Fig. 8.3) from about 5·5 m^2/g in the original material to 0·8 m^2/g at 20 GPa.

Zinc oxide forms a basis for the shock-induced synthesis of zinc ferrite (ZnFe$_3$O$_4$). The latter is synthesised explosively from a mech-anically blended, in stoichiometric ratios, mixture of ZnO and haema-tite, at pressures ranging from 7·5 to 27 GPa.[32,33] The corresponding shock-induced temperatures lie between 125 and 1110°C. At 27 GPa and 1100°C, the yield is 85%, but since less zinc oxide than haematite is engaged in the synthesis, zinc-deficient ferrite can be formed.

In the shock-synthesised material small grains are present thus indicating primary crystallisation. Depending on the pressure, detect-able quantities of spinel ferrite are found. For example, at 16 GPa and at temperatures in excess of 500°C, and again at 20 GPa and 175°C the presence of these particles is noted. At pressures above 22 GPa, spinel ferrite is present under all conditions. Two spinel phases exist, one is magnetic and the other paramagnetic. The latter is composed of very small crystallites (< 500 nm) and supermagnetic relaxation occurs.

At high pressures, the small ferrite grain is strain- and defect-free; a mildly shocked specimen exhibits heavy dislocations and plastic deformation. The ferrite phase also displays a high degree of magnetisa-tion and of high-field susceptibility. This becomes clear when a compar-ison is made between the condition of the original, ball-milled stoichiometric mixture, and the effect that the increasing pressure has on the compact. In the former case, the saturation magnetisation, and the high-field susceptibility, both at 7 K, were assessed as 0·015 emu/g and 1·14 × 10^{-5} cm^3/g. At a pressure of 22 GPa, but for two different densities of 2·15 and 3·30 g/cm^3 (bulk temperatures of 655 and 300°C respectively), the saturation magnetisation values were 30·5 and 8·15 emu/g, susceptibility rose to 7·7 × 10^{-5} and 2 × 10^{-5} cm^3/g, with the corresponding formation of the ferrite phase of 27 and less than 6%. At a higher pressure of 27 GPa, and a density of 3·38 g/cm^3 (400°C bulk temperature), the saturation level became 19·2 emu/g, the high-field susceptibility 4·04 × 10^{-5} cm^3/g, but the percentage of formed spinel fell to 15%.

8.2.14 Zirconium Composites

Of these, zirconia is of the greatest interest because of its inherent toughness, and therefore its resistance to the formation and/or propagation of cracks. In its metastable, tetragonal form and as a partly stabilised compound, it is used in metal-forming tools — although in this form it is less resistant to wear than other ceramics — as an abrasive, in the manufacture of glass cutters, and, finally, in the fabrication of refractory materials.

On average and as a solid, it has a melting point of 2700°C, a density of 5·6 g/cm³, a Knoop hardness of 1160, a Young's modulus of 207 GPa (equal to carbon steel), an UTS of 500 MPa, and a thermal conductivity of 3 W/m K. Zirconia's enhanced chemical reactivity, when shocked, was discussed in detail in Section 8.2.5, and the ease of retention of its tetragonal phase was described in connection with the fabrication of a zirconia–alumina composite (Section 8.2.1).

The compaction of zirconia mixtures has been carried out satisfactorily, using powders of an average particle size of 3 μm and of a density of 3·18 g/cm³ TD. Refractories and glasses were produced in this way; two typical examples being;[34] $ZrO_2 + Si_3N_4$ and $ZrO_2 + CaO + 2$ w/o MgO. In both cases, the initial tap density of 52·2% TD was raised to 96·3% TD after compaction. In general, special compounds containing rare earth (RE) metals are shock produced from blends of $RE_2O_3 + ZrO_2$.

8.3 NON-OXIDE CERAMIC COMPOSITES

Generally, in conventional processing these can either be hot-pressed or, more often, require sintering through the medium of high pressure, high temperature and some additives. Although sintering may still be required in, at least, some shock-consolidated materials, the conditions will be less severe and additives may be avoided altogether.

In addition to boron compounds, it is mainly carbides, nitrides and silicides of aluminium, molybdenum, nickel, niobium, silicon, tin, titanium and zirconium that are used industrially. The various tungsten compounds come under the specific name of cermets and are described in Chapter 10.

8.3.1 Aluminium Nitride — AlN

This is a corrosion resistant, high-strength, refractory structural ceramic that can be hot-pressed to 96% TD consistency, but, under standard, conventional conditions it cannot be sintered without additives. However, it responds well to shock loading.

The shock consolidation of the powder produces considerable deformation of grain structure. The relatively high temperature induced, combined with a high quenching rate, appears to be responsible for the presence of an amorphous, rich in silicon, interparticle phase that, in turn, creates good bonding conditions.[12,35]

Shock Hugoniot measurements[36] made on unsintered powder at two tap densities of 40·6 and 47·8% TD, and at low pressures of 0·25 and 1·8 GPa, showed that only 70% TD densification could be achieved with a single shock wave at 200–600 m/s impacts. Typically, the lower density sample would give 67% TD, at 0·41 GPa and a plate impact velocity of 312 m/s, rising to 70% TD at 1·76 GPa and 738 m/s. The corresponding values for the higher tap density specimens were 73% TD at 0·268 GPa and 323 m/s, and 68% TD at 1·268 GPa and 749 m/s. The reshocking of the two compacts would raise the density to 74% TD at 1·33 GPa, and to 85% TD at 3·2 GPa for the lower tap density, and to 81% TD and 86% TD at, respectively, 2·0 and 3·9 GPa for the higher tap density.

The mutually complementary work of Hoenig and Yust[12] and Gourdin *et al.*[35] at pressures ranging from 3·5 to 7·4 GPa, provided further information about the likely response of AlN. A 99% pure aluminium nitride powder, 10–300 μm particle size, densified from 50 to 60% TD tap densities to between 92 and 98·1% TD compact density, showed little or no grain growth, retaining the original crystallite size. Variations in properties across the grain were observed with 2–3 μm equiaxed, strain-free grains in the core, damaged and deformed material at half-radius, and a pseudo amorphous phase on the grain boundaries. Otherwise, the normal hexagonal structure was preserved, with a centre and edge hardness of 2050 and 1600 KHN100 respectively.

The specific surface area[14] (Fig. 8.3) increases from about 2 m²/g to 3 m²/g at 10 GPa, and then reduces to 1 m²/g at 27 GPa. A very considerable increase in the concentration of dislocations is observed.[37] In a particular, average grain size of 37 μm, polymodal (10 μm-fine, 40 μm-coarse) mixture of 93·4% AlN and 6·6% Al_2O_3 of less than 10^6/cm² dislocation concentration, shock treatment at pressure levels of 16–22 GPa produces so high a dislocation density that only the lower bound can be estimated. The initial lattice strain of near 10^{-5} reaches a saturation level[31] of 3×10^{-3} (Fig. 8.3) at pressures of between 14 and 18 GPa.

Electron spin measurements reveal a change in resonance. Whereas the starting powder displays a weak resonance of g-factor 2·0052 and a concentration of $2 \times 10^6/cm^3$, the shocked material, whilst retaining the unchanged width of 50 Oe, is characterised by a very high resonance concentration of $3 \times 10^{18}/cm^3$. The highest degree of concentration is associated with the lowest pressure since the shock-induced temperature results in annealing at high pressure levels.

Wilkins *et al.*[38] carried out an extensive investigation of the computer modelling of AlN densification using the HEMPDC2A code. Both cylindrical and plane geometries were considered (Fig. 8.12). Figure 8.12(a) represents a model for an axisymmetric, tube compacting system in which an incident shock wave S_i is induced in the powder, and is then followed by a reflected wave S_r. The section in which this wave impacts the sample is denoted by a, its distance from the detonation front determines the strength of the wave. The nearer the point is to the detonation front the higher the pressure. Figure 8.12(b) shows the corresponding model for a flat geometry developed by the authors.

Using the computer models of Fig. 8.12, Wilkins *et al.* compacted AlN powders, ranging in particle size from 40 to 250 μm, at 3600 m/s impact velocity. The 'as received' powder had a hardness of 780–1200 KH100, and a lattice strain of 0·55%. After the compaction, the strain rose to 0·75% for the 40–50 μm particle range, and was accompanied by an increase in hardness to between 1310 and 1390. For the coarser grain (210–250 μm) the strain fell to 0·48%, and so did the hardness (950–1030 HK100). A higher impact-induced temperature associated with the coarser grain, produced a higher degree of strain relief than was possible in the finer-grained material.

Final compact densities were of the order of 97% TD, increasing from 65% TD tap density.

8.3.2 Boron Composites

Since some boron compounds are used extensively in industry and boron nitride is an extremely useful substitute for diamonds in cutting tools for sintered iron composites, including steels, their fabrication by dynamic means acquires a special significance. Crystalline boron itself has been processed in this manner. Here, an explosively compacted powder of starting particle size range 44–105 μm and tap density of 38% TD, displayed a highly faulted polycrystalline post-shock structure with grains less than 1 μm in size and with mainly planar defects.[12] Good integrity, crack-free compacts 97% TD in density were obtained

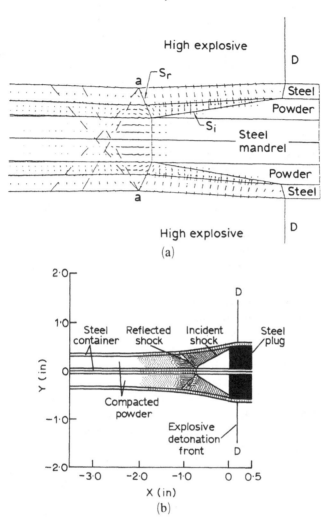

FIG. 8.12 Computer models for the explosive densification of AlN powder. (a) Axisymmetrical, tube producing system (S_i incident, S_r reflected shock waves); (b) a plane geometry system (courtesy Dr M. L. Wilkins).

at a pressure of 7·5 GPa, and $E/P = 13·8$; with porosity ranging from 1·4% in the core to 6·7% in the outer region, and the corresponding hardness ranging from 3170 to 2200 KHN100.

The response of boron carbide (B_4C) is conditioned by the initial grain characteristics and further heat treatment. Fine-grained material

(0–90 μm) explosively consolidated from 1·56 g/cm^3 (62·4% TD) reaches a final green compact density of 97·7% TD, whilst a coarse powder agglomerate (up to 300 μm), compacted from an initial density of 70% TD acquires a final density of almost that of the solid, namely 99·6%.

The early work on the effects of explosive compaction[7] showed that at lower pressures of about 0·14 GPa a final density of 61% TD is obtained, as compared with 50% in conventional processing. The same compacts after sintering for 4 h at 2200°C in an argon atmosphere, reached 87 and 71·3% TD respectively, whilst their corresponding UTS values rose to 172 and 103 MPa.

Boron nitride (BN) is normally obtainable in cubic form only. Shock loading is used in the synthesis of the material and changes the lattice[18, 39] constant of the cubic form to a graphite-like form (g-BN) which is stable at high temperatures that can transform the hexagonal, wurtzite-form to a highly dense w-BN. This is stable at high pressures, but can also exist in a metastable form at ambient temperatures. The yield of the hexagonal form depends on the green density and on the crystallinity of the starting material. The more perfectly crystalline the powder the higher the yield. Yields of over 80% are obtainable, but on average the conversion rate varies from 10–25% within a sample.

Again, as in the case of B$_4$C, the initial conditions play a decisive role in establishing the composition of the compact. A fine powder (2–4 μm) agglomerate of cubic c-BN powder, explosively compacted from a 60% TD at 2100 m/s (estimated pressures of 33 and 75 GPa in the outer and central regions), gives a compact with a final density of 89% TD, characterised by a variation in its hardness across, in this case a 5 mm thick sample of 1450–2500 HV. A coarse powder sample (10–20 μm) consolidates to 94% TD (2950–3180 HV) and shows partial inter-particle bonding.

A small amount of c-BN transforms into an amorphous phase in the coarse grain, but in the fine-grained material partly amorphous and partly graphite-like g-BN phases are present, with the latter probably crystallising out of the amorphous phase on cooling from the high, shock-induced temperature.

Even at a high pressure of 75 GPa, a minimal change in the residual lattice strain is observed.[40] In the fine powder the strain varies from 0·29 to 0·26% across the sample, and the crystallite size reduces slightly from 36 to 32 nm. In the coarser, middle range the strain remains constant at 0·19% across the sample, and the crystallite size increases slightly from 30 to 33 nm. The coarse powder (40–60 μm) compact shows a low strain range of 0·17–0·14%, with the crystallite at 31 nm.

Some interesting double material blends can also be produced by shock compaction,[19] as for instance, unseparated mixtures of BN-33W, and BN-85 Nichrome powders.

8.3.3 Molybdenum, Nickel and Niobium Compounds

Although of less practical importance, these dynamically consolidated and/or synthesised compounds serve as further examples of the versatility of these techniques.

Molybdenum disilicide ($MoSi_2$) responds well to explosive and impact densification[18,21] producing compacts of green densities that vary from 60 to 98% TD at shock pressures in the range of 0·054–2·20 GPa. In particular, a 50 μm powder was compacted from a tap density of 36·8% TD into cylindrical, high-density components.

The difficult to densify conventional mixtures of the disilicide with either boron nitride, or with zirconium diboride and boron nitride, have been compacted to densities of about 88% TD. The subsequent HIP-treatment at a relatively low pressure of 70 MPa at 1260°C (3 h) produced interparticle bonding.[21]

Good quality nichrome–8BN mixtures are also produced[18] by gas impact methods, using relatively low release pressures of between 0·19 and 1·27 GPa. The corresponding green densities range from 80 to 93% TD.

As an introduction to the fabrication of metallic and ceramic super-conducting materials (Chapter 10), the syntheses of niobium stannide[41] and niobium silicide[42] can be usefully considered.

The explosive synthesis of niobium stannide was carried out using high purity, 325 mesh, niobium and tin powders in a stoichiometric mixture of 3Nb:1Sn. At estimated pressures of some 50 GPa and temperatures exceeding 2000°C, a homogeneous, but disordered compact of intermetallic Nb_3Sn was formed, the latter having a lattice parameter of 5·287 Å. Some traces of the unreacted elemental powders were also found. The resistive temperature increased from 7·6 K at the lowest shock pressure to 16·3 K at higher pressures (the temperature of unshocked material is 9·2 K).

Niobium silicide (Nb_3Si) composite has also been explosively synthesised, and, in this condition, is characterised by a lower than normal transition temperature in comparison with that necessary to effect the change in a conventionally processed material.

8.3.4 Silicon Composites

Silicon carbide (SiC) and silicon nitride (Si_3N_4) are the two compounds that contribute greatly to the improvements in the design and manufacture of vital engineering parts and tools. The carbide is now widely used in metal-forming dies, cutting tools and in abrasive products. The more 'sophisticated' nitride, with its high resistance to creep and thermal shocks, combined with low thermal expansion, is a natural material for ceramic gas turbines and car engines.

Explosive compaction of SiC powder (solid density 3·2 g/cm^3) is carried out either to consolidate the material to the required density or to increase the activity of it with a view to further sintering. Shock compaction of the β-type material takes place without the addition of a binder and, depending on the initial particle size and tap density, it produces different results.

Although even the very early work (Fig. 6.1) indicated that an agglomerate of specific properties of ball-milled fine particles can be obtained, it is only more recently that the investigations of Sawaoka and his co-workers produced more detailed information.

A comparison between the response of a fine (average particle size of 0·28 μm, with 95·8% of grains under 1 μm and a few of up to 1·7 μm) and coarse (10 μm of α- and a small quantity of β-phase) SiC powders subjected to impact compactions indicated the existence of interesting differences.[43,44]

In the fine powder systems, shock pressures of between 9·4 and 29·9 GPa, associated with impact velocities of 1500–3000 m/s, resulted in a range of green density compacts. Depending on the tap density and impact velocity, variations in the densities, hardnesses and shock temperatures were observed. For example, at a pressure of 9·4 GPa a compact of 94·9% TD (70% tap density), 1570 HV, and 970 K shock temperature would be produced at an impact velocity of 1500 m/s. On the other hand, a higher pressure of 29·9 GPa at 3000 m/s, would give a compact of a lower green density of 89·4% TD, 2180 HV, and a temperature of 2280 K. The best results were obtained at a velocity of 2500 m/s, and at a pressure of 22·2 GPa, when the green density increased from 70% tap to 96·8% TD. The corresponding hardness was 2730 HV, and the temperature 1780 K. The green density increases with temperature to values in excess of 90% TD, reaching a peak at around 1800°C, and then reduces to below 90% TD. Partial melting and bonding are observed, with the shock temperature being the dominant process parameter. In general, the microstrain increases with the pressure and the crystallite size is reduced.

In the coarse powder agglomerates,[44] shock treated at 8 and 13 GPa respectively, the green density reaches some 98·6% TD (tap density of 60%) irrespective of the pressures and the corresponding impact velocities of 1600 and 2100 m/s. However, substantial differences in material response accompany the variation in pressure. At a lower pressure of 8 GPa, the distribution of microhardness across the sample shows a range of 15·7–22·5 GPa, with the corresponding range, at 13 GPa, being 21·6–28·2 GPa. The lower pressure produces temperatures which are too low to allow full interparticle bonding, the recrystallisation takes place only between the large particles in the area where sintering occurs. The lattice strain is increased and the crystallite size is reduced. The transgranular fracture of the compacts obtained at the higher pressures shows that recrystallisation extends to the interior of the large particles, and that in contradistinction to the effects of the low pressures, the lattice strain decreases and the crystallite size is actually increased. This unusual effect appears to be caused by annealing at the higher generated temperatures. Strong interparticle bonding is observed and points to the fact that coarser powders of SiC are more easily consolidated than the fine varieties. The final density can be raised to 99% TD by post-shock heating,[45] and a model of thermal deposition is also available.[46]

Silicon nitride powder consists normally of α- and β-crystalline phases and an amorphous residue. Because of its strong covalent properties, low diffusion rates and considerable intercrystal vacancies, the material is difficult to densify and sinter. This is particularly so in the absence of a liquid-phase because the transport across the grain boundary is a pre-requisite of sintering. The production of industrially acceptable silicon nitride depends on the ability to control the amount of impurities and doping agents present on the surfaces of the individual particles. These, if in excess, can easily reduce the resistance of the compound to high temperatures and the associated oxidation, as well as its anticreep and anticorrosive properties. It is for this reason that the effect of using, or not using, additives has been investigated in the last few years. A successful attempt at avoiding 'doping' was made by Hoenig and Yust[12] who explosively compacted a relatively coarse Si_3Ni_4 powder (10–300 μm) from a tap density of 71% TD at a detonation velocity of 4700 m/s and at a pressure of 4·3 GPa. Compacts of 92% TD green density were obtained with an E/P ratio of 8. The originally amorphous material did not change in its pattern, showed little residual stress, and its uniform Knoop hardness had a value of 690. On annealing in a nitrogen atmosphere for 1 h at 1373 K, the compact remained structurally amorphous, but the microhardness increased to 985 Knoop. Similar treatment at 1823 K resulted in total conversion to the α-phase, with some β-phase, but in a drastic reduction in hardness to KHN 441.

The shock treatment of a finer-grained powder carried out by Akashi *et al.*[47] on an amorphous spherical powder produced somewhat different results. Starting with a tap density of 44·6% TD, and using a flying impactor, they densified the powder at impact velocities of 1600 and 2100 m/s (70 and 100 GPa pressures). The originally amorphous powder crystallised to the β-phase with the crystallites forming hexagonal prisms 1 μm in diameter, and some 10–20 μm long.

The effect of the addition of Y_2O_3 was studied by Somiya *et al.*[48] on a 0·50 μm powder (94%-α-, 3%-β-phases) of 23·19 m^2/g specific surface area. A 5% (by weight) admixture of yttrium oxide would result, after HIP-treatment of the compact for 3 h at 1400°C and at 430 MPa, in a final density of 95·4% TD which compares favourably with 82·4% TD reached in the same conditions by the unshocked material. The corresponding figures for the composite HIP-ed at 100 MPa are 78% for the shocked and 70% TD for the unshocked powders. In this respect, Fig. 6.2, produced in the early investigations of this problem, is rather instructive.

The material defects induced by the shock are paramagnetic in nature. At liquid helium temperature, and at a frequency of 9·8 GHz, electron spin resonance shows that in the unshocked powder the narrow line-width resonance with g-factor = 2·0028 is absent, but it is present in the shocked material. The broader line-width resonance of g-factor of 2·0038 is increased twenty times at 20 GPa, and by a factor of a hundred at 27 GPa. At high pressures, the defect concentration increases to 1018/cm^3 and a significant reduction is observed at temperatures below 1200 K.[37]

8.3.5 Titanium Diboride, Carbide and Nitride

Although of the three titanium compounds it is the carbide that has been widely used in tool making — particularly as a cermet with steels — both the diboride and nitride are beginning to gain in importance; the former mainly because of its readiness to respond to shock treatment.

The recent work of Linse[49] carried out on 11 μm particle size TiB_2 powder, showed the pronounced dependence of the shock front velocity in the agglomerate on the generated pressure. The velocity increases to about 2100 m/s at 7·5 GPa, but it is then rapidly reduced.

The relationship can be described by an empirical equation of the form

$$V = 1475 + 0\cdot699 \, P_c^3 \qquad (8.8)$$

where P_c is the pressure in the container.

Structural changes take place on compaction of the powder. For instance, a 14·6 μm mean particle size powder — compacted from a tap density of 2·80 g/cm³ (62% TD) — reaches a bulk temperature of 150°C at 17 GPa, and 425°C at 27 GPa. This is accompanied by increases in the specific surface area of between 100 and 200% above that of the unshocked (0·6 m²/g) powder.[14] The residual lattice strain increases with pressure (Fig. 8.4), but even at a high pressure of 28 GPa, reaching a value of 0·2%, it does not show any sign of approaching saturation level.

Material defects, revealed by TEM and electron spin resonance investigations,[37] in a 99·3% pure TiB₂ power of 14·6 μm particle size, are those of resonance, dislocations and distortion. Whilst the original powder is characterised by two resonances — arising out of the presence of the rutile impurity — one of small isotropic absorption near a g-factor of 2·0029, and another of larger absorption (corresponding to trivalent titanium), the material shocked by pressures ranging from 17 to 27 GPa produces a third, broad resonance with a concentration of 3×10^{18}/cm³. Planar defects and high density dislocations are present in the original powder, but mainly deformed microstructures appear at a pressure of 22 GPa.

The response of titanium carbide to the passage of a shock wave is far less marked by changes in the specific surface area, but is more pronounced in the magnitude of the lattice strain. The compaction of 3·93 μm particle size TiC powder from tap densities of 59% TD (2·92 g/cm³) and 62% TD at 17 and 27 GPa shock pressures, resulted in a wide range of temperatures (160–425°C).[14] The specific surface area remains practically constant at 3·8 m²/g (Fig. 8.3) increasing from the original value of 2 m²/g. This particular modification parameter appears to be insensitive to both the tap density and the value of the resulting shock temperature. However, the large plastic deformation, associated with lower tap densities, results in higher lattice strain values.[31] Whereas, in general, the strain increases from $0\cdot7 \times 10^{-3}$ to $1\cdot4 \times 10^{-3}$ within the 16–27 GPa range (Fig. 8.4), for a constant pressure of 27 GPa, it reaches $1\cdot5 \times 10^{-3}$ for 45% TD tap density, and $1\cdot2 \times 10^{-3}$ for the higher value of 62%. The strain ratio along $\langle 111 \rangle$ and $\langle 100 \rangle$ is found to be less than unity in the shocked material, but it exceeds one in the milled powder. Within this particular range of shock pressures, the coherent domain size remains practically constant at just over 400 Å.

The presence of planar defects, dislocations and low-angle grain boundaries in the unshocked material and of heavily defective powder with piled-up dislocations in the shocked condition at 22 GPa is indicated by TEM analysis.[50]

A blend of 98·8% TiC and 1·16% TiO_2 (3·9 μm particle size) contains, in its original state, an isotropic resonance with a g-factor of 2·0022 (in the narrow line-width of 9 Oe) and a concentration of $3 \times 10^{15}/cm^3$. At 27 GPa the shocked material increases in defect concentration by one order of magnitude, but no new resonance is produced.[37]

Of direct application in tool manufacture is a mixture of TiC and steels. Compaction of this type of material has been accomplished using explosive systems. For instance, a 1:1 mixture of TiC and steel powders, 3 μm average particle size (6·55 g/cm^3 TD) was densified to 95·6% TD from a tap density of 55·8% TD. No separation of the constitutive powders was detected in small diameter samples at 80% TD, but in larger containers, the heavy component concentration was present in the cylindrical core. The separation did not involve transition into the liquid state.[19]

Similarly compacted, less well-known mixtures of titanium carbide and titanium nickelide powders (1:1) showed no separation at 58% TD green density.

Since the properties of titanium nitride lie somewhere in-between ionic and covalent, shock treatment of this material is likely to produce a swing to either group. The exploratory work of Sawaoka *et al.*[51] seems to indicate a preference for the ionic-type of material.

The densification of a 5 mm thick, 12 mm diameter TiN powder sample from a tap density of 67% TD (3·59 g/cm^3) to 92% TD at an impact velocity of 2500 m/s and an estimated pressure of 60 GPa, produced compacts of HV700. The broadening of the half-maximum intensity was substantial (some 50% increase), but with no change in crystallite size and an increase in microstrain, the ionic properties appear to be predominant.

REFERENCES

1. BATSANOV, S. S. In: *Proc. 6th HERF Conf.*, Haus der Technik, Essen, FRG, Paper 5.5, 1977.

2. Graham, R. A., Morosin, B., Venturini, E. L. and Carr, M. J., *Ann. Rev. Mater. Sci.* (1985).

3. Wilkins, M. L. In: *Impact Loading and Dynamic Behaviour of Materials*, Eds C. Y. Chiem, H.-D. Kunze and L. W. Meyer, DGM Informationgesselschaft Verlag, Oberursel, FRG, 1988, p. 965.

4. Graham, R. A. and Webb, D. M. In: *Shock Waves in Condensed Matter*, Eds J. R. Asay, R. A. Graham and G. K. Straub, North Holland, Amsterdam, 1984, p. 211.

5. Raybould, D. and Blazynski, T. Z. In: *Materials at High Strain Rates*, Ed. T. Z. Blazynski, Elsevier Applied Science Publishers, London, New York, 1987, p. 71.

6. Mathia, T. and Louis, F. In: *Tribology in Particulate Technology*, Eds B. J. Briscoe and M. J. Adams, Adam Hilger, Bristol, Philadelphia, 1987, p. 273.

7. Bergmann, O. R. and Barrington, J. *J. Am. Ceram. Soc.*, **49** (1966), 502.

8. Pruemmer, R. A. and Ziegler, G. In: *Proc. 5th HERF Conf.*, University of Denver, Colorado, Paper 2.4, 1975.

9. Pruemmer, R. A. and Ziegler, G. *Powder Met. Int.*, (1) (1971).

10. Hoenig, C., Holt, A., Finger, M. and Kuhl, W. In Ref. 1, Paper 6.3.

11. SRI Report No. DNA 3412 F, 1974.

12. Hoenig C. L. and Yust, C. S. *Ceram. Bull.*, **60** (1981), 1175.

13. Beauchamp, E. K., Carr, M. J. and Graham, R. A. *J. Am. Ceram. Soc.*, **68** (1985), 696.

14. Lee, Y. K., Williams, F. L., Graham, R. A. and Morosin, B. *J. Mater. Sci.*, **20** (1985), 2488.

15. Williams, F. L., Morosin, B. and Graham, R. A. In: *Metallurgical Applications of Shock-Wave and High-Strain-Rate Phenomena*, Eds L. E. Murr, K. P. Staudhammer and M. M. Meyers, Marcel Dekker, New York, Basel, 1986, p. 1013.

16. Morosin, B. and Graham, R. A. *Mater. Sci. Engng*, **66** (1984), 73.

17. Akashi, T. and Sawaoka, A. *Mater. Letters*, **3** (1984), 11.

18. Roman, O. V. and Gorobtsov, V. G. In: *Shock Waves for Industrial Applications*, Ed. L. E. Murr, Noyes Publishers, Park Ridge, NJ, 1988, p. 335.

19. Kostyukov, N. A. In: *Proc. 10th HERF Conf.*, Litostroj, Ljubljana, Yugoslavia, 1989, p. 206.

20. Murat, B., Inal, O. T. and Hellman, J. R. *J. Am. Ceram. Soc.*, **73** (1990), 346.

21. Carlson, R. J., Porembka, S. W. and Simons, C. C., *Ceram. Bull.*, **44** (1965), 266.

22. Yaziv, D., Bless, S. J. and Daudekar, D. P. In Ref. 15, p. 793.

23. Marsh, S. P. *High Temp. High Press.*, **5** (1973), 503.

24. Graham, R. A., Morosin, B., Horie, Y., Venturini, E. L., Boslough, M., Carr, M. J. and Williamson, D. L. In: *Shock Waves in Condensed Matter*, Ed. Y. M. Gupta, Plenum Press, New York, 1986.

25. Hankey, D. L., Graham, R. A., Hammeter, W. F. and Morosin, B. *J. Mater. Sci. Lett.*, **1** (1982), 445.

26. Hammeter, W. F., Hankey, D. L. and Dosch, R. G. *J. Am. Ceram. Soc.*, **68** (1985).

27. Gibbons, R. *Bull. Am. Phys. Soc.*, **17** (1973), 1106.

28. ADADUROV, G. A., BREUSOV, O. N., DREMIN, A. N. and DROBYSHEV, V. N. *Russian J. Inorganic Chem.*, **16** (1971), 1073.
29. BATSANOV, S. S. In: *Explosive Welding, Forming, Plugging and Compaction*, Eds I. Berman and J. W. Schroeder, ASME, New York, 1980, p. 59.
30. STAVER, A. M. In Ref. 8, Paper 2.1.
31. MOROSIN, B. and GRAHAM, R. A. In Ref. 15, p. 1037.
32. VENTURINI, E. L., MOROSIN, B. and GRAHAM, R. A. *Mater. Lett.*, **3** (1985), 349.
33. VENTURINI, E. L., MOROSIN, B. and GRAHAM, R. A. *J. Appl. Phys.*, **57** (1985), 3814.
34. PRUEMMER, R. A. In: *Explosive Welding, Forming and Compaction*, Ed. T. Z. Blazynski, Applied Science Publishers, London, New York, 1983, p. 369.
35. GOURDIN, W. H., ECHER, C. J., CLINE, C. F. and TANNER, L. E. In: *Proc. 7th HERF Conf.*, Ed. T. Z. Blazynski, University of Leeds, 1983, p. 233.
36. GOURDIN, W. H. and WEINLAND, S. L. *J. Am. Ceram. Soc.*, **68** (1985), 674.
37. JACK, K. H. In: *Nitrogen Ceramics*, NATO Seminar, University of Kent, Ed. F. C. Riley, Noordhoff, Leyden, 1977, p. 257.
38. WILKINS, M. L., KUSUBOV, A. S. and CLINE, C. F. In Ref. 15, p. 57.
39. AKASHI, T. and SAWAOKA, A. B. In Ref. 24.
40. MEYERS, M. A., THADDANI, N. N. and LI-HSING, YU In Ref. 18, p. 265.
41. OTTO, G., REECE, O. Y. and ROY, U. *Appl. Phys. Lett.*, **18** (1971), 418.
42. HUGHES, D. D. and LINSE, VONNE D. *J. Appl. Phys.*, **50** (1979), 3500.
43. KONDO, K. I., SOGA, S., SAWAOKA, A. and ARAKI, M. *J. Mater. Sci.*, **20** (1985), 1033.
44. AKASHI, T., LOTRICH, V., SAWAOKA, A. and BEAUCHAMP, E. K. *J. Am. Ceram. Soc.*, **68** (1985), C-322.
45. SAWAOKA, A. In Ref. 18, p. 380.
46. LOTRICH, V. F., AKASHI, T. and SAWAOKA, A. In Ref. 15, p. 277.
47. AKASHI, T., SAWAOKA, A. and BEAUCHAMP, E. K. In Ref. 24.
48. SOMIYA, S., YOSHIMURE, M., FUJIWARA, S., KONDO, K., SAWAOKA, A., HATTORI, T. and MOHRI, J. *Comm. Am. Ceram. Soc.*, (1984), C-51.
49. LINSE, VONNE D. In Ref. 15, p. 29.
50. KURODA, D. and HEUER, A. H. In: Second Report of the DARRA Dynamic Synthesis and Consolidation Process, Lawrence Livermore Laboratory, Rep. UCID-19 663-83-1.
51. SAWAOKA, A., SOGA, S. and KONDO, K. *J. Mater. Sci. Lett.*, **1** (1982), 347.

Chapter 9

Shock Consolidation of Polymeric Powders

9.1 INTRODUCTION

Although structural synthesis of polymeric powders can and does take place when suitable shock conditions are established, the main interest in the use of this kind of treatment lies in the possibility of the bulk manufacture of new materials. Densification of either single polymers, or of their mixtures and blends has increased in its importance with the volume and range of specialised polymeric–matrix composites required for a variety of industrial applications.

Unlike compacts of pure ceramics or cermets, plastics, whether moulded or compacted, have had relatively little utilisation as machine or plant components. Because of their generally low thermal conductivity, ranging from 0·165 to 0·340 W m/m²°C, and low resistivity of the order of 10^{10}–10^{16} Ω cm, they have traditionally been regarded as thermal and electrical insulators rather than potential, industrially viable materials. In this latter role, their utilisation suffers because of, for instance, overheating of parts such as plastic housings and similar components, or because of the degradation of properties that occur in moulding and extrusion and are caused, again, by excessive local heating.

These limitations have been recognised and there have been attempts to widen the scope of materials like polyesters, polycarbonates, polyethylenes and polypropylenes by blending them with other non-metallic or metallic materials either by purely mechanical means or by direct incorporation into the polymer chain. The shock processing of particulate either 'pure' polymeric matter or of a blend leads not only to a more 'homogenised' material being produced, but also to a relatively strong one since the interparticle bonding is much easier to attain here than in either purely metallic or ceramic agglomerates.

An attraction, additional to consolidation, is the low adiabatic heat generated which is of particular interest in the forming of a number of thermoplastics. The existing evidence suggests that thermal degradation, often associated with conventional moulding and extrusion processes, is absent. Additionally, polymeric materials, obtainable generally only in powder form and susceptible to temperature effects, viz. polypropylene, can be easily consolidated and therefore the utilisation of often cheap but otherwise difficult to produce in bulk materials may be possible on a commercial scale.

The needs of the rapidly evolving electronics industry include the manufacture of metalloplastics, which possess electromagnetic shielding properties, are capable of either discharging static electricity or converting mechanical to electrical signals, and can possibly be used in electrical heating installations. Successful, conventional production techniques involving, for example, PMMA/copper composites,[1] polyester resin filled with aluminium powder,[2] and polypropylene with high purity aluminium fibres[3] have been developed. This type of approach to the manufacture of sophisticated materials is very conveniently extended to dynamic processing, which, as already observed, will be characterised by strong bonding properties.

A novel application of shock pressures is that of pulses of a few microseconds duration of PVF_2 material. These induce ferroelectric properties, including polarisation of the material and also of VF_2/C_2F_3H copolymer. Consequently, a new range of possibilities is thus created.[4]

Conventional P/M processing of polymeric particulate matter is still making use, where appropriate or indeed possible, of sintering in order to increase structural strength. The hot forging of porous sintered preforms of ultra-high molecular weight polyethylene powder is a typical example of this particulate route,[5] although a more refined method of ultrasonic moulding has also been used. With the constantly increasing interest in direct semi-fabricate processing of polymeric materials, e.g. cold compaction of powders in tapered dies,[6] or tube drawing (from solid) to produce oriented material,[7] attention is more and more focused on the production of materials that are mechanically sound, and quickly obtainable in reasonable quantities. In this respect, the wear resistant, but low friction blends, and at the other extreme, abrasive materials, are particularly important. Since, however, in all of these applications the properties of the starting materials decide the outcome of the consolidating operation, the response of the 'elemental' or simple polymers forms a good guide to the shock reactivity of mixtures formed from them. Consequently, as in the case of metals (Chapter 7), a brief review of the results of dynamic consolidation of some basic polymeric powders precedes the main discussion.

9.2 DENSIFICATION OF SINGLE POLYMERS

The two techniques employed here are those of indirect (gas-gun or explosively activated sub-press) or direct, usually cylindrical, implosive systems. The generation of a conical-shaped front is of even greater importance here than in the case of metallic or ceramic powders because overcompaction and the consequent core burning or melting are common occurrences (Fig. 9.1).

Since the shock processing of polymeric powders is a relatively recent development, the volume of the available literature is still limited. The experimental work of Blazynski *et al.* has been confined mainly to either a selection of PVC materials (in view of the existence of comparable information about their response to conventional treatment) or a feasibility study of a few other polymers.

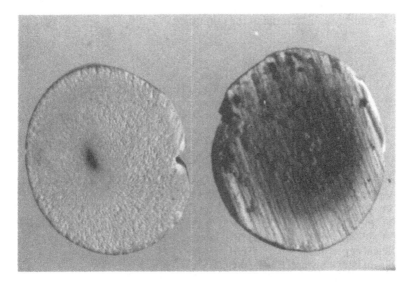

FIG. 9.1   An example of overcompaction of a polymer powder compact: (left) core burning; (right) core melting.

In the detailed investigations, three ICI, Corvic-type homopolymers and the ICI Welvic-type copolymer were used. Two suspension homopolymers (S57/116 and S67/111), and one paste-making homopolymer (P67/579) were employed. The copolymer PC7/314 was included to represent the composition type of a PVC material. The relevant pre- and post-compacting physical data are given in Table 9.1. These values refer to mixtures of average particle size, but in each case the actual size band extends from -150 to $+212$ μm. On average, an 's'-type polymer mixture contains some $40\% + 212$, $35\% - 212$ and the remainder of -150 μm particles.

Additionally, a diluent homopolymer and a diluent copolymer powders, together with polypropylene granulated material were used (Table 9.1).

The direct densification used initially, produced solid compacts 60 mm long and 25 mm in diameter, whereas the indirect technique resulted in smaller 20 mm long and 15 mm diameter specimens.

In all the materials concerned, the success of the compacting operation depends on a number of basic parameters. These are:

- initial particle size,
- tap density,
- level of energy,
- geometry and material of the container.

TABLE 9.1
Shock Compaction of Polymer Powders and Granules

Material	Average initial particle size (μm)	Density		Temperature (°C)	
		Tap (g/cm³)	Max. final (%)	Glass transition	Decomposition
PVC P67/579 (P)	75	0·53	94	82	242
PVC S57/116 (P)	160	0·66	97	82	237
PVC PC7/314 (P)	140	0·60	93	82	247
PVC S67/111 (P)	170	0·60	95	87	287
Diluent copolymer (P)	166	0·49	85		
Diluent homopolymer (P)	200	0·40	85		
Polypropylene (G)	3460	0·29	70		

P, powder; G, granule.

The information provided in the following section is based on Refs 8–16.

9.2.1 Direct Compaction

9.2.1.1 Morphology

A set of representative micrographs showing the pre- and post-compacting surface morphologies of the PVC and diluent copolymer are given in Fig. 9.2. A high degree of densification, low porosity, and considerable plastic deformation are evident in each case. The micrographs refer to optimal shock pressure values of about 11 GPa.

The effect of the initial particle size was described, with reference to Fig. 6.9, in Section 6.1.2 on the example of the copolymer PC7/314. A general conclusion that can be drawn is that the degree of cohesion is larger in the small particle range than in the band $> 212\ \mu$m where some porosity is discernible. An increase in the particle size results in a reduction in impact velocities — for the same level of energy supplied — and consequently it produces both the lower surface temperatures and much lower plastic grain deformations. The 'as supplied' mixture, by providing a wide range of impact conditions, offers a good chance of satisfactory bonding. This, in turn, influences the green density and compressive strength to fracture of the compact. Figure 9.3 quantifies the effects discussed by indicating that the levels of both these parameters are the highest for the mixture and the lowest for the large-particle bond. Both these curves show the same pattern of initial increase in the value of the parameter considered to a maximum, followed by a rapid reduction. The characteristic three compacting zones are present and represent the same structural changes in the polymeric aggregate as those discussed earlier in connection with the metallic and ceramic compacts.

The level of energy dissipated with the consequent value of the shock pressure, and the profile of the shock front have an additional effect on the properties of the compact. This is demonstrated in Fig. 9.4, with reference to the same PC7/314 copolymer. The fracture surface of a compact obtained at a low level of impact energy of about 30 J/cm^3 is of a brittle nature, indicating a degree of undercompaction (Fig. 9.4(a)). A higher energy level of about 60 J/cm^3 results in a ductile fracture surface (Fig. 9.4(b)), but an excess of energy (> 80 J/cm^3) affects the basic structure of the material revealing stretched polymer chains (Fig. 9.4(c)).

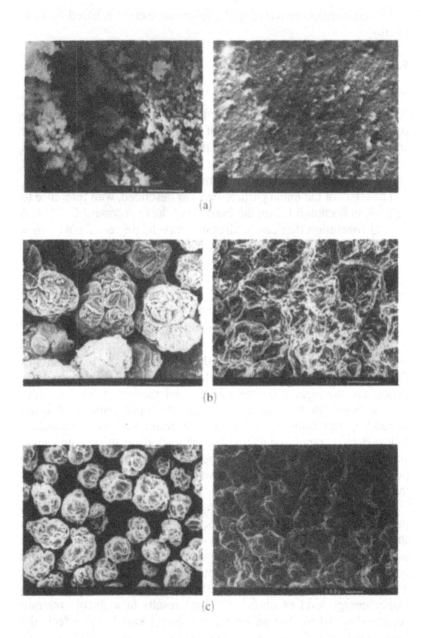

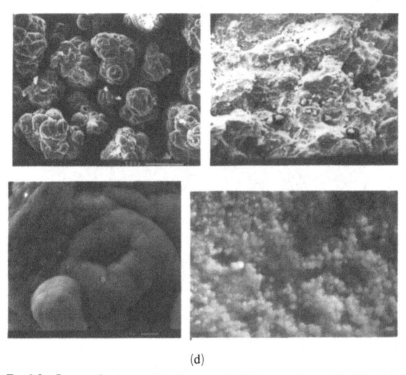

(d)

FIG. 9.2 Pre- and post-compacting morphologies at about 11 GPa. (a) P67/579; (b) S57/116; (c) PC7/314; (d) S67/111.

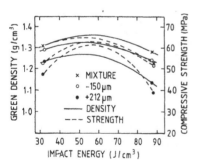

FIG. 9.3 The effect of particle size on the green density and compressive strength of a S57/116 compact.

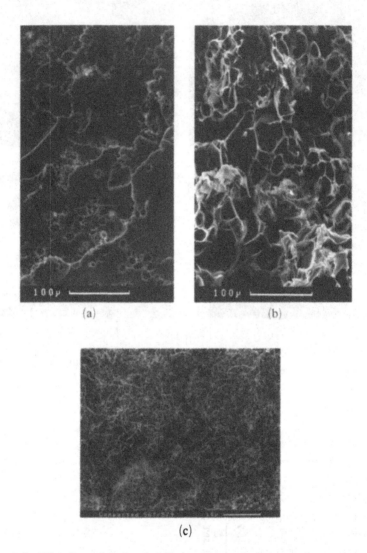

Fɪɢ. 9.4   The effect of the level of dissipated impact energy on the structure of
a PC7/314 copolymer compact. (a) Brittle fracture surface — 30 J/cm³;
(b) ductile fracture surface — 60 J/cm³; (c) stretching of polymeric chains above
80 J/cm³.

9.2.1.2 Mechanical properties

As long as sufficient interparticle 'stand-off' distances are maintained, the tap density should be as high as possible. The pre-shock consolidation of the powder results in considerable saving on energy that can be more usefully utilised in producing intergranular bonding and substantial reduction in porosity. Again, the corresponding improvements in the green density and compressive strength will follow. This supposition is borne out by Fig. 9.5 which shows that irrespective of the actual grain size, the effect of apparent, and by implication of the initial tap, density is to increase the levels of the final green density and of the strength of the compact, providing that the degree of the original powder density is sufficiently high. The higher the initial, apparent density the higher the level of the final green density, although, again, the actual value depends on the impact energy. Similarly, the compressive strength to fracture increases with an increase in the apparent density, but varies with the energy dissipated. The shapes of the respective curves follow the same pattern as that of Fig. 9.4 and thus reflect the influence of the three zones of shock compaction.

The general effect of the stiffness of the container, whether cylindrical or planar, was discussed in detail in Chapter 6. The conclusions reached there apply fully to the compaction of polymeric particulate matter.

The determination of the optimal E/P or E/M values can be based on the already discussed square of the detonation velocity, or on the impact pressure (IP) parameters. The relationship between the E/P ratio and the square of the velocity on the one hand, and impact energy, on the other, is shown in Fig. 9.6. The figure also indicates the corresponding dependence of the green compact density on these quan-

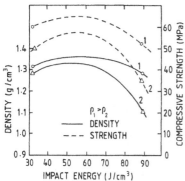

FIG. 9.5 The effect of tap density on the green density and compressive strength of a S57/116 compact.

tities. It is seen that, for a given material, there exists a definite value of E/P which, in conjunction with a specific impact energy (or square of the velocity), gives higher values of density and strength. For the S57/116 material, an explosive with 2900 m/s detonation velocity and 1·62 E/P ratio will produce satisfactory compacts. The corresponding impact energy required amounts to 56 J/cm^3. Similar relationships must be established for each individual material and used in production.

The degree of densification of finer-grained powders is high with a green density of over 90% TD (average theoretical density is about 1·4 g/cm^3), but is reduced for coarse or granulated materials (Table 9.1). A comparison between the implosively compacted, solid cylindrical samples (7–13 GPa) and conventionally densified specimens (350 MPa) is provided by Fig. 9.7. This shows that although the final density is not

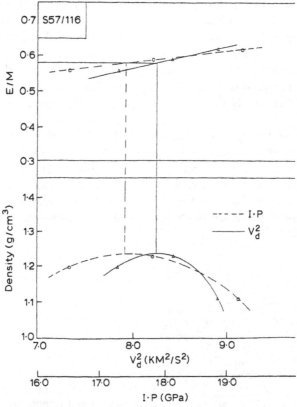

Fig. 9.6   The relationships between E/P (E/M) ratio and green density, the shock parameters of detonation velocity V_D, and impact pressure (IP).

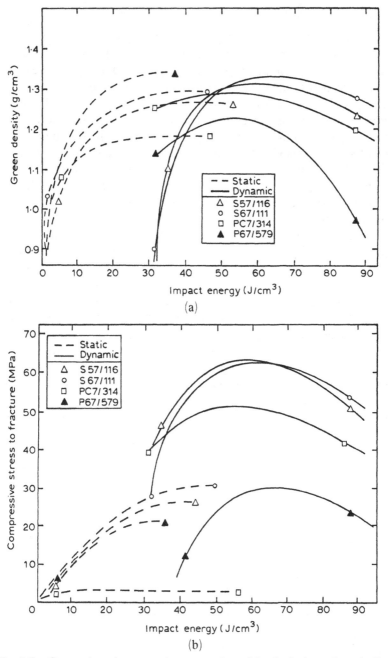

FIG. 9.7 Comparison between the properties of implosively and statically compacted PVC powder compacts. (a) Green density as a function of impact energy; (b) compressive stress to fracture as a function of impact energy. Pressure ranges: 7–13 GPa, 350 MPa.

affected by the method of compaction (Fig. 9.7(a)), the pattern of the green density/energy level relation is very different in each case. In static compaction the density increases initially with energy, but reaches a plateau appropriate to the material. In explosive compaction, up to an energy level of some 40 J/cm^3 only a rearrangement of particles takes place — as in a static process — with a minimum of plastic deformation and lattice distortion. In the 45–65 J/cm^3 range optimal conditions are reached, with a high level of plastic deformation and particle comminution present. Above this energy level, however, excessive fragmentation and damage, caused by the release wave, reduce the maximum degree of density attained earlier. Degradation of properties begins to take place rapidly.

The strength of the compact, as represented, for example, by compressive stress to fracture (Fig. 9.7(b)), shows the same pattern of behaviour, but static and dynamic conditions produce substantially different levels of strength. In the static process, because of the lack of adequate bonding and the consequent relaxation, the strength can be very low, viz. PC7/314 material, but with interparticle bonding present in the explosive operation, high levels of stress are attained, even in the case of this copolymer. Generally, in optimal conditions, the strength of the respective polymer compacts increases by 150–200% when explosively consolidated. In terms of the actual numerical values, the compressive strength to fracture of the explosively consolidated PVC powders ranges from some 30 to 64 MPa, whereas the corresponding static values remain in the 20 MPa range (Table 9.2).

Since the deformation of plastics is dependent upon time and temperature, the post-shock creep properties of the compacts are of special

TABLE 9.2
Post-Shock Properties

Material	Theoretical density (%)			Maximum compressive strength (MPa)	
	Tap	Maximum final static	Explosive	Static at 380 MPa	Explosive
P67/579	37·8	95·0	92·0	29·5	30·9
S57/116	47·1	97·0	98·5	23·3	64·2
PC7/314	42·8	90·5	96·2	0·5	52·0
S67/111	42·8	92·0	98·5	28·0	63·0

interest. An indication of this is provided by Fig. 9.8 which shows that under a load corresponding to a stress of 20 MPa, the elastic strain increases with the temperature, but the slopes of the strain/time curves do not differ much. All experiments were carried out at temperatures lower than the glass transition range (Table 9.1). Very low strain levels of no more than 2% are reached at temperatures of up to 55°C, but a significant change takes place at temperatures exceeding 70°C, with the strain increasing to between 15 and 20%. Conversely,[12] at a constant temperature of 20°C, the creep strain increases with stress. A constant 2% strain is maintained for 24 h at 20 MPa, an almost constant strain of about 7% at 40 MPa, but a strain of 12% at 50 MPa results in fracture after about 4 h.

9.2.2 Indirect Compaction

The effectiveness of this method depends very much on the generated impact velocities of the press rams or gas-gun flyer-plates. The experimentation carried out on P67/579 powder, compacted in an explosively activated press, shows the difference in material response associated with the variation in impact velocities. For instance, at 150 m/s (400 J/cm^3) a compact of 92·3% TD is obtained, whereas the density increases to 95% TD at 165 m/s (420 J/cm^3). The corresponding compressive strengths are, however, only 9·7 and 11·8 MPa, and are very much below the values attained by shock-consolidated samples.

Although interparticle boundaries are not detectable on fracture surfaces (Fig. 9.9), the level of bonding, if any, is clearly very low.

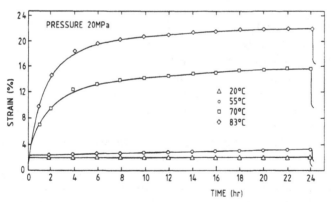

Fig. 9.8 Development of creep strain at a stress of 20 MPa for shock densified S57/116 compacts.

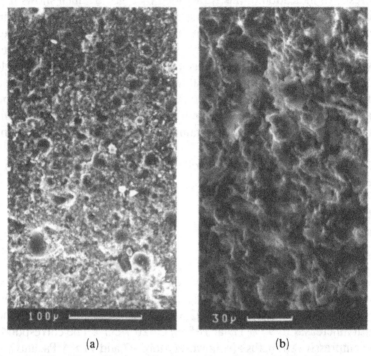

(a)   (b)

FIG. 9.9   The effect of impact velocity on the structure of indirectly compacted
P67/579 PVC powder. (a) $V = 150$ m/s; (b) $V = 165$ m/s.

9.3 POLYMERIC BLENDS

The structural and mechanical properties of compacts can be considerably enhanced by the blending of two, or possibly more, polymeric
powders of different characteristics. As long as sufficiently strong
interparticle bonds are obtained, a highly 'homogenised' mixture is
produced in either cylindrical or planar geometries.

Direct implosive compactions of the single polymers, discussed in
Section 9.2, were successfully performed using different volume
fractions of these materials. The relevant data about post-shock
morphologies and the physical and mechanical properties of the
composites were obtained and were reported.[8,11,12,15] In particular,
mixtures of S67/111 and PC7/314 with diluent copolymer, those of
S57/116 with S67/111 and P67/579, and of S67/111 with P67/579
were consolidated.

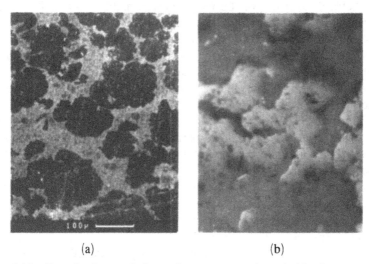

(a)   (b)

FIG. 9.10   Post-shock morphology of one-to-one polymeric blend compacts.
(a) Mixture of S57/116 (dark) and P67/579 PVC powders; (b) mixture of
S57/116 (dark) and S67/111 PVC powders (× 100).

The structural characteristics of two composite materials are indicated in Fig. 9.10. Figure 9.10(a) refers to a one-to-one blend of S57/
116 and P67/579 PVC powders. A considerable reduction in the
particle size of both materials, in comparison with the original sizes
(Table 9.1), and the change in shape of both powders are evident. In
their original state, the particles of P67/579 resembled quite closely the
morphology of the uncompacted S57/116 showing a number of small,
randomly grouped nodules on the surfaces. Substantial integration of
the latter material has taken place, partly as the result of plastic
deformation and partly as a consequence of the comminution of individual particles. The S57/116 material appears to be less affected and
maintains fairly uniform dispersion in the mixture. Interparticle bonding
is achieved and a voidless structure obtained. The individual grains of
S57/116 have been reduced, on average, to about 100 μm at a pressure
of 6 GPa, and those of P67/579 to about 55 μm.

Undercompaction of this particular mixture is associated with shock
pressures of under 4 GPa. Sound compacts are obtained at pressures of
about 5–6 GPa.

A fracture surface of the S57/116–S67/111 mixture (Fig. 9.10(b))
reveals a voidless structure of well mixed and bonded powders. The

geometrical similarity of the two materials makes it difficult to distinguish between them on a b/w photograph, but the actual SEM examination confirms this statement. Although no fragmentation is observed at a pressure of about 6·5 GPa, the plastic straining has reduced individual grains of both materials to about 120 μm.

Although in both cases there is a suggestion of the existence of wavy interfaces, the actual mechanism of interparticle bonding is not obvious. Fusion bonding was observed in the earlier experimentation of Blazynski *et al.*[8,11] between the particles of single polymers, and, no doubt, excess of energy dissipated locally may well produce the same result in the mixture.

In general, the post-compacting density is effected by the volume fraction of the constituent materials and so is the compressive stress to fracture. The nature of these relationships depends, however, on the initial properties of the agglomerate.

Mixtures of the P67/579 material with either of the other two S-type powders are characterised by a decrease in both the density and strength with increasing volume fraction of P67/579. The reason for this is likely to be the smaller particle size and lower strength of the latter. The rates of energy absorption and of pressure attenuation increase directly with a decrease in the density of the powder.

With all three powders having the same theoretical density, it is observed that the densities of the respective mixtures reduce from 97 to 88% and 98·5 to 91·4% TD for S57/116–P67/579 (Fig. 9.11(a)), and S67/111–P67/579 (Fig. 9.11(b)) compacts respectively within the 3:1–1:3 volume fraction range. The strengths of these compacts reduce correspondingly and equally from 65 to 35 MPa. The closeness of these figures is linked with the basic physical similarity of the two S-type powders. The properties of these two powders (Fig. 9.11(c)) — with S67/111 showing lower tap density and slightly larger average particle size — are considerably affected by the composition. The addition of S67/111 results in a decrease in the density from 98·5% TD for a 3:1 ratio of S57/116–S67/111 mixture to 89% TD for a ratio of 1:1, but a further increase in the volume fraction of this material restores the 3:1 properties at the reverse ratio of 1:3. Similarly, the compressive stress to fracture reduces from 65 to 35 MPa and increases again to 65 MPa. This pattern of property distributions is probably due to the fact that the slight size advantage of the S67/111 material is offset by the lower initial density. It reasserts itself, however, as the proportion of the powder is increased. The results obtained when compacting the two

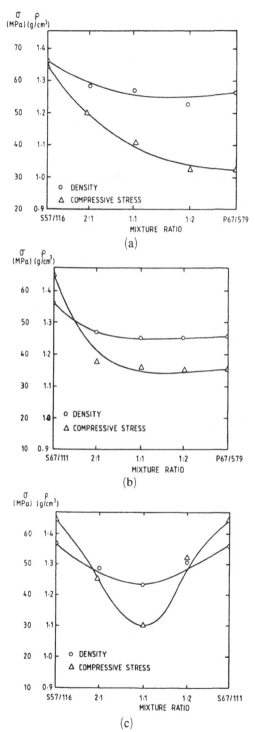

Fig. 9.11 The effect of volume fraction on the levels of density and compressive strength to fracture of PVC polymeric, implosively consolidated composites obtained at pressures of between 5 and 6 GPa. (a) S57/116–P67/579 mixture; (b) S67/116–P67/579 mixture; (c) S57/116–S67/111 mixture.

materials singly confirm this supposition since in that case, for a given consolidating pressure, the green density and the strength of S67/111 tend to be slightly lower than those of S57/116 powder.

The wide variety of distribution patterns provides clear evidence of both the versatility of the implosive technique and of the range of attainable physical and mechanical proeprties of purely polymeric blends.

9.4 POLYMER–CERAMIC COMPOSITES

9.4.1 PVC–Alumina

A good insight into the polymer matrix–alumina filler composite properties is afforded by the recent work of Muhanna and Blazynski.[16-19] The S57/116 PVC polymer powder already discussed was used as the matrix and three particle size alumina powders were admixed mechanically in a range of volume fractions. The volume fraction, f, used in the following discussion refers always to the filler. The pre-compaction properties of these materials are listed in Table 9.3. The densification of the composites was carried out in an implosive, cylindrical system using ICI Trimonite explosive. Impact pressures generated were in the 12–22 GPa range.

Optical microscope and SEM fractographs for compacts containing 53, 150 and 300 μm alumina fillers at $f = 0.25$, are shown in Fig. 9.12. These serve as examples of satisfactory production of composites. The optical micrograph (Fig. 9.12(a)) indicates good mixing and blending of the matrix and the filler and, consequently, a satisfactory degree of homogenisation of the structure; this in spite of some porosity (dark

TABLE 9.3
Pre-compaction Properties of Alumina/PVC Mixtures

Particle size	53 μm			150 μm			300 μm			
Filler/matrix	0·25	0·54	1·00	0·25	0·54	1·00	0·33	1·00	1·85	3·00
Apparent density (g/cm³)	0·71	0·93	1·15	0·78	0·92	1·08	0·86	1·00	1·44	1·62
Tap density (g/cm³)	1·00	1·30	1·47	0·98	1·24	1·45	1·04	1·42	1·68	1·94
Theoretical density (g/cm³)	1·95	2·42	2·65	2·01	2·56	2·77	2·12	2·80	3·16	3·30

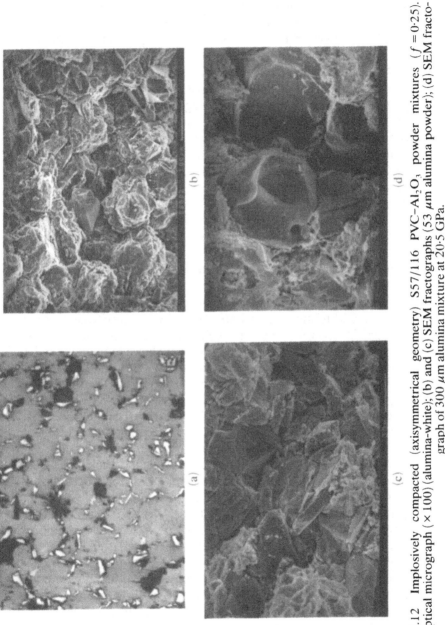

FIG. 9.12 Implosively compacted (axisymmetrical geometry) S57/116 PVC–Al$_2$O$_3$ powder mixtures ($f = 0.25$). (a) Optical micrograph ($\times 100$) (alumina-white); (b) and (c) SEM fractographs (53 μm alumina powder); (d) SEM fractograph of 300 μm alumina mixture at 20·5 GPa.

areas). Interparticle meshing, combined with plastic deformation, is likely to result in bonding. The evidence of this is provided by Figs 9.12(b, c), which refer to the specimen of Fig. 9.12(a), and by the density, strength and hardness distribution of Fig. 9.13.

At the other extreme of the investigated filler particle size (300 μm) (Fig. 9.12(d)), the situation is different. The disparity between the particles of the matrix and filler is too high to be fully accommodated at the lower levels of pressure (20·5 GPa) and therefore the porosity of the compact is higher.

In common with other dynamically compacted particulate matter, the PVC/alumina mixtures respond to the passage of a shock wave in accordance with their relative initial density — defined to a significant degree by the relative particle sizes — volume ratio, and the level of energy delivered.

The effect of these parameters, singly or jointly, depends ultimately on the impact pressure available (or energy dissipated) and consequently variations of these parameters with the pressure show some basic characteristics.

The optimal compacting conditions can be gauged again from an E/M or E/P relationship and are shown, for the specific example of $f = 0.25$, 53 μm alumina mixture, in Fig. 9.14. The effect of particle size of the filler, for the same volume fraction, is shown in Fig. 9.13 and is compared with the properties of the matrix.

The presence of a specific volume of the filler results in an increase in compact density of between 6 and 12% (peak) in comparison with that of the pure polymer (Fig. 9.13(a)). The respective peaks of the curves are displaced with respect to that of the polymer, thus indicating that a higher level of pressure is required for small alumina compounds. The highest green density is of the order of 95%. On the whole, for a given impact pressure, the degree of cohesion in the small particle range is higher than for the larger sizes. This is because an increase in the particle size causes a reduction in impact velocities and, therefore, lowers surface temperatures and produces smaller surface deformations. In consequence, the compact density is reduced.

As regards the strength (Fig. 9.13(b)), the pattern of stress dependence on pressure is, naturally, the same as that of density, with the possible exception of the 300 μm material. The slight difference in the volume ratios is insignificant here. The maximum strength of this particulate composite is the same as that of pure S57/116, but it is reached at a lower pressure. The highest attainable compressive stress of 45 MPa occurs for 53 μm filler particle size compound at a pressure of 19·5 GPa.

Rockwell hardness of the low and medium sized alumina mixtures (Fig. 9.13(c)) shows the same distribution tendencies as those of the compressive strength. The influence of the filler-matrix ratio on the properties of the composite is demonstrated in the example provided by Fig. 9.14(d). This refers to 53 μm alumina powder, or to the optimal mixture composition. The relative density and compressive strength to fracture reduce with an increase in the polymer content for a given operational impact pressure. This is associated with the comminution, degradation and/or overcompaction of the matrix. The bending strength of the composite (Fig. 9.13(e)) increases initially with decreasing alumina content, reaches its maximum at $f = 0·53$, and then falls rapidly. This pattern results from the lack of tensile strength — brought into play by the bending test — of the large volume of unreinforced matrix. Although the actual values of these parameters clearly depend on the particle size, the basic trends in their variations with the impact pressure, are similar to those displayed by the mixture containing 53 μm alumina material. It is, however, evident that the optimal compact properties are obtained with the small-grained 53 μm filler.

Wear and frictional properties of the compacts containing 53 μm filler were assessed by tests carried out on an Amsler machine. EN31 steel discs of 60 C Rockwell hardness (0·05 and 0·18 μm roughness) were used at a constant load of 15 N and a sliding velocity of 0·24 m/s. The load was applied in dry conditions over a sliding distance of approximately 104 km. The specimen pins were machined out of the compacted material to a diameter of 12·5 mm with the flat pin ends reduced to 8 mm, and polished to a surface finish of 3 μm.

Wear tests produced surfaces of the type shown in Fig. 9.15(a, b), but, with surface melting of the polymer occurring as a consequence of the rise in frictional temperature, certain areas of the optical micrographs were affected.

As already observed, an increase in alumina content to, say, $f = 0·54$, results in a lower level of adhesion (Fig. 9.15(c)) and consequently in a less satisfactory set of properties. Again, wear tests, as exemplified by Fig. 9.15(b) produce some melting of the matrix.

Both the wear rate (Fig. 9.16(a)) and the coefficient of friction (Fig. 9.16(b)) as functions of the volume ratio, depend also on the finish of the disc and their respective levels are reduced when better finish of disc surfaces is achieved. However, the main observation is that of

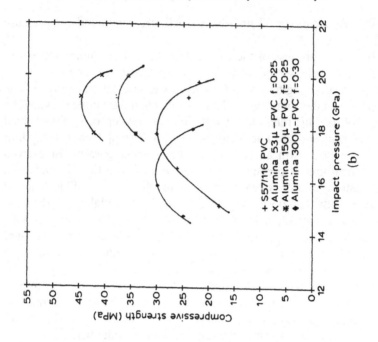

(b)

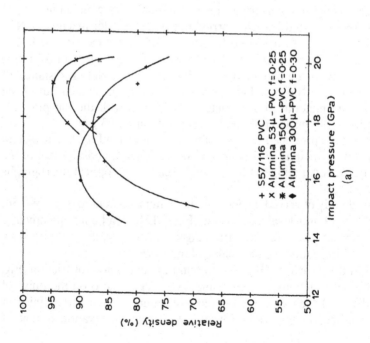

(a)

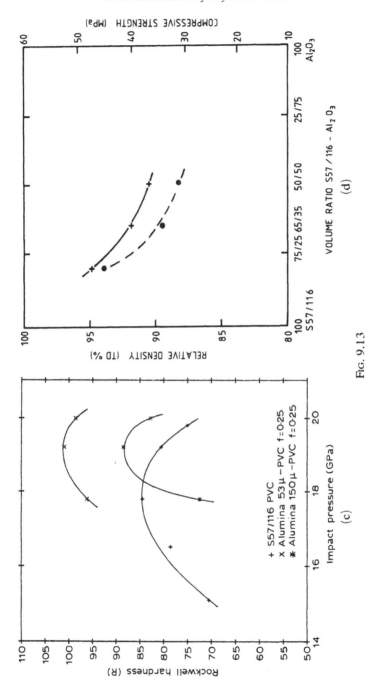

Fig. 9.13

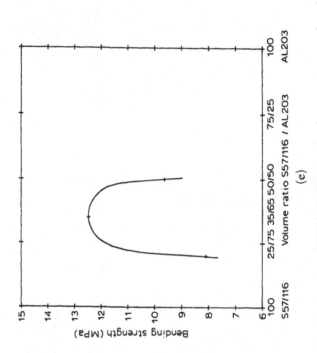

FIG. 9.13   Physical and mechanical properties of S57/116–53 μm alumina compacts as functions of filler particle size, impact pressure, and volume ratio. (a) Variation of density with impact pressure; (b) variation of compressive stress to fracture with impact pressure; (c) distribution of Rockwell hardness values; (d) effect of volume fraction and compressive strength; (e) effect of volume fraction on bending stress.

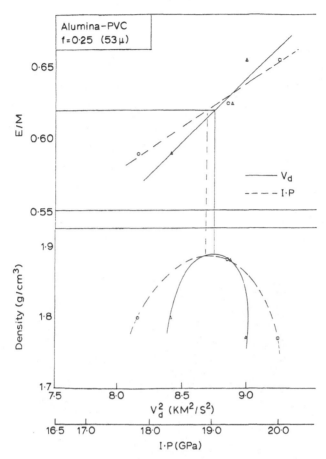

FIG. 9.14   Determination of the optimal conditions for the compaction of 53 μm, $f = 0.25$, alumina mixture.

increasing rate of wear, and of the coefficient of friction with increasing alumina content. This is associated not only with the natural abrasiveness of alumina, but also with the fast generation of debris which changes the distribution of the now variable size particles of the filler. This particular pattern of behaviour is directly related to the degree of interparticle bonding existing between the two materials and therefore to the degree of densification of the original compact. When the matrix content predominates, the higher degree of coalescence present reduces the dislodgement of the particles from the pin surfaces. The converse of this obtains when the ratio of the polymer is significantly reduced.

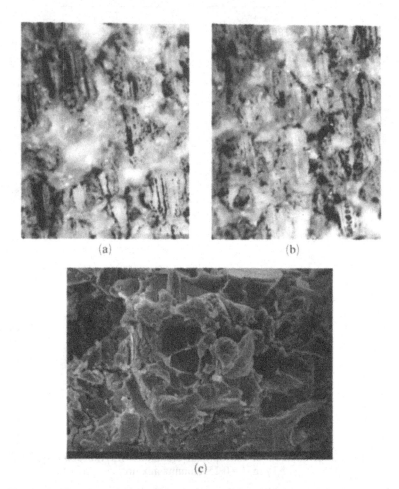

FIG. 9.15   Post-wear test morphology of S57/116-53 μm alumina compacts.
(a) Optical micrograph, $f = 0.25$ ($\times 100$); (b) optical micrograph, $f = 0.54$
($\times 100$); (c) SEM fractograph, $f = 0.54$.

The effect of the disc surface conditions is emphasised in Fig. 9.16(a),
which shows that for, say, $f = 1$, the highest rate of wear produced by the
'rougher' disc falls from some 11×10^{-6} to 5×10^{-6} mm^3/N m
associated with the 'smoother' disc surface. Within the investigated
f-range, the 'lower' rate of wear varies from 0.8×10^{-6} to 5×10^{-6}
mm^3/N m. The corresponding coefficients of friction range from 0
to 0.58.

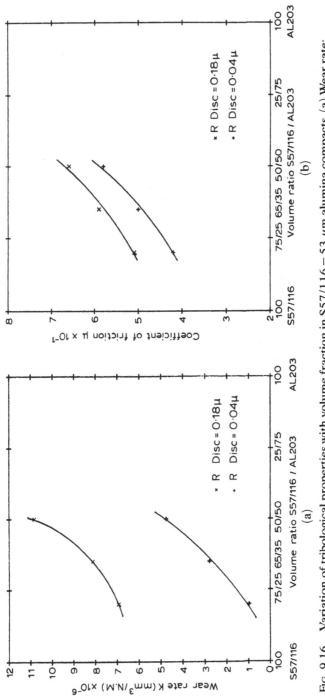

FIG. 9.16   Variation of tribological properties with volume fraction in S57/116 − 53 μm alumina compacts. (a) Wear rate; (b) coefficient of friction.

9.4.2 PVC–Carbon Fibre

Reinforcement of a weak polymer matrix by the incorporation of pre-orientated fibres is the ultimate target of designers of this kind of composite. However, because of the possibly severe anisotropy developing in a system in which fibre reinforcements are unidirectional, the preliminary investigation of Muhanna[16] was concerned with short carbon fibres randomly incorporated into S57/116 PVC powder matrix. Examination of the properties was mainly confined to a mixture of the fibre volume fraction $f = 0.30$, and the compaction was carried out in an implosive, axisymmetrical system. The E/M (E/P) ratio corresponding to the optimal processing conditions was about 0.62 (Fig. 9.17). The pre-compaction mixture properties were:

- apparent density 0.58 g/cm^3,
- tap density 0.74 g/cm^3,
- theoretical density 1.53 g/cm^3.

The morphology of the fracture surface of a composite is shown in Fig. 9.18(a). The degree of bonding between the adjacent polymer particles is satisfactory, but between the polymer and fibre is not as good. Some porosity is evident, due mainly to a relatively low plastic deformation of the fibre and its insufficient comminution in the 16–19 GPa range of impact pressures.

A lower level of intergranular bonding than that recorded in the case of polymer/polymer composites is responsible for a reduction in the strength of the compacted mixture in comparison with the S57/116 polymer matrix itself. The peak of the latter is about 43 MPa as compared with the maximum of 30 MPa of the mixture. An improvement in strength is achieved when a small reduction in the fibre content is made and results in an increase in the degree of bonding. Even a small reduction to $f = 0.25$ produces a combined strength of the same magnitude as that of the polymer, but at a cost of a higher operational pressure.

A comparison between the relative densities of the pure polymer and the mixture (Fig. 9.18(b)) shows that the same level of density of about 90% TD is reached at a pressure of 18 GPa for the latter, but a higher pressure of 19.5 GPa is required for the former. The degree of comminution of the fibre, although low, is sufficient to reduce the size of the interparticle voids. The graph exhibits the usual 'three zone' characteristics, as does that of the compressive strength.

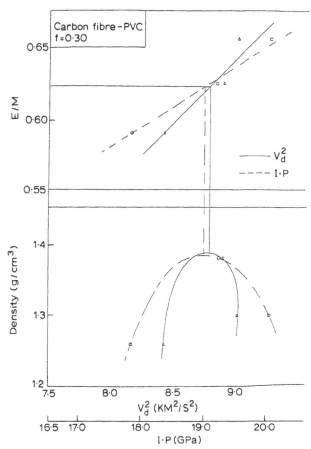

FIG. 9.17 Optimal conditions for implosive densification of S57/116 PVC powder and carbon fibre ($f = 0.30$).

9.4.3 Polyethylene–Calcium Hydroxyapatite

The search for particulate reinforced composites characterised by high toughness and moduli of elasticity drew the attention of Abram *et al.*[20] to blends of a high molecular weight, high density polyethylene (PE/8410) matrix and calcium hydroxyapatite [$Ca_{10}(PO_4)_6(OH)_2$] filler powders. Their investigation indicated clearly the dependence of toughness on the processing route, although the tensile modulus itself appeared to be unaffected.

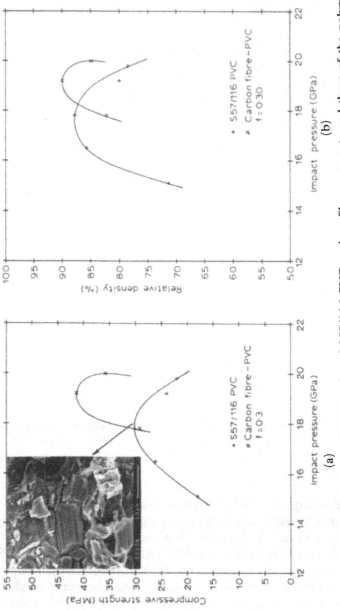

FIG. 9.18   Comparison between the properties of S57/116 PVC–carbon fibre compacts and those of the polymeric matrix. (a) Compressive strength to fracture; (b) relative density.

To explore these relationships more fully, an implosive, axisymmetrical compacting system was used by Blazynski and Muhanna[21] in their feasibility studies. The investigation was concerned with a specific mixture of hydroxyapatite of volume fraction 0·6. The filler particle size was less than 30 μm and the particles were approximately spherical $(\rho = 3·16$ g/cm$^3)$. The high density PE material had the theoretical density of 0·945 g/cm^3.

The densification of the mixture was carried out from a tap density of 0·97 g/cm^3, at impact pressures of 18 and 19 GPa. These correspond to impact energies of 300 and 342 J/cm^3 and container impact velocities of 388 and 414 m/s.

A satisfactory measure of consolidation and a virtual elimination of porosity were achieved at the lower pressure (Fig. 9.19), but even a slight increase in the pressure would cause structural damage and a consequent reduction in both the density and strength of the compact.

FIG. 9.19   Implosive   compact   of   polyethylene–calcium   hydroxyapatite $(f = 0·6)$ composite implosively consolidated at 18 GPa.

9.4.4 PVC–Silica

Experimentation with mixtures of silicon dioxide and PVC-type polymers arose out of an industrial interest in frictional and good wear-resistance properties of these agglomerates. In view of their generally satisfactory self-lubricating tendencies, their incorporation into the bearing design is a logical step, but requires more detailed knowledge of both the likely behaviour of SiO_2 and that of its effect on a polymer powder. Because of its well-established characteristics the PVC S57/116 powder was selected as the matrix and a very fine silica powder, average particle size 15 μm, as the filler. The pre-compact properties are listed in Table 9.4. Implosive compaction in cylindrical geometries was carried out in the 12–25 GPa range, varying the filler/matrix volume fraction from 0·3 to 3·0. Wear and frictional properties of the compacts were assessed, as in the case of alumina composites (Section 9.4.1), but EN58J, 10C Rockwell hardness steel discs, polished to either 0·13 or 0·036 μm, were used. The optimal operational conditions were assessed on the basis of the E/P relationships[16] of the type shown in Fig. 9.20 for a specific volume ratio.

The morphologies of the two extreme silica volume fractions of 0·3 and 3·0 (reflecting the optimal conditions) are shown in Fig. 9.21. The optical micrographs (Figs 9.21(a, d)) show a high degree of interparticle cohesion and bonding, particularly in the polymer, and a relative absence of porosity. The SEM fractographs (Fig. 9.21(b, e)) indicate considerable plastic deformation and interparticle integration. Figures 9.21(c, f) refer to the post-wear, originally polished, worn surfaces.

The respective effects of the silica volume fraction and impact pressure, on the one hand, and of the volume fraction at a constant operational pressure, on the other, are indicated in Figs 9.22 and 9.23. The tri-zonal distribution, relative to impact pressure, is again present in all the cases[12-14] with the highest degree of densification (Fig. 9.22(a)), and that of compressive strength (Fig. 9.23(a)) occurring for the lowest *f*-

TABLE 9.4
Pre-compaction Properties of SiO_2–S57/116 Mixture

Filler/matrix ratios	0·3	1·00	1·85	3·00
Apparent density (g/cm³)	0·56	0·46	0·28	0·23
Tap density (g/cm³)	0·60	0·65	0·44	0·36
Theoretical density (g/cm³)	1·43	1·48	1·55	1·61

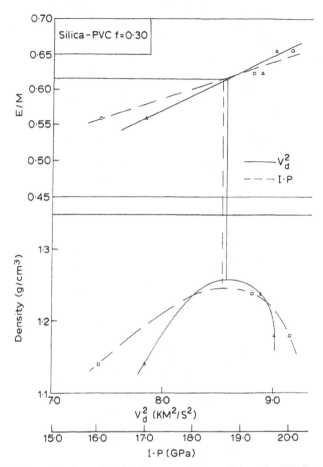

FIG. 9.20 Determination of optimal compaction conditions for S57/116 PVC-silica $(f = 0.30)$ compacts.

value of 0·33. Any increase in the filler content requires an increase in impact pressure and consequently the range of the 'optimal' pressure level extends from some 19 to 22 GPa for f lying between 0 and 3·0.

The highest relative compact density of 87% TD $(f = 0.3)$ is lower than the comparable value of the alumina-type composite. This is mainly due to the very fine silica powder which, although filling well the voids in the matrix structure, reduces the respective interparticle stand-off distances and thus affects adversely the conditions of individual impacts.[14,19]

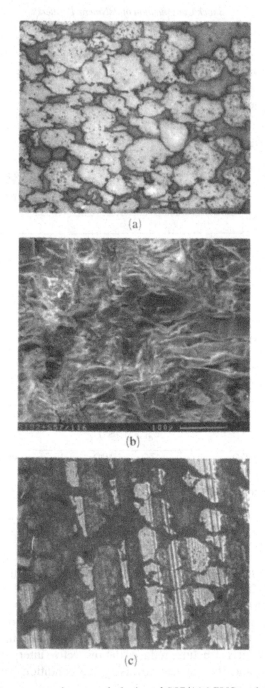

FIG. 9.21 Post-compaction morphologies of S57/116 PVC and 15 μm silica powders. (a) Optical micrograph ($f = 0.3$) ($\times 100$); (b) SEM fractograph ($f = 0.3$) ($\times 100$); (c) worn surface ($f = 0.3$); (d) optical micrograph ($f = 3.0$); (e) SEM fractograph ($f = 3.0$); (f) worn surface ($f = 3.0$). Operational pressures: 19 GPa for $f = 0.3$, 23 GPa for $f = 3.0$.

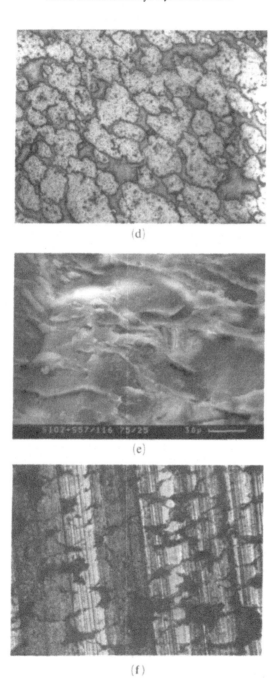

(d)

(e)

(f)

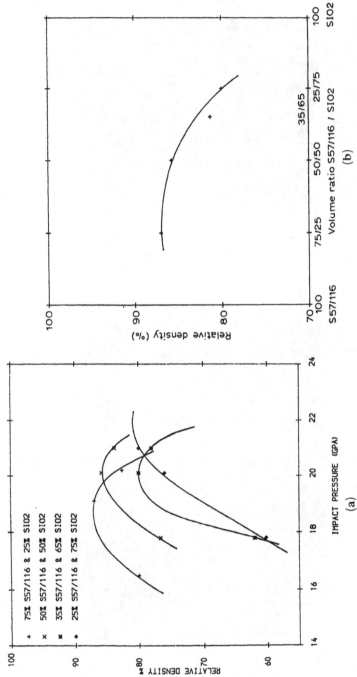

Fig. 9.22   Variation of compact density: (a) with silica volume fraction and impact pressure; (b) with volume fraction at a pressure of 19 GPa.

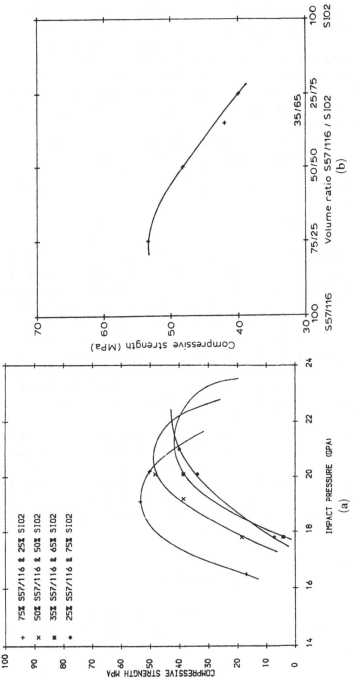

FIG. 9.23   Variation of compressive stress to fracture: (a) with silica volume fraction and impact pressure; (b) with volume fraction at a pressure of 19 GPa.

The compressive strength to fracture reaches slightly higher levels, but, on the whole, is comparable with the alumina-filler composites.

The dependence of both the compact density and strength on the volume ratio is shown in Figs 9.22 and 9.23. However, for a given value of the impact pressure, any increase in the silica content will reduce the numerical values of the two parameters. The density is reduced by some 9% (Fig. 9.22(b)) and strength by 35% (Fig. 9.23(b)) within the 0·33–3·0 range of f-values.

The rate of wear (Fig. 9.24(a)), which depends on the volume fraction and, naturally, on the condition of the disc, is almost twice that of alumina-filler composites, but the numerical differences between the roughness levels of the disc are less pronounced. The rate varies from between 8×10^{-6} and 11×10^{-6} mm^3/N M, at the higher matrix content to around 18×10^{-6} mm^3/N M at higher silica content.

The wear of the surfaces of the composites is due mainly to the abrasive action of the metal asperities and silica debris. The penetration of the tool asperities into the soft polymer matrix, and the presence of debris result in the loss of the material. The silica debris appears to be produced as a result of the increase in frictional temperature and the consequent enhanced viscosity of the polymer which, in turn, facilitates the dislodging of silica particles. Some surface melting is also present.

Both the wear rate and coefficient of friction (Fig. 9.24(b)) are significantly affected by the silica content. The two parameters increase with the increasing f-value. In the lower range of this, the coefficient of friction lies between 0·32 and 0·40, but it increases to about 0·5 as the influence of the matrix is reduced.

From a practical point of view, the importance of matching particle sizes of the matrix and the filler has to be emphasised. Inadequate mechanical mixing may well result in a localised concentration of one material and a loss of the uniformity of structural properties. An example of this is provided by a bush machined out of a compact, but containing a large, bonded insert of the polymer matrix that weakens the structure of the element (Fig. 9.25).

9.5 METALLOPLASTICS

The main interest in metalloplastics arises out of their electrical properties and, in particular, their electrical resistivity. The early work of, among others, Malliaris and Turner,[22] Aharoni[23] and De *et al.*[24] on

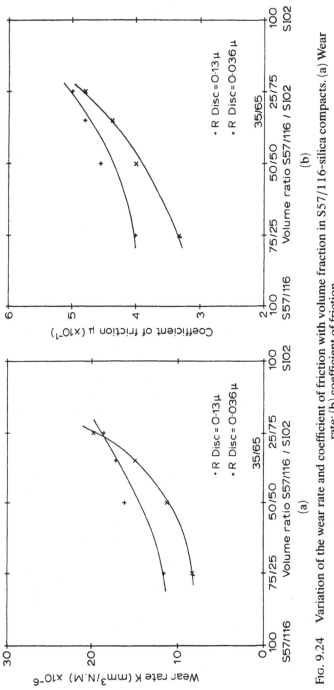

FIG. 9.24   Variation of the wear rate and coefficient of friction with volume fraction in S57/116-silica compacts. (a) Wear rate; (b) coefficient of friction.

Fig. 9.25 Inclusion of a large, bonded element of polymeric matrix in the structure of a bush machined out of a PVC–silica compact.

conventionally compacted or prepared polymer–metallic mixtures, revealed reasonably well-defined tendencies. For instance, a mixture of a 70 μm PE and 4–50 μm nickel powders, compacted at room temperature at 98 MPa, displayed a resistivity of only 10^{-4} Ω cm; the low value was attributed to the higher metal content of $> 30\%$. However, a similar result could be produced by a 6% nickel-content mixture when large PVC particle powders were mixed with small iron or nickel powders varying in particle size from 4 to 7 μm. The effect of particle size was thus clearly demonstrated.[23] A similar conclusion was reached on examination of the properties of an iron powder (about 10 μm) blend in a polyimide–amide solution. Again, a drop in resistivity was observed.[23]

A different aspect was emphasised by the evaluation of the resistance to corrosion of metalloplastics. A mixture of PMMA (50–150 μm) and of 20% copper powder of the same particle size, was compacted at 14°C at 9·8 MPa. The resistivity of the dry compact was found to be 198 Ω cm, but whereas after immersion in nitric acid the corroded part of the composite showed a resistivity of $6·07 \times 10^{10}$ Ω cm, that of the internal section was only slightly lower than the original value of 187 Ω cm.

Since, quite clearly, this particular parameter depends not only on the actual blend of the materials used, but also on the physical properties of the constituents, and on the interparticle bonding, the enhancement of the latter by implosive compaction should be examined.

The following sections contain information about the mechanical and physical properties and the PVC (matrix) and steel, iron and copper filler composites.

9.5.1 Polymer–Armco Iron

This particular series of investigations[12,15,25] was concerned with polyvinyl chloride powders of S57/116 and P67/579 serving as matrices. The characteristics of the two polymers are given in Table 9.1. The particle size of the Armco iron employed was within a 100–210 μm range, and the iron powder volume fraction varied from 0·2 to 0·5, with the theoretical densities of these mixtures ranging from 3·0 to 4·9 g/cm^3.

The importance of correct pre-compaction mixing is demonstrated by the compact morphologies. Figure 9.26(a) shows separation of heavier iron powder layers from the lighter polymer matrix and the resulting loss of uniformity of distribution throughout the compact. In a correctly mixed sample, however, the distribution is uniform (Fig. 9.26(b)) and macroporosity is practically eliminated. A SEM fractograph (Fig. 9.26(c)) shows that, nevertheless, interparticle bonding is very localised and areas of macroporosity and of purely mechanical compacting still exist at pressures lower than optimal.

However, at higher values of impact pressures a certain degree of comminution occurs and may be detrimental if too much debris accumulates and thus reduces the stand-off distances between individual particles.

Both the post-compaction density (Fig. 9.27(a)) and compressive strength to fracture (Fig. 9.27(b)) increase with reduction in polymer content. In both cases, the respective curves have the tri-zonal characteristics, but their maxima do not reflect the compositions of the individual mixtures. For instance, the difference between the maximum densities of $f = 0·25$ and $f = 0·42$ compacts is only of the order of 12%, whereas that between the volume fractions of $f = 0·42$ and $f = 1·0$ amounts to some 46%. This would indicate that better bonding, and therefore lesser porosity, is reached at higher levels of metallic powder content.

A similar situation arises in the case of strength properties, but the difference between the 'medium' and 'high' iron compacts is less pronounced and amounts to some 25%, with the 'lower' and 'medium' composites still separated by 12%. The best results can only be

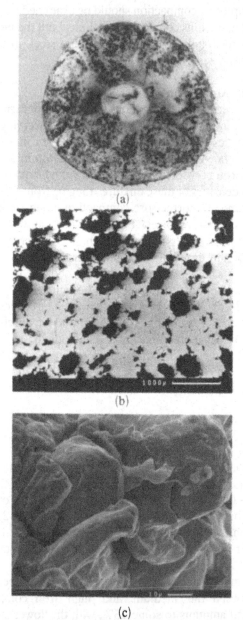

FIG. 9.26   Morphology of S57/116 PVC and Armco iron powder compacts. (a) Segregation of the constituents in incorrectly mixed blend. (b) Correctly mixed one-to-one blend. (c) SEM fractograph of 30% iron blend at 13 GPa.

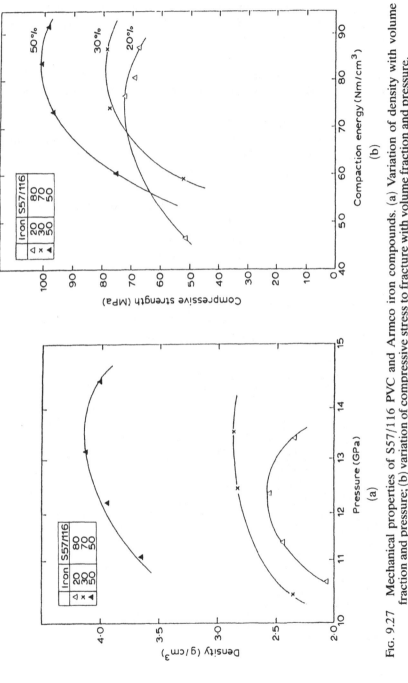

FIG. 9.27 Mechanical properties of S57/116 PVC and Armco iron compounds. (a) Variation of density with volume fraction and pressure; (b) variation of compressive stress to fracture with volume fraction and pressure.

expected at the optimal pressure values which for this particular material combination range from 12·5 GPa for $f = 0.25$, through 13·5 GPa for $f = 0.43$, to 13·8 GPa for $f = 1.0$.

The maximum Shore D hardness of 82 ($f = 1.0$), 80 and 74 respectively is associated with these optimal impact pressures,[26] and its distribution, related to the pressure, again displays tri-zonal characteristics.

9.5.2 Polymer–Stainless Steel

The 304 stainless steel powder ($+ 100–325$ μm) was used as a filler in S57/116 PVC powder. The volume fraction of the steel (SS) varied from 0·1 to 0·5. An implosive, cylindrical configuration system compaction[10, 11, 14] produced well densified specimens, but, at the same time, a fair level of porosity was present. This is due mainly to the basic shape characteristics of steel particles which differ considerably from the essentially spherical particles of the polymer. A thorough mixing of the aggregate is thus made more difficult. It follows that the interparticle collision conditions are less uniform and a full range of bonding possibilities is available.

Fracture surfaces of compacted aggregates (Fig. 9.28) show good dispersion of the constituent particles and indicate satisfactory bonding, in particular in the $f = 1.0$ mixture (Fig. 9.29(b)).

Fig. 9.28   SEM fractograph of a S57/116 PVC and 304L stainless steel powder compact ($f = 1.0$).

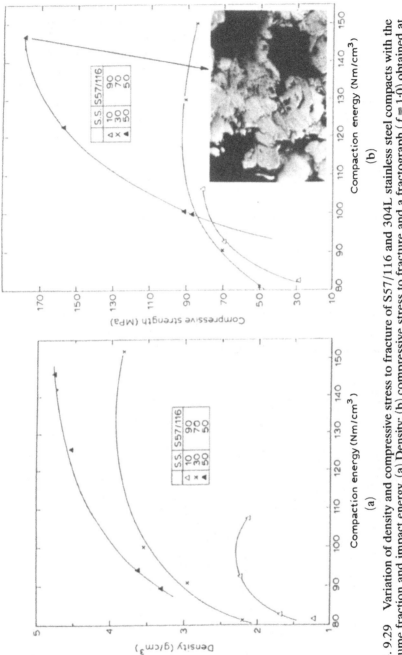

FIG. 9.29 Variation of density and compressive stress to fracture of S57/116 and 304L stainless steel compacts with the volume fraction and impact energy. (a) Density; (b) compressive stress to fracture and a fractograph ($f = 1.0$) obtained at optimal conditions ($\times 260$). (Polymer–light area.)

The variation of density and strength with process parameters is shown in Fig. 9.29. At a relatively high energy level of some 150 J/cm^3, as compared with 90 J/cm^3 for an iron blend, the optimal conditions for $f = 1.0$ have just been reached (Fig. 9.29(a)). The maxima of obtainable densities range from 2.25 to 4.75 g/cm^3 for f lying between 0.11 and 1.0. The energy required is, however, 70% in excess of that needed for the manufacture of comparable polymer/iron composites.

Again, as in the case of iron-filler composites, the compressive strength to fracture of the 'lowest' and 'medium' metallic content compacts is of the same order of magnitude of between 80 and 90 MPa, but that of $f = 1.0$ composites rises to 170 MPa for the same level of energy expended (Fig. 9.29(b)).

Hardness distribution varies, as in the case of iron compounds, with the basic process parameters. The highest value of 88 is reached in the $f = 0.5$ compact, and the lowest of 75 in the $f = 0.1$ mixture. This is slightly higher than the maximum hardness of a $f = 0.2$ iron compact.[25]

9.5.3 Polymer–Copper

Compacts of S57/116 powder and of both electrolytic (dendritic) and irregular/spherical copper powders were implosively compacted by Hegazy and Elbially.[26] The basic characteristics of these composites

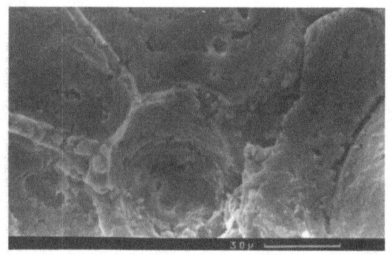

FIG. 9.30   Fractograph of S57/116 PVC and copper ($f = 1.0$) powders (courtesy Dr A. A. Hegazy).

closely resemble those of stainless steel and iron in that the relationship between the impact pressure and green density is, again, trizonal. A high degree of consolidation is attained at a pressure of 17·4 GPa when the green density reaches 97% TD for a volume fraction $f = 1·0$ (Fig. 9.30). However, an excess of energy dissipated at a higher pressure of 20 GPa results in interparticle melting and resolidification, with the layers of melt approaching 10 μm in thickness. The likelihood of brittleness setting in, or, at least, of some loss of ductility will exist, but the plastic deformation and comminution of the polymer are also responsible for a certain reduction in the value of the compact density.[26] At a pressure of 20 GPa the density reaches only 90% TD.

REFERENCES

1. MUKHOPADHYAY, R., DE, S. K. and BASU, S. *J. Appl. Polym. Sci.*, **20** (1976), 2575.
2. BERGER, M. A. and McCULLOUGH, R. L. *Compos. Sci. Technol.*, **22** (1985), 81.
3. BIGG, D. M. *Composites*, **4** (1979), 95.
4. BAUER, F. In: *Metallurgical Applications of Shock-Wave and High-Strain-Rate Phenomena*, Eds L. E. Murr, K. P. Staudhammer and M. A. Meyers, Marcel Dekker, New York, 1986, p. 1071.
5. HALLDIN, G. W. and KAMEL, L. L. *Polym. Engng Sci.*, **17** (1977), 21.
6. CRAWFORD, R. J. and SPREVAK, D. *Eur. Polym. J.*, **20** (1984), 441.
7. CRAGGS, G. *Proc. Inst. Mech. Engrs*, **204** (1990), 43.
8. BLAZYNSKI, T. Z., HEGAZY, A. A. and EL-SOBKY, H. In: *High Energy Rate Fabrication — 1984*, Eds I. Berman and J. W. Schroeder, ASME, New York, 1984, p. 113.
9. BLAZYNSKI, T. Z. In Ref. 4, p. 189.
10. BLAZYNSKI, T. Z. and HEGAZY, A. A. *Z. Werkstofftechnik*, **17** (1986), 363.
11. RAYBOULD, D. and BLAZYNSKI, T. Z. In: *Materials at High Strain Rates*, Ed. T. Z. Blazynski, Elsevier Applied Science Publishers, London, New York, 1987, p. 71.
12. HEGAZY, A. A. and BLAZYNSKI, T. Z. *J. Mater. Sci.*, **22** (1987), 3321.
13. BLAZYNSKI, T. Z. In: *Tribology in Particulate Technology*, Eds J. B. Briscoe and M. J. Adams, Adam Hilger, Bristol, 1987, p. 303.
14. BLAZYNSKI, T. Z. In: *Shock Waves for Industrial Applications*, Ed. L. E. Murr, Noyes Publishers, NJ, 1988, p. 406.
15. HEGAZY, A. A. PhD thesis, University of Leeds, 1985.
16. MUHANNA, A. H. PhD thesis, University of Leeds, 1990.
17. MUHANNA, A. H. and BLAZYNSKI, T. Z. *Compos. Sci. Technol.*, **33** (1988), 121.
18. MUHANNA, A. H. and BLAZYNSKI, T. Z. In: *Proc. 10th HERF Conf.*, Litostroj, Ljubljana, Yugoslavia, 1989, p. 58.

19. MUHANNA, A. H. and BLAZYNSKI, T. Z. *Materialwissenschaft u. Werkstofftechnik*, **21** (1990), 305.
20. ABRAM, J., BOWMAN, J., BEHIRI, J. C. and BONFIELD, W. *Plast. Rubber Process. Applicat.*, **4** (1984), 261.
21. BLAZYNSKI, T. Z. and MUHANNA, A. H. Interim Report, Department of Mechanical Engineering, University of Leeds, 1987.
22. MALLIARIS, A. and TURNER, D. T. *J. Appl. Phys.*, **42** (1971), 614.
23. AHARONI, S. M. *J. Appl. Phys.*, **43** (1972), 2463.
24. DE, S. K., MUKHOPADHYAY, R. and BASU, S. *Composites*, **7** (1976), 159.
25. MARDANI, H. A. and BLAZYNSKI, T. Z. *J. Mech. Work. Technol.*, **18** (1989), 315.
26. HEGAZY, A. A. and ELBIALLY, B. H. In Ref. 18, p. 67.

Chapter 10

Cermets, Metallic Glasses and Superconductors

10.1 INTRODUCTION

In addition to the well-defined, but general areas of application of shock waves, discussed in the previous chapters, there exists a range of highly specialised and, sometimes, novel ways of utilising their effects. Although the main representative composites of this kind are cermets, metglasses (amorphous composites), and, more recently, super-conductors — most of these involving some form of powder processing — other possibilities also exist. One such example is the consolidation of pharmaceutical powders by means of gas guns.[1] The high strain ratios attained in this case are sufficiently high to approximate to those associated with explosive techniques. A general tendency to an increase in the resistance to compaction at strain rates of up to 10^5/s is observed in, for instance, the strain-rate sensitive NaCl, Di-pac-sugar, lactose, Avicel, KBr, $Ca_3(PO_4)_2$ and $CuSO_4$ powders. The exception, in this range, is the well-known drug Paracetamol which, because of its particle morphology and composition, softens at strain rates below 10^3/s, but at higher strain rates displays the same tendency as other materials.

Another, more specialised application is that of the bonding of metallic elements to alumina. Examples of this exist in the welding of metal foils to this material.[2] Here, pure metal foils of aluminium, niobium, tantalum, titanium and zirconium, $2\cdot5-17\cdot8$ μm thick, were welded to high-alumina ceramic substrates $0\cdot76 \times 4\cdot83$ mm^2 in area. Typical explosive welding jetting was observed and resulted in both chemical bonding and mechanical interlocking.

These disparate examples illustrate well the versatility and potential of the technique, but form, of course, only a very small proportion of the actual investigative and production work done. A more detailed review of the state of knowledge in the field of main experimentation is given in the following sections.

10.2 CERMETS

These are basically ceramic materials bonded with metals with a view to producing high-strength, tough composites that can work in severe stress conditions — particularly in tooling. It is usually the carbides and/ or nitrides of high quality, metallic bases that are used. An introduction to cermets has already been given in Section 8.3.5 when the relatively simple titanium-based composites were discussed. However, a number of more complex materials has been produced by shock-processing and is now industrially employed.

A brief review of the effects of shock treatment in the area of tungsten composites serves as a useful introduction to the fabrication of cermets. Two well-documented examples are those of the carbides[3] and mixtures with uranium dioxide.[4]

The explosive synthesis of stoichiometric mixtures of tungsten and carbon powders $(E/P = 5)$ yields some 90% of WC, or α-W_2C. The synthesised material has a microhardness in excess of HV4000, which compares favourably with HV2000 of the conventionally, cold densified compacts.

The compaction of a W–30UO$_2$ powder mixture (150 μm mean particle size) from a 30·3% TD (solid theoretical density is 15·6 g/cm^3) to 96·8% TD in the form of 20 mm in diameter, 150 mm long rods and tubes is easily achieved[5] with either a fine or coarse powder.

The fine powder green compacts, on heat processing for 3 h at 1150°C and 69 MPa, show sintering, interparticle bonding and particle growth. In coarse powder compacts, the same results in UO$_2$ particles remaining unchanged in the tungsten matrix and showing no sign of rupture.

On the other hand, zircon bromide and zirconia can form bases for a different set of compacted mixtures. $(ZrB_2)_{80}(MoSi_2)$ explosively compacted from a mixture of 49 μm initial particle size and 56·8% TD, forms cylindrical rod compacts of 95·6% TD, whereas a $(ZrB_2)_{86}(MoSi_2)_{12}(BN)_2$ similarly formed mixture results in a compound of 88% TD.[4]

Compacts of zirconia with silicon nitride and also with calcium and manganese oxides are of high quality. For instance, $(ZrO_2) + (Si_2N_4)$ of

3 μm and 2·85 g/cm³ original size and density, shows an increase from a 53% TD tap to 95·5% final density when explosively shocked. A compound of $(ZrO_2)(CaO)(2$ w/0 MgO), also 3 μm, but 1·66 g/cm³ density, is compacted from 52·3% to 96·3% TD.[5]

A variety of cermets was obtained by gas-gun compaction in the USSR.[6] For instance, Nichrome–8BN composites were formed at pressures ranging from 195 MPa to 1·28 GPa with corresponding densities of 80–93% TD. Lanthanum-based materials and, in particular, $(LaCr)_2$–40Cr of 98% TD density were obtained and possessed heat transfer coefficients and resistance to oxidation (up to 1350°C in air) twice those associated with statically compacted composites.

Of further and considerable interest are two-layer seals made in 20Cr–80Ni + BN, and 23Cr–18Ni + BN powder compacts. These are reported to show a high degree of resistivity at temperatures of between 400 and 900 K. The seals are particularly suitable for working in weak acidic conditions, and in nitrogen oxide media.[7]

10.3 METALLIC GLASSES

Metal alloys lacking in long range atomic order (Fig. 10.1) form amorphous metals of glass-like structure. For this reason they are often referred to as 'metallic glasses' and proprietary, registered alloys are known under the name of Metglas.

The standard production methods are based on rapid solidification of the metal from liquid or gas at cooling rates of the order of 10^6 K/s. This is usually achieved by the casting of a stream of liquid metal onto a rotating wheel at speeds approaching 100 km/h and results in the

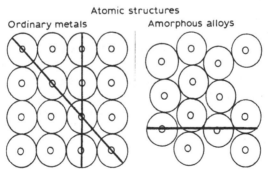

FIG. 10.1 Schematic representation of atomic structures.

production of a thin foil or ribbon. To attain high cooling rates, the thickness of the molten layer must not exceed a critical value which depends on the thermal and mechanical properties of the processed alloy. Although the mass fraction of the necessary molten material increases with the specific surface area of the particles, there exists a limit beyond which difficulties arise because excessive melting leads to resolidification and recrystallisation.

The usefulness of the metallic glasses lies, on the one hand, in their very high strength, approaching 1·7 GPa and combined with a hardness of up to 1000 HV (whilst retaining a degree of flexibility that allows bending by 180°), and, on the other, in their enhanced electromagnetic properties. It is this characteristic of the amorphous structure that makes metallic glasses particularly important in the electrical industry. However, although the use of these materials in the foil form in, say, electrical transformers — where their presence reduces core losses to a quarter of the conventional materials — creates no difficulties, the manufacture of '3-D' parts, involving the bulk of the material, presents considerable problems. Bulk-form manufacture requires the employ-ment of powder-consolidating techniques and the amorphous alloy powders are notoriously difficult to process in this form. It is here there-fore, that a dynamic compacting technique offers a solution, because the very high rates of cooling, necessary to induce or maintain amorphous structures, are easily attained in a gas-gun or in implosive compaction. The compacts possess high strength, unlike those produced by warm pressing methods, and are usually characterised by a satisfactory degree of interparticle bonding.

An additional, but important, application of dynamic treatment is the bonding of conventionally produced foil to a metal substratum. The technique employed here is explosive welding and, as such, does not necessarily imply shock treatment.

10.3.1 Bonding of Foil

Explosive welding of 40–45 μm thick metallic glass foils onto metal bases was successfully carried out by Prümmer.[8] Some 25×0.04 mm samples of $Fe_{80}B_{20}$, $Fe_{40}Ni_{40}B_{20}$, $Ni_{78}Si_8B_{14}$ and $Co_{48}Ni_{10}Fe_5Si_{11}B_{16}$ alloys were attached to St37 and stainless steels, as well as to brass, fully annealed aluminium, and Kanthal. Collision point velocities of up to 3400 m/s were used. At velocities lower than 2300 m/s only the shear band mode of deformation was observed and resulted in a number of interfacial fractures. However, at higher collision point velocities a homogeneous type of deformation prevailed with the interface being welded in a wavy form.

The collision point velocity was directly linked to both the type of weld and to its strength. For instance, at velocities lower than 2300 m/s only line welds (Fig. 10.2) of a relatively low strength were obtained and bonding was present along a 4 mm length only. The debonding, shown in Fig. 10.2(a), was the result of the initial distortion that occurred during the assembly of the specimen and is thus indicative of the degree of care necessary in manufacture. In the 2300–3200 m/s velocity range, crack-free wavy interfaces are obtained with bonded surfaces extending

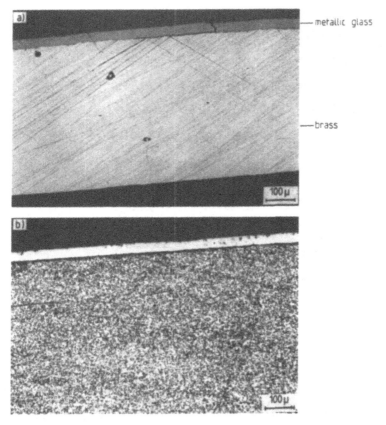

FIG. 10.2 Micrograph of metallic glass foil explosively bonded to metallic base. (a) $Ni_{78}Si_8B_{14}$-brass, at 2200 m/s; (b) $Ni_{78}Si_8B_{14}$-Kanthal Al, at 2050 m/s (courtesy Professor R. A. Prümmer).

over 20–25 mm lengths (Figs 10.3 and 10.4). At velocities exceeding 3240 m/s, the wave amplitude is of the order of foil thickness and conditions of both spalling of the foil and fracture of the weld develop.

The explosive welding technique thus finds an important field of application in allowing strong connections to be made between the very hard and thin metallic glass foil and a metallic base.

10.3.2 Dynamic Powder Compaction

A substantial volume of investigative work, covering almost a decade, is now in existence. A relatively early investigation of Vreeland *et al.*,[9]

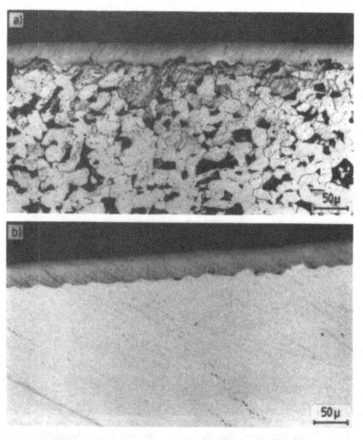

FIG. 10.3   Interface between the explosively bonded metallic glass and metal base. (a) $Fe_{40}Ni_{40}B_{20}$–St37 steel, at 2800 m/s; (b) $Co_{58}Ni_{10}Fe_5Si_{11}B_{16}$-Kanthal Al, at 2800 m/s (courtesy Professor R. A. Prümmer).

concentrating on the relationship between and the effect of the ratio of shock/melting point energies and shock pressure, has produced an insight into this complex matter.

The experimentation was concerned with both the microcrystalline, iron based Markomet 3.11 ($Fe_{BAL}W_{5.75}Mo_{4.5}Cr_{4.25}V_{4.0}C_{1.38}B_{0.65}$) and the amorphous, nickel-based Markomet 1064 ($Ni_{52.5}Mo_{38}Cr_8B_{1.5}$). Shock consolidation of approximately 2 μs duration was used in both

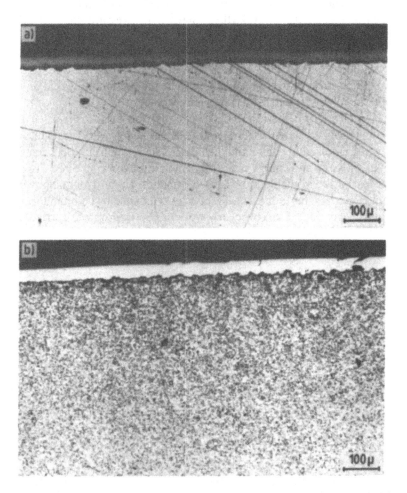

FIG. 10.4   Wavy interfaces between metallic glass foil and metallic base. (a) $Ni_{78}Si_8B_{14}$-stainless steel, at 2800 m/s; (b) $Ni_{78}Si_8B_{14}$-stainless steel, at 2700 m/s (courtesy Professor R. A. Prümmer).

cases and was carried out in a 40 μm Hg vacuum at 22°C. The two powders were obtained by foil comminution. In general, a decrease in the amorphous content of the structure was noted with increasing shock energy. In particular, the microcrystalline Marko 3.11 ($<$38 μm particle size), which originally possessed amorphous layers formed on its particle surfaces (the result of the melting and rapid solidification), reacted strongly to the value of the shock pressure. At a low pressure of 5·2 GPa a mixture of amorphous and crystalline structures was observed, but at a slightly higher pressure of 7·2 GPa the amorphous material recrystallised. A further increase in pressure to 13·8 GPa produced a number of bubbles.

The amorphous Marko 1064 material suffered transformation to crystalline at about 600°C. At a low pressure of 4 GPa only a few crystalline particles were formed, but their proportion increased at 7·7 GPa, and almost full recrystallisation took place at 12 GPa.

It is clear from these authors' work that the balance between the energy supplied and the temperature generated is very delicate and must be assessed in each individual case. The relationship between the shock pressure and energy ratio is shown in Fig. 10.5.

Most of the available published information about dynamic powder compaction relates to the iron-based alloys obtained by comminution of metallic glass foils. Extensive early work on these alloys was carried out

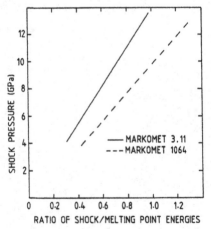

FIG. 10.5   Experimentally established relationship between the shock pressure and the ratio of shock/melting point energies for shock consolidated Markomet 3.11 and 1064 powders (based on the data of Ref. 9).

in the USSR where Roman *et al.*,[10] for instance, reported on the explosive compaction of $Fe_{40}Ni_{40}P_{16}B_4$ and $Fe_{48}Ni_{48}B_4$. Shock treatment resulted here in magnetic properties of a good coercive force of 0·0375–0·05 Oe, but a low saturation induction of 0·58–0·64 Ti. A similar experimentation[11] with $Fe_{40}Ni_{40}P_{14}B_6$ and $Fe_{70}Cr_{10}P_{13}C_7$ produced discs 15 mm in diameter and 1–5 mm thick with densities between 96 and 99% TD. Although two different techniques of shock loading were used, and three starting temperatures of 77, 293 and 523 K, the variation in temperature did not affect the final density, but both types of shock-wave loading, and the starting temperature had an effect on the ordering/alignment of the platelet shaped powder. The starting temperature did not influence the response of the structure to shock treatment. Specimens were amorphous when compacted at pressures below 7 GPa, but showed recrystallisation at a pressure of 11 GPa, with 6–30% crystallisation being detected. Some work on annealing at 673 K in air, and 753 K in vacuum was also reported.

The paramagnetic properties of the iron-based alloys were examined by, among others, Murr *et al.*,[12] Caliguri *et al.*,[13] and Hasegawa and Cline[14] and Raybould and Hasegawa.[15]

An examination of the properties of explosively compacted Metglas 2605-S2 ($Fe_{78}B_{13}Si_9$) showed[12] that when compared with the properties of ferrites, the paramagnetic properties of cores made of the amorphous material result in considerably smaller total power loss. The hysteresis loops of the two materials are completely different. The Metglas compact hysteresis is also different (at 80 Hz) from that of the corresponding foil-type material. Further, whereas the resistivity of the compact is 220 $\mu\Omega$ cm, that of the foil is only 130 $\mu\Omega$ cm.

The work on the $Fe_{78}B_{13}Si_9$ compact showed that after annealing, a coercivity of 1·60 Oe and a B_{10} of 0·8 T were obtained. The a.c. properties were typified by a permeability of around 300 at 1 kHz and 0·1 T. Raybould and Hasegawa[15] showed that using a longer anneal could produce properties superior to those of the, otherwise, identical warm-pressed compacts; the permeability being around 600 at 0·1 T and 1 kHz, compared with 300 for the die-pressed compact.

Further work was carried out on Metglas 2605-S2 by Page and Raybould,[16] who compacted both the amorphous and crystalline powders by means of a gas-gun. In both cases, particle sizes varied from 100 to 200 μm.

The amorphous material, when compacted in the 0·372–1·86 GPa (446–1550 m/s) range, produced green densities of 4·9 and 7·0 g/cm^3

respectively. The hardness of 950 HV of the original material was reduced to between 920 and 940 HV. No evidence of interparticle melting was found, but it was noticed that the interparticle boundaries would disappear and good bonding would be present. Also, no evidence of crystallisation below the critical tempreature of 750°C was reported.

The crystalline material, after heating for 6 h in vacuum, had a fine microstructure and its original hardness of 750 HV increased to between 800 and 825 HV after compaction. At 1·61 GPa (310 m/s) the density of the compact was 6·01 g/cm^3, rising to 7·1 g/cm^3 at 1·65 GPa and 1430 m/s impact velocity. After the compaction, a fine-grained material containing hard borides was obtained. No melting was detected, but no conclusion about possible interparticle bonding could be reached because the boundaries remained visible. There was no evidence of twinning.

An example of the difficulties that can be experienced in obtaining well-homogenised structures is given by Prümmer[17] in Fig. 10.6. In an explosive compaction of $Fe_{40}Ni_{40}B_{20}$ alloy powder, three different types of structure were observed within a distance of 100 μm from each

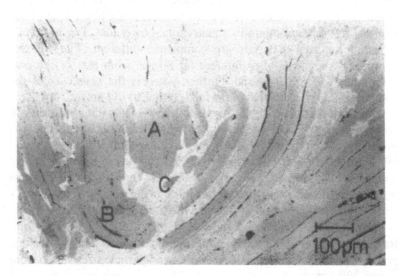

FIG. 10.6   Changes in structural properties within an explosive compact of $Fe_{40}Ni_{40}B_{20}$ alloy powder. (A) Well-bonded areas; (B) unbonded areas; (C) recrystallised material (courtesy Professor R. A. Prümmer).

other. For instance, well-bonded areas exist at A, with unbonded areas at B, and recrystallised material at C.

Of non-iron based alloys, nickel, cobalt and aluminium metallic glasses have been processed. An explosively compacted $Ni_{89}P_{11}$ material[17] can be consolidated either in powder or foil form. In powder form, its green density is closely related to the compacting pressure (Fig. 10.7). At a low pressure of 3·7 GPa full densification is not achieved and the compact is porous. At 4·4 GPa, however, satisfactory compaction takes place, accompanied by full densification. Any further increase in pressure is counterproductive, and at, say, 5·1 GPa excessive melting produces recrystallisation. To avoid the onset of it, Prümmer suggests limiting the particle size to below 250 μm.

When 40 μm thick foils are used in compaction, densities of close to 100% TD can be reached, but it is difficult to obtain interface bonding whilst preventing the formation of molten pockets and their subsequent crystallisation.

Cobalt-based alloys such as, for example, $Co_{50}Ni_{10}Fe_5B_{16}Si_{11}$ were consolidated by impact to form rods 10 mm in diameter, and tubes 16 mm in diameter with 4 mm thick walls.[11] Extensive work was carried out to determine any structural changes in the amorphous material that could not be detected by X-ray or DSC. Differences were found in the local anisotropy (Ha) and correlation radius (re).

An aluminium-based material $(Al_{BAL}Si_6)$ was investigated by Smugeresky and Gourdin.[18] Monosize powder, produced by ultrasonic

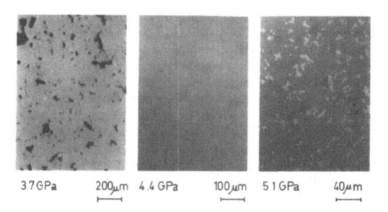

3·7 GPa   200μm   4·4 GPa   100μm   5·1 GPa   40μm

Fig. 10.7   Explosively compacted metallic glass $Ni_{89}P_{11}$ powder (courtesy Professor R. A. Prümmer).

gas atomisation, in a range of sizes $(-45+38 \ \mu m, \ -105+74 \ \mu m, -205+210 \ \mu m)$ was compacted at pressures of about 2·2 GPa in cylindrical tubes 31·8 mm in diameter and 6 mm thick. The compaction was achieved by the use of a projectile with an impact velocity of 840 m/s. At pressures lower than 1·5 GPa, full densification was impossible, but microstructure modification increases in the 1·5–4·0 GPa range. Melting and resolidification appear to occur as a result of deformation rather than any enhanced recovery and recrystallisation. The mechanical properties, especially the ductility, exceed those of cast Al–Si390 alloy (used in car engines). In fact, the medium particle-sized material compact reaches some 277 MPa in UTS, 240 MPa in yield stress, and displays 2% elongation.

10.4 SUPERCONDUCTORS

10.4.1 Introduction
Superconductivity or zero electrical resistance $(R = 0)$ is associated with the effects of the supercooling of matter. Diamagnetic shielding, together with the induced Meissner effect result in a substance which is highly desirable from the point of view of its electrical properties. The degree of supercooling necessary to achieve this objective depends on the processed material. The elemental, single metals usually require very low temperatures of the order of 4·2 K (liquid helium temperature), but the superconducting transition temperatures (T_c) of many alloys are well above those of the elemental metals. Most of the metallic oxides of lead, bismuth and particularly niobium acquire superconductivity within the 13–23 K range. In particular, the Nb_3Sn superconducting compound of crystalline A15 structure already discussed (Section 8.3.3) can be shock synthesised from 99·9% niobium and 99·99% pure tin powders in a stoichiometric ratio of 3:1 at pressures of 50 GPa, and at generated temperatures of over 2000°C to give the intermetallic Nb_3Sn phase only, with the onset of superconductivity at around 16·7 K. Similar T_c values are recorded for niobium titanide, Nb_3Ge and NbAlGe (23 K). The recent introduction of ceramic superconductors, operational at a nitrogen T_c of 77 K, has opened a new range of possibilities in this area. The oxide-based materials such as, for instance, $LaBaCuO_x$ (30 K), $YBa_2Cu_3O_x$ $(>90 K)$, $Bi_2Sr_2CaCu_2O_x$ $(>85 K)$ and $Tl_2Ca_2Ba_2Cu_3O_x$ $(>120 K)$ are of particular interest and their successful bulk manufacture and large part shaping engage the attention of both researchers and production engineers.[19-21]

It is precisely the manufacturing problems of the successful integration of component materials of these more complex systems that has focused attention on the explosive, or other dynamic means of processing. The brittleness of these materials, on the one hand, and the necessity to produce a superconducting crystal phase (A15) or a cubic β-tungsten crystal-type structure for isotropic superconductivity, limit the choice of manufacturing methods. In addition to a direct powder synthesis and compaction, Murr and his collaborators successfully use a bulk system of an array or a series of slots machined out of an aluminium or copper monolith. A mixture of the constituent powders, contained within, can be synthesised and consolidated explosively, and an orthorhombic, superconducting, compacted material is obtained.[19] The consolidation is due mainly to the presence of shock-induced dislocations which, in turn, guarantee a sufficient degree of plasticity to allow the basically brittle material to undergo compaction without cracking.

The success of any such operation depends heavily on the strength of the magnetic field, field current density, and transition temperature T_c. The characteristics of this 3-D system depend, in turn, on the material involved in the operation. Single, elemental systems (referred to as Type I) are characterised by a critical value of the magnetic field below which the particular material is superconductive, but above which $R > 0$ and the material reverts to its normal state. The interplay between the three parameters of state is determined by plotting the respective values in field–current–T_c space coordinates and thus producing a well-defined and unique superconducting surface.

The situation becomes much more complicated in the case of alloys and ceramics, or Type II, superconductors. Here, there exist two critical field values, the lower and upper, and these produce, in turn, not a single superconducting system, but a two-stage system. Real superconductivity, with $R = 0$, exists below the lower critical field value. Between the lower and upper magnetic field levels there exists a mixed state containing partially trapped flux, and it is, finally, above the higher critical boundary that the material reverts, once again, to its normal ($R > 0$) state. It is particularly interesting to note that a Type II superconductor can sustain fields higher than any elemental Type I system.

The explosive manufacture of either system, but particularly of Type II, generally increases the field current and its density, since, as discussed earlier, the accompanying deformation of the material modifies its structure and properties.[19]

10.4.2 Shock Synthesis and Modification

The effects of shock synthesis and/or modification of structure have been recently investigated, with reference to the copper oxide-type of superconducting material, by Murr et al.,[19,23] Venturini et al.[20,21] and Staudhammer.[22] Shock-induced chemical synthesis, if successful, is likely to be of greater practical importance than any conventional method since both the time element and elaborate procedures would be avoided. Shock modification is of interest if it can lead to the production of stable structural defects that would be expected to enhance the interparticle magnetic flux pinning. The two effects could, ideally, be combined, but in both cases the problem of the possible degradation of superconducting properties, rather than the improvement, has to be investigated. The main objectives of any such process are therefore the conversion of a large part of the considered material to an orthorhombic superconducting phase, and retention of the oxygen content to preserve magnetic shielding values.

Some degradation in superconducting properties in 'as shocked' materials is attributable to the loss of oxygen just mentioned and partial conversion to a non-superconducting tetragonal phase. The latter effect is associated with the induced bulk shock temperatures. An additional, recently observed, parameter is that of the actual volume of the sample of the shocked material. The larger the sample the lower the content of the produced skin superconducting phase.[22]

Shock synthesis was studied on the lanthanum–strontium–copper, and yttrium–barium–copper systems.[21] A stoichiometric ($La_{1.85}Sr_{0.15}$-CuO_4) mixture of lanthanum oxycarbonate, lanthanum and strontium oxides, and of some 5–10% of a material of K_2NiF_4-type structure was shocked in gas tight recovery capsules. The compact density ranged from 1·5 to 1·77 g/cm^3 at the corresponding pressures of 17–22 GPa. The bulk temperatures varied from 800 to 900°C, and the essential for a superconduction K_2NiF_4-type phase was produced in a large volume of 85–90% at a pressure of 22 GPa, but only as 60% at 17 GPa. In the 'as received' material loss of superconducting properties was observed, but oxygen annealing at around 400°C restored the properties.

The yttrium–barium–copper (1:2:3) mixture was synthesised from initial densities of 1·75–2·3 g/cm^3 at 22 GPa. The bulk temperatures ranged from 600 to 800°C. As in the case of lanthanum system, postshock oxygen annealing was necessary.

Shock modification effects were investigated[20-22] on high purity (90%) starting powder materials of $Bi_2CaSr_2Cu_2O_x$, ($Tl_2Ca_2Ba_2Cu_3O_x$

(Tl-2223), and $YBa_2Cu_3O_7$ (Y-1237). The consolidation was again carried out in gas-tight copper recovery capsules at shock pressures ranging from 7·5 to 27 GPa.

In general, a substantial decrease in both the diamagnetic shielding and Meissner effect was observed for Bi-2122 and Tl-2223 compacts in the 150–850°C temperature range. However, a 5 K (122 K) increase in the onset of superconductivity of the thallium material was observed (no change in the bismuth material), and a threefold increase in the post-shock flux pinning energy at low tempreatures was noted. Tl-2223 powder was compacted from a density of 4·08 g/cm^3 (60% TD) at pressures of between 20 and 27 GPa with corresponding bulk temperatures of 150–450°C. The oxygen stoichiometry appeared to be maintained in the shock process and substantial Meissner and diamagnetic shielding were present. For instance, the Meissner effect measured in a magnetic field of 2·5 mT, varied with temperature from some $10^{-1·8}$ emu/g at 94 K to $10^{-4·6}$ emu/g at 122 K. The corresponding values of the 'as supplied' material were 10^{-1} to 94 K, and $10^{-4·8}$ at 118 K.

Bi-2122 powder was consolidated at a pressure of 7·5 GPa from a tap density of 4·8 g/cm^3 (60% TD) giving a bulk temperature of 200°C. The removal of the 100 K phase resulted from this treatment, possibly accompanied by the beginning of the formation of glassy phases.[20] In the shock-modified material, the onset of the Meissner effect was associated with a temperature of 86 K, as compared with 85 K for the 'as supplied' powder.

The yttrium-type powder (Y-123) was compacted from a tap density of 4·03 g/cm^3 (63% TD) at pressures varying from 7·5 to 20 GPa that produced an average bulk temperature of 150°C. A partial conversion from the orthorhombic, superconducting phase to the tetragonal would result, but oxygen annealing could restore most of the original shielding properties, while retaining the enhanced magnetic flux pinning at temperatures near the transition (T_c) temperature.

It is mainly in connection with the $YBa_2Cu_3O_{7-x}$ explosively processed materials that Staudhammer's investigation[22] of the volume effect indicates a possible reason for the degradation or even loss of superconductivity after shock treatment. It appears that a second order transformation, present in the Y-123 orthorhombic phase, produces two superconducting phases at 60 and 90 K respectively. Incorrect heat treatment can therefore easily lead to the preponderance of the wrong phase and a consequent degradation in properties. Staudhammer has shown that small quantities of the correctly cooled material (from a

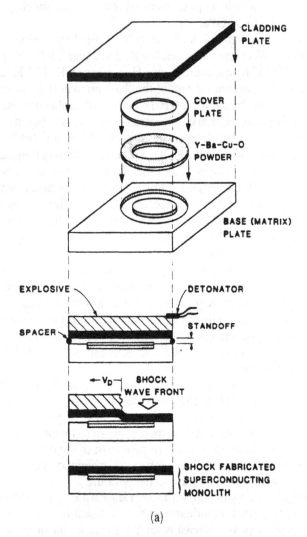

(a)

FIG. 10.8   Schematic representation of bulk-monolith superconducting explosive systems. (a) Fabrication of a ring of ceramic, initially powdered, material in a metal matrix; (b) fabrication in a linear system (courtesy Professor L. E. Murr).

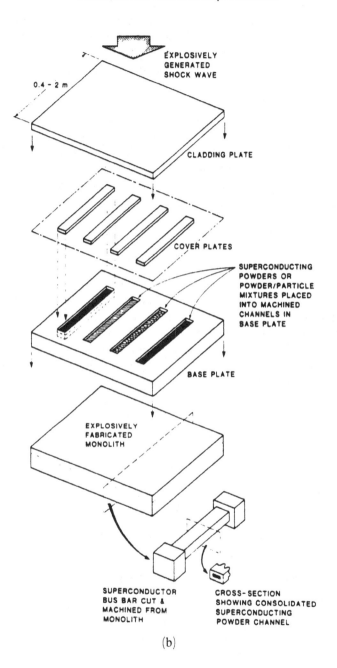

(b)

homogenisation temperature of 1073 K to between 673 and 723 K) produce higher volume fractions of the desired 90 K phase, whereas large-bulk samples produce little or no 90 K phase.

A similar situation may, of course, arise in other materials and may provide at least a partial explanation for some of the observed phenomena.

A general conclusion that can be drawn from these and similar experiments is that although the correct, for superconductivity, structure can be created in shock processes, the superconductivity itself may not necessarily be present in the synthesised or compacted material, but may have to be restored, perhaps incompletely, by a post-shock treatment, e.g. oxygen annealing.

10.4.3 Monolith Fabrication

Whereas the small-scale tests described above are sufficient from the point of view of investigations into the likely pattern of behaviour of the processed superconductor, their practical usefulness is limited because of the relatively small volumes of material involved. Bulk fabrication of suitably pre-shaped superconductors is of importance in any manu-facturing process and is therefore briefly considered here.

Since the oxide-type ceramics are generally brittle and therefore difficult to consolidate integrally, a method of avoiding this problem, as well as that of post-compacting fracturing, had to be devised. An

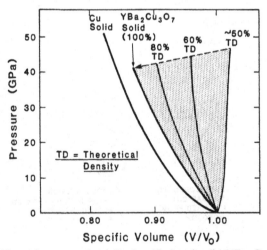

FIG. 10.9   Hugoniot curves for fully densified and solid $YBa_2Cu_3O_7$ material, and, also, solid copper (courtesy Professor L. E. Murr).

interesting solution, involving the elements of both explosive welding and powder compaction, was proposed by Murr *et al.*[19,23-26] The developed system consists essentially of a flyer and base metal matrices (aluminium, copper or silver) which on the detonation of an explosive charge encapsulate and compact the superconducting powder, placed originally in conformal channels. The plates weld to each other and on completion of the operation a superconducting monolith, containing geometrically pre-shaped high-temperature superconducting ceramic powder, is formed. Two such basic systems, e.g. ring and linear, are shown in Fig. 10.8. Multilaminate systems, based on the basic arrangements, are also produced.

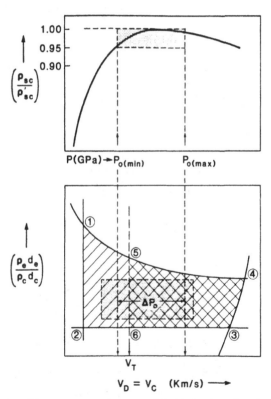

FIG. 10.10   Welding parameters for monolith fabrication, showing the optimal welding window (shaded) (ρ, density; d, thickness; ρ'_{sc}, average of starting bulk and superconductor TD densities; subscripts; c, cladding, e, explosive; sc, superconductor) (courtesy Professor L. E. Murr).

The monoliths fabricated in the manner described, exhibit a strong Meissner effect and are therefore likely to be of practical use where magnetic field shielding is required.

Although by the consolidation of suitable mixtures fabrication of new superconducting materials is possible, the bulk of the investigative work carried out so far has been concerned with $YBa_2Cu_3O_7$ material. Its

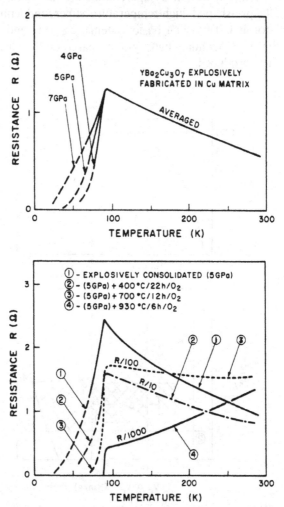

FIG. 10.11   The effect of impact pressure, temperature and oxygen annealing on the superconducting properties of $YBa_2Cu_3O_7$ monolith (courtesy Professor L. E. Murr).

Hugoniot curves — for different theoretical densities — are shown in Fig. 10.9, where a curve for solid copper (as a matrix) is included for comparison. The degradation of properties of superconductors after shock treatment already discussed (see preceding section) has been confirmed by Murr in connection with the reduction in superconductivity of the monoliths. Whereas the suppression of T_c, sometimes to < 77 K, can be moderated by the use of highly refined starting powder, the effect of impact pressure is less easy to manipulate since the welding conditions limit freedom of choice (Fig. 10.10). Some representative results obtained with copper–matrix monoliths are shown in Fig. 10.11 and indicate the fact that microstructural defects must have been created which cannot be annealed out even at elevated temperatures. Murr suggests that major crystallographic changes must occur, in addition to simple oxygen loss, and that they are associated with the passage of the shock front. The problem appears to be related to the operational pressures in that some degradation of the residual superconductivity is observed even at low peak shock pressures of 4 GPa, but becomes more pronounced above 6 GPa. The high pressures (20 GPa or so) used by Morosin et al.[20,21] in their experiments (Section 10.4.2) would, naturally, accentuate this problem.

The continuing development of the monolith systems augurs well for their successful integration into the normal production routines.

REFERENCES

1. AL-HASSANI, S. T. S. and ES-SAHEB In: *Mechanical Properties at High Rates of Strain*, Ed. J. Harding, Institute of Physics, Bristol, London, 1984, p. 421.
2. SHAFFER, J. W., CRANSTON, B. H. and KRAUSS, G. *Proc. 5th HERF Conf.*, Ed. R. Wittman, University of Denver, Colorado, 1975, Paper 4.12.
3. STAVER, A. M. In Ref. 2, Paper 2.1.
4. CARLSON, R. J., POREMBKA, S. W. and SIMONS, C. C. *Ceram. Bull.*, **44** (1965), 266.
5. BARRINGTON, J. and BERGMANN, O. R. US Patent 3,337,766.
6. ROMAN, O. V. and GOROBTSOV, V. G. In *Shock Waves for Industrial Applications*, Ed. L. E. Murr, Noyes Publishers, NJ, 1988, p. 335.
7. ROMAN, O. V. and GOROBTSOV, V. G. In: *Powder Metallurgy of Aerospace Materials, Proc. Int. Metal Powder Conf.*, Berne, Switzerland, 1985, p. 1.
8. PRÜMMER, R. A. *Z. Werkstofftechnik*, **13** (1982), 44.
9. VREELAND, T. Jr, KASIRAJ, P., AHRENS, T. J. and SCHWARZ, R. B. *Proc. Materials Soc. Annual Meeting*, Boston, MA, 1984.

10. ROMAN, O. V. *et al.* 6th Collection — Powder Metallurgy, Higher School, Minsk, USSR, 1982 (in Russian).

11. ROMAN, O. V. *et al. Sov. J. Powder Metall.*, (1984), 345.

12. MURR, L. E., SHAUKAR, S., HARE, A. W. and STAUDHAMMER, K. P. *Scripta Metall.*, **17** (1983), 1353.

13. CALIGURI, R. D., DECARP, P. S., CURRAN, D. P. and SHOCKEY, D. A. *Explosive Compaction of Ferromagnetic Metal Powders*, ERRI Contract Ts-82-610, Final Report, 1983.

14. HASEGAWA, R. and CLINE, C. In: *Proc. 5th Int. Conf. on Rapidly Quenched Metals*, Wurtzburg, FRG, 1984.

15. RAYBOULD, D. and HASEGAWA, R. *Powder Rep.*, **39** (10) (1984).

16. PAGE, N. W. and RAYBOULD, D. *Mater. Sci. Engng*, **A118** (1989), 179.

17. PRÜMMER, R. A. *Mater. Sci. Engng*, **98** (1988), 461.

18. SMUGERESKY, J. E. and GOURDIN, W. H. In: *Metallurgical Applications of Shock-Wave and High-Strain-Rate Phenomena*, Eds L. E. Murr, K. P. Staudhammer and M. A. Meyers, Marcel Dekker, New York, Basel, 1986, p. 107.

19. MURR, L. E., HARE, A. W. and EROR, N. G. In Ref. 6, p. 473.

20. VENTURINI, E. L., GRAHAM, R. A., GINLEY, D. S. and MOROSIN, B. In: *Shock Compression of Condensed Matter — 1989*, Eds S. C. Schmidt, J. N. Johnson and L. W. Davison, Elsevier Science Publishers, BV, Amsterdam, 1990, p. 583.

21. MOROSIN, B., VENTURINI, E. L., GRAHAM, R. A. and GINLEY, D. S. *Synthetic Metals*, **33** (1989), 185.

22. STAUDHAMMER, K. P. In: *Proc. 10th HERF Conf.*, Litostroj, Ljubljana, Yugoslavia, 1989, p. 1.

23. MURR, L. E., NIOU, C. S. and PRADHAN, M. Ibid., p. 21.

24. MURR, L. E. In: *Proc. 28th NATO Defense Res. Group Seminar on Novel Materials for Impact Loading*, Bremen, FRG, 1988.

25. MURR, L. E. and EROR, N. G. *Mater. Manufact. Process.*, **4** (1989), 177.

26. MURR, L. E. and EROR, N. G. In: *2nd Workshop on Industrial Application Feasibility of Dynamic Compaction Technology*, Tokyo Institute for Technology, Tokyo, 1988.

Index

Printed in the United States
By Bookmasters